Birkhäuser

Static & Dynamic Game Theory: Foundations & Applications

For further volumes:
www.springer.com/series/10200

Lacra Pavel

Game Theory for Control of Optical Networks

Birkhäuser

Lacra Pavel
Department of Electrical and
Computer Engineering
University of Toronto
Toronto, ON
Canada

ISBN 978-0-8176-8321-4 ISBN 978-0-8176-8322-1 (eBook)
DOI 10.1007/978-0-8176-8322-1
Springer New York Heidelberg Dordrecht London

Library of Congress Control Number: 2012938417

Mathematics Subject Classification: 91-02, 93-02, 94-02

Printed on acid-free paper

Springer is part of Springer Science+Business Media (www.birkhauser-science.com)

Preface

Game theory has recently been enjoying increased popularity in the research community, as it provides a new perspective on optimization, networking, and distributed control problems. It incorporates paradigms such as Nash equilibrium and incentive compatibility, it can help quantify individual preferences of decision-makers, and it has an inherently distributed nature. Consequently, game theoretic models have been applied to a variety of problem domains ranging from economics to computer science, networking, and security. Game theoretic models provide not only a basis for analysis but also for design of network protocols and decentralized control schemes. This makes it attractive as a theoretical framework for the design of networks and in particular communication networks. Applications range from power control in wireless communications to congestion control for the Internet, as well as sensor or ad-hoc networks.

As the backbone of the Internet, optical networks provide huge bandwidth and interconnect countries and continents. Unlike conventional networks, which are well-established and even standardized, the optical networks field is much younger, and in fact still evolving. Designing and setting up networking applications within the optical context is inherently more difficult than in conventional wired or wireless networks, in part due to the more complex physical-layer effects, and in part due to the lack of automatic methodologies developed for optical networks. But in these respects, the optical networking field has seen a tremendous growth in recent years.

From a game theory and control perspective, there are a multitude of problems to be tackled in the optical networks area and the field is still in its infancy. Aside from the fact that the area is much younger, research work requires an unusual blend of interdisciplinary expertise. In this spirit, the present monograph draws the author's research background in control theory, practical industrial experience in optical networks, and more than five years of research in game theory and control in optical communications.

The book is focused on mathematical models, methodologies, and game theory for optical networks from a control perspective. The general setup is that, given an optical communication network and some performance measures to be optimized among many networks units/players/channels, one must design an algorithm that achieves as good a performance as possible for each channel. The algorithm should be decentralized and provably convergent, and should use localized and decentralized feedback. By regarding channels in the network as players in a game this multi-objective optimization problem fits within the setup of game theory.

The first theoretical question of interest is how to formulate meaningful and tractable game theoretical problems that can be used as a basis for developing such algorithms, taking into account the various mathematical characteristics arising naturally in optical networks. These mathematical characteristics are imposed by physical constraints and by network specific features and topologies. They translate into constraints on the game formulation (separable versus coupled action space), constraints on player's interaction (localized versus global interaction, i.e., one shot game versus stage or partitioned game), and constraints on players' actions (global constraints versus propagated/modified constraints).

Thus, one of the characteristics of optical networks is that game theoretic formulations and results cannot be transferred directly from other application domains (e.g., wireless networks or congestion control). Due to inherent physical complexities, the optical area uncovers new theoretical problems, such as games with coupled utilities and coupled constraints. Moreover, in the networking setup these constraints are modified across the network links and a framework for dealing with network games has to be developed. This makes game theory in optical networks an excellent starting point that could open new research problems and that could be generalized to other classes of games in networks.

This monograph has a two-fold aim. Its first goal is to provide researchers in the control and game theoretic community with background on the rich problems and the initial results in this area. There is a broad scope for fundamental control and game theoretical research; the hope is that the book will provide background material such that non-specialists in optical networks can approach these research problems within the optical networking domain. The book's second goal is to provide researchers in the networking and optical community with game theoretical methodologies that could be used to solve optical networking problems.

The following topics are covered. In the first part of the book, two chapters present non-cooperative game theory background and some new mathematical results for Nash equilibria computation in games with coupled constraints. In the second part, background and mathematical models for optical networks are presented, followed by game theory formulations developed for various topologies. The basic game considered is a power control game in the class of games with continuous action spaces, coupled utilities in normal form. The book in fact introduces in a systematic, gradual way the different types of game theoretical problems: first games with no coupled constraints in normal form, then games with coupled constraints; all-to-all interaction versus localized player interaction, which leads to games in ladder-nested form; multi-link (single-sink) and finally mesh topologies and how to deal with games in such scenarios, while building on the simpler cases and results. The third part considers issues such as robustness and time-delay effects, as well as other types of game problems in optical networks, including routing and path coloring games.

Much of the work in game theory applications has been within the framework of networking or computer science fields. Essentially, this book uses a mathematical approach developed from a control theoretical perspective. This is different from a networking approach, which is typically application-specific and focused on particular protocols, and unlike algorithmic-only approaches as typically used in computer science. Game theoretical problems are mathematically formulated herein in a systemic manner, with analytic conditions derived for existence and uniqueness of Nash equilibrium. Based on these, iterative algorithms with provable convergence properties to Nash equilibria are developed. The control theoretic approach to games as taken here allows for the treatment of important issues such as stability and time-delay effects in a dynamic system context. The abstract mathematical models and results could be applied to other application domains.

I wish to acknowledge and express my thanks to the many people who influenced and shaped my work over the past several years and thus indirectly helped me in writing this book: Tamer Başar, Tansu Alpcan, Peter Caines, Roland Malhamé, and Eitan Altman from the game theoretical and control community; and Stewart Aitchison, Li Qian, and Dan Kilper from the optical communications community. I am indebted to my graduate students, Yan Pan, Nem Stefanovic, and Quanyan Zhu, with whom many of the research work was done and without whom this book would not have been possible.

Toronto Lacra Pavel

Contents

List of Acronyms

ABC	Automatic power control
APC	Automatic power control
ASE	Amplified spontaneous emission noise
BER	Bit-error rate
CW	Continuous wave
DSL	Digital subscriber line
GA	Gradient algorithm
LHS	Left-hand side
LS	Light source
NE	Nash equilibrium
OA	Optical amplifier
OADM	Optical add/drop multiplexer
OLT	Optical line terminal
ONTS	Optical network test system
OSA	Optical spectrum analyzer
OSC	Optical service channel
OSNR	Optical signal-to-noise ratio
OXC	Optical cross-connect
PUA	Parallel update algorithm
RHS	Right-hand side
Rx	Receiver
SIR	Signal to interference ratio
Tx	Transmitter
VOA	Variable optical attenuator
VOL	Virtual optical link
WDM	Wavelength division multiplexing

Chapter 1
Introduction

Abstract This chapter provides an introduction and overview of the monograph, which is aimed at understanding how control algorithms can be designed for optical networks from a game-theoretic perspective. The first section gives a review of work in game theory for networks, followed by a brief introduction to the area of optical networks. The last section presents the scope of the monograph, followed by a chapter by chapter description of the monograph.

1.1 Game Theory in Networks

Our dependence on communication networks to distribute information is growing at an exponential rate. Different communication networks have evolved to provide a variety of services to end users. These communication networks include radio networks, satellite networks, and optical networks to name a few. In the past decade, game theory and system control have been utilized to optimize the performance of these networks, where heuristics were previously used [68].

Game theory, an area initiated more than fifty years ago [96, 159], has been of interest for researchers working in a broad range of areas from economics [16], computer science [73, 97, 129], social studies, and engineering and communication networks [8, 10, 13, 14, 41, 64, 78, 79, 83, 118, 133, 160, 163]. The recent popularity it has been enjoying in engineering has to do with the fact that game theory brings new perspectives to optimization and control of distributed systems and networks. It incorporates paradigms such as Nash equilibrium and incentive compatibility, which can help quantifying individual preferences of decision-making agents. In fact game theory provides a rigorous mathematical framework for modeling actions of individual selfish or cooperating agents/players and interactions among players. Furthermore, it has an inherently distributed nature and provides a foundation for developing distributed algorithms for dynamic resource allocation. Moreover, recent interest has been on extending the standard game setup in various ways, some having to do with computation of equilibria [36, 40, 120] others being concerned with the inherent static nature of a traditional game setup, how to extend it to a dynamic process by which the equilibria is to be achieved [84, 137] or to address equilibria efficiency [1, 60, 63, 123].

Game-theoretic applications in communication networks can involve either cooperative or noncooperative games, static or repeated games, finite or continuous

L. Pavel, *Game Theory for Control of Optical Networks*,
Static & Dynamic Game Theory: Foundations & Applications,
DOI 10.1007/978-0-8176-8322-1_1,

strategy games [14]. Applications include power control problems in different multiuser environments [8, 41, 61, 72, 118, 133], routing [12, 99] or congestion control [4, 11, 13, 21, 76, 129, 139, 160], extending the system-based optimization approaches [68, 80–82, 87, 153]. The many "players" interact within the network and make (sequential) decisions, i.e., play a game. For example, in a noncooperative (Nash) game framework the natural players/agents can be the Internet service providers (ISP) or domain operators [79, 83], the routers [12, 21, 99, 162], or even the users themselves in an access network application with dedicated wavelengths [112]. As another example in wired networks, there could be two sets of players: telecom firms/ISP and end users. Both sets of players have different objectives and non-negligible interaction across players exists. In wireless networks there could be wireless LANs where users/players communicate with a fixed access point, or wireless ad-hoc networks where users/players communicate with each other in the absence of any fixed infrastructure support.

In a noncooperative (Nash) game [20, 96, 128] each player pursues the maximization of its own utility, or equivalently the minimization of its own cost function, in response to the actions of all other players. The stable outcomes of the interactions of noncooperative selfish agents correspond to Nash equilibria. On the other hand, in a cooperative game framework the natural players/agents can be the network nodes, routers or switches (as software agents), or users [9]. These players/agents cooperate to redistribute the network resources (bandwidth, wavelength capacity, power).

Why game theory? Consider the case of a multiple access network problem; most optimization-based approaches find the optimal multiple access control (MAC) and routing parameters that optimize network throughput, lifetime, delay etc. and assume all nodes in the network use these parameters. But there is no reason to believe that nodes will adhere to the actions that optimize network performance. Cheaters may deviate in order to increase their payoffs which in turn affects other users. In effect any scenario where there is some strategic interaction among self-interested players is best modeled via a game-theoretic model. Game theory helps to capture this interaction, the effect of actions of rational players on the performance of the network. Although the selfish behavior of players causes system performance loss in a Nash game [1, 38, 64, 73, 122], it has been shown in [64, 134] that proper selection of network pricing mechanism can help preventing the degradation of network system performance. In the context of evolution, a game captures the essential features where strategic interactions occur. The basic idea is that actions which are more "fit", given the current distribution of actions, tend over time to displace less fit actions [47, 49, 50].

1.2 Optical Networks

Optical networks represent a typical example of complex communication networks and the Internet relies on them for its backbone infrastructure. This book is aimed at understanding how such control algorithms can be designed for optical networks from a game-theoretic perspective.

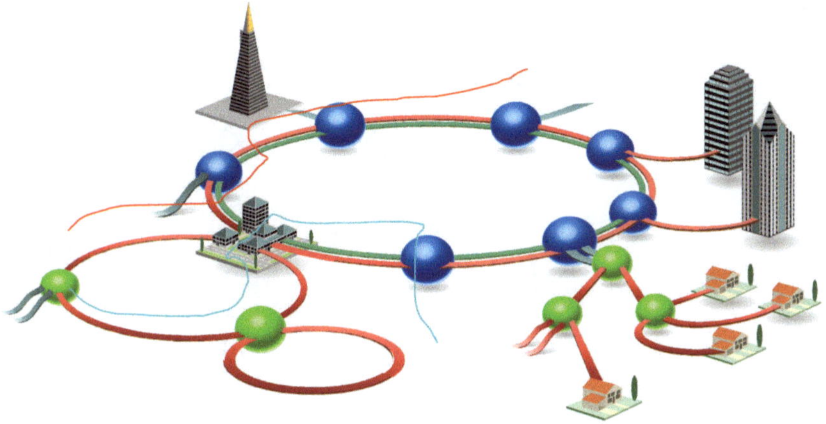

Fig. 1.1 Network deployment example

Unlike conventional networks which are well-established and even standardized, the optical networks area is much younger, in fact is still an evolving area. Designing and setting up networking applications within the optical context is inherently more difficult than in conventional wired or wireless networks in part due to the more complex physical-layer effects, in part due to the lack of automatic methodologies developed for optical networks. The combination of (a) need for dynamism—rapid provisioning of links; or allowing large but fairly brief bursts of data to travel through the network from source to destination, (b) need for transparency—the desire not to convert from optical to electrical to optical at switches, but to have continuous optical flow of signals, means that one needs to create a dynamic network using optical components—which are themselves continuing to evolve with new capabilities. This involves reconfigurable lightpaths, a dynamic optical layer with dynamic virtual topologies and optimized, adaptive transport layer (see Fig. 1.1). One of the aims of this book is to address this issue.

In optical networks information transmission is realized by transmitting several intensity-modulated wavelengths of light over the same optical fiber. After many years of delay, communication via the optical fiber is now being accepted in a wide range of military, aerospace, commercial, and security applications. About 85 percent of all global communications traverse optical fibers, including cable and telephone signals.

Optical networks are comprised of a set of fiber optic links that connect nodes which facilitate adding, dropping, and rerouting of signal channels in a dynamical manner. Laser-based transmitters emit continuous wave (CW) light with wavelengths around 1550 nm (optical frequencies of around 190 THz). This continuous wave light is modulated in intensity by the information data and represents the transmitted optical signal. Each channel is associated with a single wavelength (frequency) in the beam of light. Signals (channels) are multiplexed into a single beam of light using wavelength division multiplexing (WDM) before being transmitted through the network. Signals are demultiplexed at their destination points (Rx)

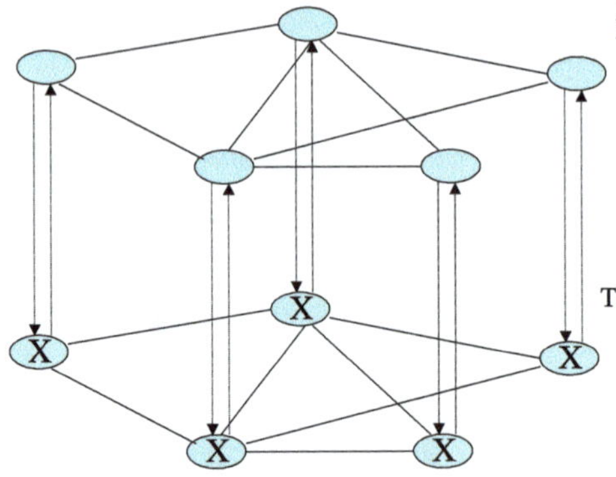

Fig. 1.2 Network layers

to be processed outside the optical domain. Optical line terminals (OLT) are network elements that interface with the optical network clients (SONET, IP, ATM) and are comprised of transponders (Tx/Rx pairs), as well as optical multiplexers/demultiplexers. Individual powers of the channels can be adjusted at the signal sources (Tx) as well as at some intermediary nodes. Optical amplifiers (OAs) are used to amplify the channel powers every 80 to 100 kilometers in an optical link. Recent years have witnessed the evolution from point-to-point WDM fiber links to *switched optical networks* (also referred to as *wavelength-routed optical networks*). Channels can be rerouted dynamically through different links within the network using optical cross-connects (OXCs) or optical add/drop multiplexers (OADMs). See the List of acronyms for commonly used ones. By using direct photonic links and optical-domain connectivity unnecessary and expensive optical to electrical to optical (o/e/o) conversion is avoided.

In static optical networks, lightpaths are statically provisioned up front. The operational expense of reconfiguring the network is quite high and such reconfiguration is time-consuming in the sense that any change in the virtual topology, i.e., the topology needed for a set of transmission requests, needs manual reconfiguration and additional equipments. Evolution of emerging tunable network element (such as reconfigurable OADMs, OXCs, and tunable lasers) make it possible to realize *reconfigurable optical networks* [131, 152, 156, 166]. Specific for these is that network reconfiguration can be quickly performed under software control for arbitrary virtual topologies (control plane layer, see Fig. 1.2). As important is the fact that the length of fiber links and the number of optical amplifiers deployed along a link can be changing. Different channels may travel on different fiber links and arbitrary virtual topologies can be formed over the transport layer (see Fig. 1.2).

Then the interesting question is how to deploy reconfigurable optical networks in which quick reconfiguration is realized under software control and at the same time, the network stability and channel transmission performance are maintained. Typical optical network topologies are mesh, consisting of interconnection of multiple links, which complicates network performance analysis and network control. This book is aimed at understanding how such control algorithms can be designed for optical networks from a game-theoretic perspective.

1.3 Scope of This Monograph

From a game theory and control perspective, there are a multitude of problems to be tackled in optical networks and the field is still in its infancy. An important problem is how to control the channel transmission performance in optical networks while incorporating physical-layer impairments. More precisely, transmitted signals degrade in quality due to physical-layer impairments (attenuation, noise, dispersion, nonlinear effects). Channel transmission performance at the physical layer (transport layer) is characterized through the bit-error rate (BER), which is defined as the average probability of incorrect bit identification. Performance degradation is associated with an increase in BER, which in turn depends on optical signal-to-noise ratio (OSNR), dispersion, and various nonlinear effects [2]. While a channel traverses a number of optical elements (e.g., OAs and OXCs) through the network (see Fig. 1.3), it may degrade in quality due to these impairments. Moreover, signals of new established channels affect the quality of existing channels on their shared fiber links. An optical network is considered as a dynamic multiuser environment, if the signal of a channel is regarded as interfering for all others on the same fiber link. Power control is a key issue in designing an interference-limited multiuser communication network system. A major portion of this monograph will be focused on this first type of problem by using game-theoretic approaches. Another part of the book will discuss game-theoretic approaches for other challenging problems such as routing and wavelength-assignment in optical networks [99, 125, 165] in the context of reconfigurable optical networks.

For simplicity, we use "optical networks" to generally refer "reconfigurable switched optical networks". We also use "channel (input) power" to refer to "channel (input) signal power". Our approach is primarily based on game theory. Throughout the book the focus is on analysis of dynamic systems arising from game formulations with non-separable player utilities and with coupled constraints as well as propagated (modified) constraints, in the class of continuous strategy games. It turns out this is the case for a class of games in optical networks, where the utility of each player is affected by decisions of other players, and action sets are no longer orthogonal (Chaps. 7 and 8).

The monograph is organized as follows. Chapter 2 provides an overview of basic concepts in game theory, mostly focused on noncooperative (Nash) game theory. Chapter 3 focuses on matrix games, while Chap. 4 on games with continuous action sets and cost functions. Chapter 5 presents some relatively new theoretical results for Nash games with coupled constraints, i.e., coupled action sets. Chapter 6 provides an overview of basic background on transmission in optical networks and on general topologies to be studied. The concepts of optical signal-to-noise ratio (OSNR) and coupled power constraints in an optical link are introduced, and the various models are presented. In Chap. 7 the basic formulation of a game framework towards maximizing channel OSNR is presented. We restrict our attention to single point-to-point optical links, as the simplest network topology. A Nash game played among channels is set up towards maximizing OSNR, in this first instance without coupled power constraints. Thereafter, two approaches are considered in order to

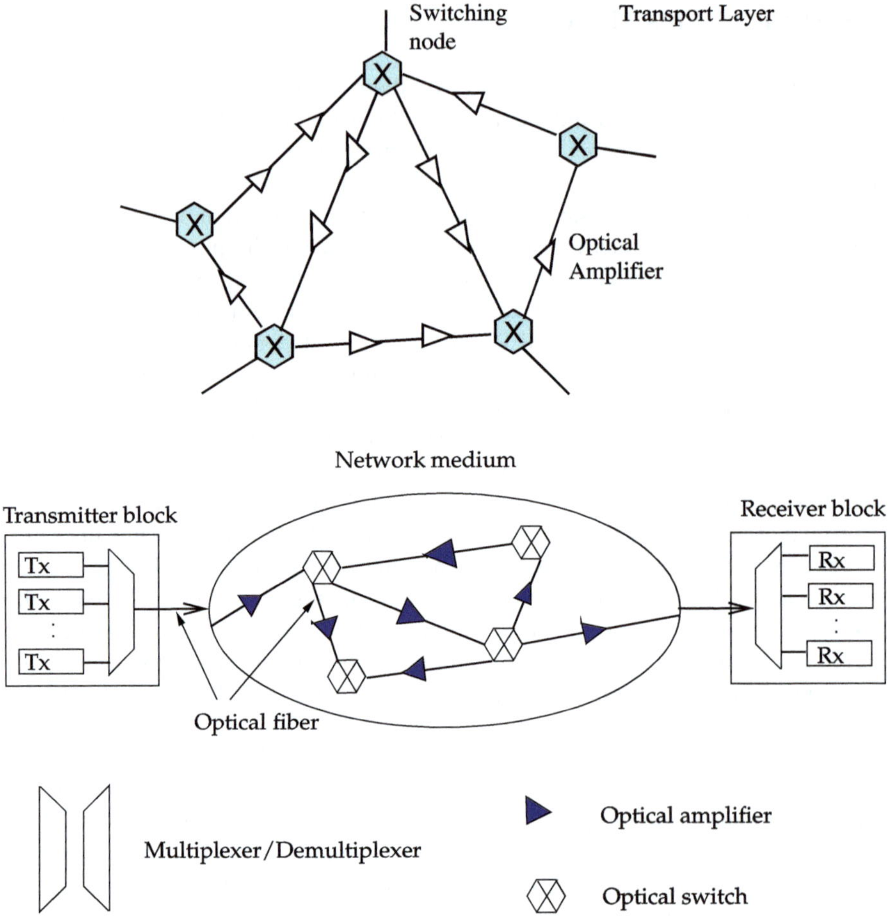

Fig. 1.3 Block diagram of an optical WDM network system

incorporate the coupled power constraint: an indirect approach (penalty-based) and a direct one, based on the theoretical results in Chap. 5. In network configurations, coupled power constraints are propagated along fiber links and these functions become intricate from the end-to-end point of view. Approaches on how to deal with this case of games in more complicated network topologies are studied in Chap. 8. Various cases of increasing complexity are considered: multi-link adjustable topologies, single-sink topologies and finally mesh topologies. Chapter 9 addresses some efficiency issues of such a Nash game. Simulations and some practical experimental implementations are presented in Chap. 10. Network control schemes and game-theoretic inspired algorithms need to consider the effects of propagation delay. This is because this propagation delay may not just impact the performance of the algorithms, but it may destabilize the network as a whole. Chapter 11 addresses the impact of time-delay in optical network control algorithms derived from game-theoretic formulations. Stability properties of game and pricing dynamics are investigated by using Lyapunov theory and a singular perturbation approach, under the assumption of time-scale separation and considering time-delay effects. Other game formulations for wavelength-routing and wavelength-assignment (routing and path-coloring) fall into the class of finite strategy games and are typically treated as extension of integer-linear programming (ILP) heuristics. These are covered in Chap. 12.

Techniques developed herein and game-theoretic formulations as exemplified for optical networks could be well extended to other cases of more general classes of games with propagated constraints. There is a lot of scope for fundamental control and game-theoretic research. The hope is that this book will provide background material such that non-specialists in optical networks can approach such research problems within the optical networking domain. Its second goal is to provide researchers in the networking and optical community with game-theoretic methodologies that could be used to solve optical networking problems.

Part I
Game Theory Essentials

Chapter 2
Basics of Game Theory

Abstract This chapter provides a brief overview of basic concepts in game theory. These include game formulations and classifications, games in extensive vs. in normal form, games with continuous action (strategy) sets vs. finite strategy sets, mixed vs. pure strategies, and games with uncoupled (orthogonal) vs. coupled action sets. The next section reviews basic solution concepts, among them Nash equilibria being of most relevance. The chapter is concluded with some remarks on the rationality assumption and learning in classical games. The following chapters will introduce these concepts formally.

2.1 Introduction

Game theory is a branch of applied mathematics concerned with the study of situations involving conflicting interests. Specifically game theory aims to mathematically capture behavior in strategic situations, in which an individual's success in making choices depends on the choices of others. The field was born with the work of John von Neumann and Oskar Morgenstern [159], although the theory was developed extensively in the 1950s by many among whom John Nash [95, 96].

In this chapter we shall introduce the game-theoretic notions in simplest terms. Our goal will be later on to study and formalize mathematically various game problems, by which we understand problems of conflict with common strategic features. While initially developed to analyze competitions in which one individual does better at another's expense (zero-sum games), it has been expanded to treat a wide class of interactions, which are classified according to several criteria, one of these being cooperative versus noncooperative interactions. Typical classical games are used to model and predict the outcome of a wide variety of scenarios involving a finite number of players (or agents) that seek to optimize some individual objective. Noncooperative game theory studies the strategic interaction among self-interested players.

Historically, game theory developments were motivated by studies in economics, but many interesting game theory applications have emerged in fields as diverse as biology [141], computer science [54], social science and engineering [74]. In engineering, the interest in noncooperative game theory is motivated by the possibility of designing large scale systems that globally regulate their performance in a dis-

L. Pavel, *Game Theory for Control of Optical Networks*,
Static & Dynamic Game Theory: Foundations & Applications,
DOI 10.1007/978-0-8176-8322-1_2, © Springer Science+Business Media New York 2012

tributed, and decentralized manner. Modeling a problem within a game-theoretic setting is particularly relevant to any practical application consisting of separate subsystems that compete for the use of some limited resource. Examples of such applications include most notably congestion control in network traffic (i.e. Internet, or transportation), problems of optimal routing [11, 13, 14], power allocation in wireless communications and optical networks [118, 133].

The chapter is organized as follows. We present basic concepts and game formulations, then we review some classifications: games in extensive vs. in normal form, games with continuous (infinite) action (strategy) set vs. finite action (finite strategy games), mixed vs. pure strategy games, and games with uncoupled (orthogonal) vs. coupled action sets. We follow with a discussion of basic solution concepts, among them Nash equilibria being of most relevance herein. We conclude with some remarks on the rationality assumption and learning in classical games.

2.2 Game Formulations

Game theory involves multiple decision-makers and sees participators as competitors (players). In each game players have a sequence of personal moves; at each move, each player has a number of choices from among several possibilities, also possible is the chance or random move. At the end of the game there is some payoff to be gained (cost to be paid) by the players which depends on how the game was played. Noncooperative game theory [96] studies the strategic interaction among self-interested players. A game is called *noncooperative* if each player pursues its own interests which are partly conflicting with others'. It is assumed that each player acts independently without collaboration or communication with any of the others [96].

This is in contrast to a standard optimization where there is only one decision-maker who aims to minimize an objective function by choosing values of variables from a constrained set such that the system performance is optimized (see the appendix for a review of basic results, mainly drawn from [24]).

So far we have mentioned three elements: alternation of moves (individual or random (chance)), a possible lack of knowledge and a payoff or cost function. A game $\mathcal{G}$ consists of a set of players (agents) $\mathcal{M} = \{1, \dots, m\}$, an action set denoted Ω_i (also referred to as a set of strategies S_i) available for each player i and an individual payoff (utility) $\mathcal{U}_i$ or cost function J_i for each player $i \in \mathcal{M}$.

In a game, each player individually takes an optimal action which optimizes its own objective function and each player's success in making decisions depends on the decisions of the others. We define a noncooperative game $\mathcal{G}$ as an object specified by the triplet $(\mathcal{M}, S, \Omega, \mathbf{J})$, where

$$S = S_1 \times S_2 \times \cdots \times S_m$$

is known as the strategy space,

$$\Omega = \Omega_1 \times \Omega_2 \times \cdots \times \Omega_m$$

is the action space, and $\mathbf{J} : \Omega \to R^m$, defined as

$$\mathbf{J}(\mathbf{u}) = \left[J_1(\mathbf{u}), \ldots, J_m(\mathbf{u})\right]^T, \quad \mathbf{u} \in \Omega$$

is the vector of objective functions associated to each of the m players, or agents participating in the game. In some cases (see Chap. 12) a graph notation might be more appropriate than the set $\mathcal{M}$ notation. Conventionally $\mathbf{J}$ represents a vector of cost functions to be minimized by the agents. An alternative formulation is to use a vector of utility functions (payoff functions) $\mathcal{U}$ to be maximized by the agents. Without loss of generality we assume that each agent aims to minimize its cost so we will use $\mathbf{J}$ throughout. Since usually we shall identify each player's action set and cost function we shall use the notation $\mathcal{G}(\mathcal{M}, S_i, \Omega_i, J_i)$, where the subscript is associated to each player i, $i \in \mathcal{M}$. In some cases we shall identify strategies with actions (one-shot games) and drop one of the arguments. The properties of the sets Ω_i and the functions $J_i(\mathbf{u})$, $i \in \mathcal{M}$ depend on the modeling scenario, and hence the type of game under consideration.

Each player's success in making decisions depends on the decisions of the others. Let Ω_i denote the set of actions available to player i, which can be finite or infinite. This leads to either finite actions set games, also known as matrix games, or infinite (continuous action set) games. In the latter case each player can choose its action from a continuum of (possibly vector-valued) alternatives. A strategy can be regarded as a rule for choosing an action, depending on external conditions. Once such a condition is observed, the strategy is implemented as an action. In the case of *mixed strategies*, this external condition is the result of some randomization process. Briefly, a *mixed strategy* for agent i is a probability distribution x^i over its action set Ω_i. In some cases actions are *pure*, or independent of any external conditions, and the strategy space coincides with the action space. In discussing games in *pure strategies* we shall use the term "strategy" and "action" interchangeably to refer to some $\mathbf{u} \in \Omega$, and the game $\mathcal{G}$ can simply be specified as the pair $\mathcal{G}(\mathcal{M}, \Omega_i, J_i)$.

In the introduction we have already distinguished between cooperative and non-cooperative games. There are numerous other classifications of games, but only a few are relevant to our purposes in this monograph. We will briefly review the distinctions between these types of game and introduce these concepts for possible forms of a game as well as what we understand by various solution concepts.

2.3 Games in Extensive Form

The extensive form of a game amounts to a translation of all the rules into technical terms of a formal system designed to describe all games.

Extensive-form games generally involve several acts or stages, and each player chooses a strategy at each stage. The game's information structure, i.e., how much information is revealed to which players concerning the game's outcomes and their opponents' actions in the previous stages, significantly affects the analysis of such games. Extensive-form games are generally represented using a tree graph. Each node (called a decision node) represents every possible state of play of the game as it is played [20]. Play begins at a unique initial node, and flows through the tree along a path determined by the players until a terminal node is reached, where play ends and cost s are assigned to all players. Each non-terminal node belongs to a player; that player chooses among the possible moves at that node, each possible move is an edge leading from that node to another node. Their analysis becomes difficult with increasing numbers of players and game stages.

A formal definition is as follows.

Definition 2.1 An *m-player game $\mathcal{G}$ in an extensive form* is defined as a graph theoretic tree of vertices (states) connected by edges (decisions or choices) with certain properties:

1. $\mathcal{G}$ has a specific vertex called the *starting point* of the game,
2. a function called the *cost function* which assigns an m-vector (tuple) $J_1, \dots, J_m$ to each terminal vertex (outcome) of the game $\mathcal{G}$, where J_i denotes the cost of player i, $\mathcal{M} = \{1, \dots, m\}$,
3. each non-terminal vertex of $\mathcal{G}$ is partitioned into $m + 1$ possible sets, $\mathcal{S}_0, \mathcal{S}_1, \dots, \mathcal{S}_m$ called the player sets, where $\mathcal{S}_0$ stands for the choice of chance (nature),
4. each vertex of $\mathcal{S}_0$ has a probability distribution over the edges leading from it,
5. the vertices of each player $\mathcal{S}_i$, $i = 1, \dots, m$ are partitioned into disjoint subsets known as *information sets*, $\mathcal{S}_{i,j}$, such that two vertices in the same information set have the same number of immediate (choices/edges) followers and no vertex can follow another vertex in the same information set.

As a consequence of (5) a player knows which information set he is in but not which vertex of the information set.

A player i is said to have *perfect information* in a game $\mathcal{G}$ is each information set for this player consists of one element. The game $\mathcal{G}$ in extensive form is said to have *perfect information* if every player has perfect information. A *pure strategy* for player i denoted by u_i is defined as a function which assigns to each of player's i information sets $\mathcal{S}_{i,j}$, one of the edges leading from a representative vertex in this set $\mathcal{S}_{i,j}$. As before we denote by Ω_i the set of all pure strategies of player i, $u_i \in \Omega_i$ and by $\mathbf{u} = (u_1, \dots, u_m)$ the m-tuple of all players strategies, with $\mathbf{u} \in \Omega = \Omega_1 \times \cdots \times \Omega_m$. A game in extensive form is *finite* if it has a finite number of vertices, hence each player has only a finite number of strategies. Let us look at a couple of examples.

Example 2.2 In the game of Matching Pennies (see Fig. 2.1) player 1 chooses "heads" (H) or "tails" (T), player 2, not knowing this choice, also chooses between H or T. If the two choose alike (matching) than player 2 wins 1 cent from player 1 (hence $+1$ for player 2 and -1 for player 1); else player 1 wins 1 cent from player 2 (reverse case). The game tree is shown below with vectors at the terminal vertices indicating the cost function, while the number near vertices denote the player to

Fig. 2.1 Game of matching pennies in extensive form

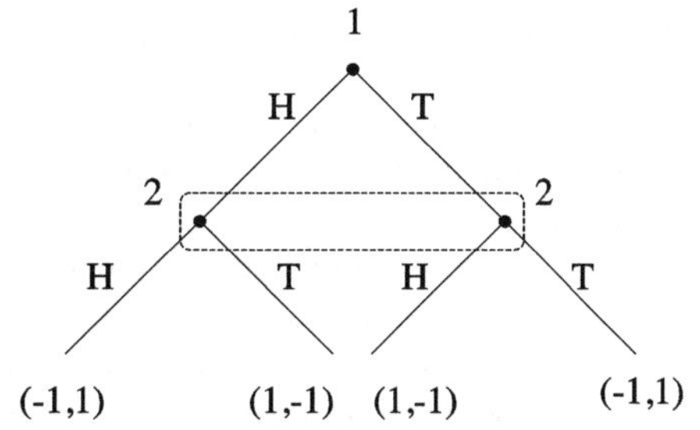

Fig. 2.2 Game of matching pennies in extensive form

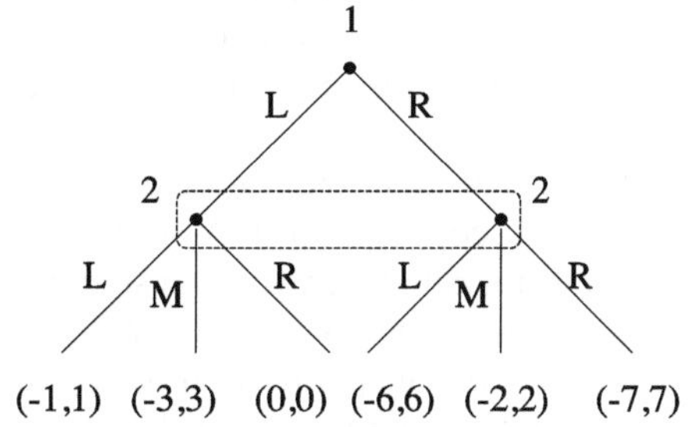

Fig. 2.3 Game of matching pennies in extensive form

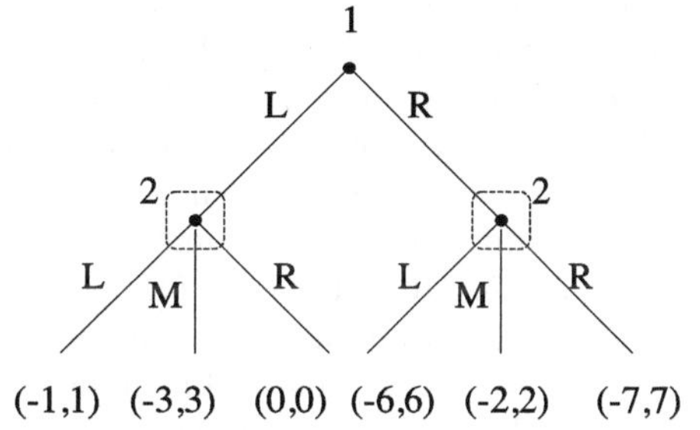

whom the move corresponds. The dotted (shaded) area indicates moves in the same information set.

The next two figures show other two zero-sum game examples which differ by the information available to player 2 at the time of its play (information set), denoted by the shaded area (dotted). In the first case, Fig. 2.2, the two possible nodes of player 2 are in the same information set, implying that even though player 1 acts before player 2 does, player 2 does not have access to it s opponent decision. This means that at the time of its play, player 2 does not know at which node (vertex) he is. This is as saying that both players act simultaneously. The extensive form in the second case Fig. 2.3, admits a different matrix game in normal form. In this case each node of player 2 is included in a separate information set, i.e., has perfect information as to which branch of the tree player 1 has chosen.

2.4 Games in Normal Form

Games in normal form (strategic form) model scenarios in which two or more players must make a one-time decision simultaneously. These games are sometimes referred to a one-shot game, simultaneous move games. The *normal form* is a more condensed form of the game, stripped of all features but the choice of each player's pure strategies, and it is more convenient to analyze. The fact that all players make their choice of strategy simultaneously has nothing to do with a temporal constraint, but rather with a constraint on the information structure particular to this type of game. The information structure of a game is a specification of how much each player knows at the time he chooses his strategy. For example, in Stackelberg games [20], where there are leaders and followers, some players (followers) choose their strategies only after the strategic choices made by the leaders have already been revealed.

In order to describe a normal-form game we need to specify players' strategy spaces and cost functions. A strategy space for a player is the set of all strategies available to that player, where a strategy is a complete plan of action for every stage of the game, regardless of whether that stage actually arises in play. A *cost function* of a player is a mapping from the cross-product of players' strategy spaces to that player's set of costs (normally the set of real numbers). We will be mostly concerned with these type of normal-form games herein.

For any strategy profile $\mathbf{u} \in \Omega$, where $\mathbf{u} = (u_1, \dots, u_m) \in \Omega$ is the m-tuple of players' pure strategies and $\Omega = \Omega_1 \times \cdots \times \Omega_m$ is the overall *pure strategy space*, let $J_i(\mathbf{u}) \in R$ denote the associated cost for player i that depends on all players' strategies.These costs depend on the context: in economics they represent a firm's profits or a consumer's (von Neumann–Morgenstern) utility, while in biology they represent the fitness (expected number of surviving offspring). All these real numbers $J_i(\mathbf{u})$, $i \in \mathcal{M}$, form the *combined pure strategy* vector cost function of the game, $\mathbf{J} : \Omega \to R_m$. A normal-form game $\mathcal{G}(\mathcal{M}, \Omega_i, J_i)$ is defined by specifying Ω_i and J_i. It is possible to tabulate functions J_i for all possible values of $u_1, \dots, u_m \in \Omega$ either in the form of a relation (easier for continuous or infinite games when Ω is a continuous set), or, as an m-dimensional array (table) in the case of finite games (when Ω is finite set). In this latter case and when $m = 2$ this reduces to a matrix whose size is given by the number of available choices for the two players and whose elements are pairs of real numbers corresponding to outcomes (costs) for the two players. Let us look at a few examples for $m = 2$, where we shall use rows for player 1 as the columns for player 2. Hence entry (j, k) indicates the outcome of player 1 using the j pure strategy and player 2 using k strategy.

Example 2.3 (*Matching Pennies*) Consider the game of Matching Pennies above, where each player has two strategies "Heads" (H) or "Tail" (T). The normal form of this game is described by the matrix

		player 2 H	T
player 1	H	$(-1, 1)$	$(1, -1)$
	T	$(1, -1)$	$(-1, 1)$

or given as the matrix

$$M = \begin{bmatrix} (-1,1) & (1,-1) \\ (1,-1) & (-1,1) \end{bmatrix}$$

Most of the times instead of M we shall use a pair of cost matrices (A, B) to indicate the outcome for each player separately, matrix A for player 1 and matrix B for player 2. For the above game this simply means the pair of matrices (A, B) where

$$A = \begin{bmatrix} -1 & 1 \\ 1 & -1 \end{bmatrix}, \qquad B = \begin{bmatrix} 1 & -1 \\ -1 & 1 \end{bmatrix}$$

One can transform any game in extensive form into an equivalent game in normal form, so we shall restrict most of our theoretical development to games in normal form only.

2.5 Game Features

Depending of various features of the game one could classify them in different categories. Below we briefly discuss such classification depending on the competitive nature of the game, the knowledge/information available to the players, and the number of times the game is repeated.

2.5.1 Strategy Space: Matrix vs. Continuous Games

In a matrix game with m players, each player i has a finite number of discrete options to choose from, i.e., there are n_i possible actions, so that the set of its actions is simply identified with a set of indices $M_i = \{1, \ldots, n_i\}$ corresponding to these possible actions. Then one considers the action $u_i \in \Omega_i$ with the action sets being defined as $\Omega_i = \{\mathbf{e}_1, \ldots, \mathbf{e}_{n_i}\}$, and $\mathbf{e}_j$ being the jth unit vector in $\mathbb{R}^{n_i}$. Given the action $\mathbf{u} \in \Omega$ chosen by all players, player i has a cost matrix $A_i \in \mathbb{R}^{n_1 \times \cdots \times n_m}$, $\forall i \in \mathcal{M}$, that defines his cost by

$$J_i(\mathbf{u}) = [A_i]_{u_1,\ldots,u_m} \in \mathbb{R}, \quad i \in \mathcal{M}$$

This is easiest seen in the case of a matrix game when we can explicitly write the cost functions as

$$J_1(\mathbf{u}) = u_1^T A u_2, \qquad J_2(\mathbf{u}) = u_1^T B u_2 \tag{2.1}$$

Such as two-player matrix game is the Matching Pennies game in Example 2.3. A symmetric game is a game where the payoffs for playing a particular strategy depend only on the other strategies employed, not on who is playing them. If the identities of the players can be changed without changing the payoff to the strategies, then a game is symmetric and this corresponds to $B = A^T$. Many of the commonly studied 2×2 games are symmetric. The standard representations of the

Chicken game, Prisoner's Dilemma game, and the Stag Hunt game are all symmetric games [20].

Unlike matrix games, where players have a finite set of actions, in a continuous game each player can choose its action from a continuum of (possibly vector-valued) alternatives, that is, $\Omega_i \subset \mathbb{R}^{n_i}$. We shall review results for both matrix games and continuous games in the next two chapters, but most of the games we shall consider afterwards are continuous games.

2.5.2 Mixed vs. Pure Strategy Games

A strategy can be regarded as a rule for choosing an action, depending on external conditions. Once such a condition is observed, the strategy is implemented as an action. In the case of mixed strategies, this external condition is the outcome of some randomization process. Consider an m-player matrix game and denote by $x_{i,j}$ the probability that player i will choose action j from n_i his available alternatives in Ω_i. Then a mixed strategy x_i is defined as the vector composed of the probabilities associated with available actions, i.e., $x_i = [x_{i,j}]$, $j = 1, \dots, n_i$, $x_i \in \Delta_i$, $i \in \mathcal{M}$ where

$$\Delta_i := \left\{ x_i \in \mathbb{R}^{n_i} \mid x_{i,j} \geq 0, \quad \sum_{j=1}^{n_i} x_{i,j} = 1 \right\}, \quad \forall i \in \mathcal{M}$$

is a simplex. In some cases actions are pure, or independent of any external conditions, and the strategy space coincides with the action space. In discussing games in pure strategies we shall use the term "strategy" and "action" interchangeably to refer to some $\mathbf{u} \in \Omega$, and the game $\mathcal{G}$ can simply be specified as the pair $\mathcal{G}(\mathcal{M}, \Omega_i, J_i)$. This will be the case considered throughout most of the monograph.

2.5.3 Competitive Versus Cooperative

A *cooperative game* is one in which there can be cooperation between the players and/or they have the same cost (also called team games). A *noncooperative game* is one where an element of competition exists and among these we can mention coordination games, constant-sum games, and games of conflicting interests. We give below a few such examples.

2.5.3.1 Coordination Games

In coordination games, what is good for one player is good for all players. An example coordination game in normal form is described by

$$M = \begin{bmatrix} (-3, -3) & (0, 0) \\ (0, 0) & (-4, -4) \end{bmatrix}$$

In this game, players try to coordinate their actions. The joint action $(j,k) = (2,2)$ is the most desirable (least cost), but the joint action $(j,k) = (1,1)$ also produces negative costs to the players. This particular game is called a pure coordination game since the players always receive the same payoff.

Other coordination games move more toward the domain of games of conflicting interest. For example, consider the Stag Hunt game: stag hare (we shall come back to this example)

$$M = \begin{bmatrix} (-4,-4) & (0,-1) \\ (-1,0) & (-1,-1) \end{bmatrix}$$

In this game, each player can choose to hunt stag (first row or first column) or hare (second row or second column). In order to catch a stag (the biggest animal, hence the bigger payoff or lowest cost of −4), both players must choose to hunt the stag. However, a hunter does not need help to catch a hare, which yields a cost of -1. Thus, in general, it is best for the hunters to coordinate their efforts to hunt stag, but there is considerable risk in doing so (if the other player decides to hunt hare). In this game, the costs (payoffs) are the same for both players when they coordinate their actions, but their costs are not equal when they do no coordinate their actions.

2.5.3.2 Constant-Sum Games

Constant-sum games are games in which the sum of the players' payoffs sum to the same number. These games are games of pure competition of the type "my gain is your loss". *Zero-sum games* are particular example of these games, which in terms of the two-players cost matrices can be described by $B = -A$. An example of such game is the Rock, Paper, and Scissors game with the matrix form

$$M = \begin{bmatrix} (0,0) & (1,-1) & (-1,1) \\ (-1,1) & (0,0) & (1,-1) \\ (1,-1) & (-1,1) & (0,0) \end{bmatrix}$$

2.5.3.3 Games of Conflicting Interests

These fall in between constant-sum games and coordination games and cover a large class, whereby the players have somewhat opposing interests, but all players can benefit from making certain compromises. One can say that people (and learning algorithms) are often tempted to play competitively in these games (both in the real world and in games), though they can often hurt themselves by doing so. However, on the other hand, taking an extreme cooperative approach (same actions) can lead to similarly bad (or worse) payoffs (high costs). One of the most celebrated games of this type is the Prisoners' Dilemma game, with the choices of to "Confess" (cooperate) or "Don't Confess" (defect) as the actions of two prisoners (players) put in separate cells. If they both confess each they each receive 3 years in prison. If only one confesses, he will be freed, used as witness, and the other will be convicted and receive 8 years in prison. If neither confesses they will be convicted of a minor offense and receive each only 1 year. The normal form (strategic form) of this game is described by the matrix

$$M = \begin{bmatrix} (-3,-3) & (0,-8) \\ (-8,0) & (-1,-1) \end{bmatrix}$$

or, as pair of cost matrices (A, B) for the two players,

$$A = \begin{bmatrix} -3 & 0 \\ -8 & -1 \end{bmatrix}, \qquad B = \begin{bmatrix} -3 & -8 \\ 0 & -1 \end{bmatrix}$$

2.5.4 Repetition

Any of the previously mentioned kinds of game can be played any number of times between the same players, and the game can be the same at each play or can be state-dependent.

2.5.4.1 One-Shot Games

In *one-shot games*, players interact for only a single round (or stage). Thus, in these situations there is no possible way for players to reciprocate (by inflicting punishment or rewards) thereafter.

2.5.4.2 Repeated Games

In *repeated games*, players interact with each other for multiple rounds (playing the same game). In such situations, players have opportunities to adapt to each others' behaviors (i.e., "learn") in order to try to become more successful. There can be finite-horizon repeated games where the same game is repeated a fixed number of times by the same players, or infinite-horizon games in which the play is repeated indefinitely.

2.5.4.3 Dynamic Games

The case where the game changes when players interact repeatedly is what can be called a *repeated dynamic game*, characterized by a state. These are also called differential games. Unlike a repeated game where the agents play the same game every time, in a dynamic game the state of the game influences the play and the outcome. Important in this class are the so called *stochastic games*, which are extensions of Markov decision processes to the scenario with m multiple players, where probabilistic transitions are modeled. We shall not cover these types of game in this monograph.

2.5.5 *Knowledge Information*

Depending on the amount of information a player has different plays and outcomes may be possible. For example, does an player know the costs (or preference orderings) of other players? Does the player know its own cost (payoff) matrix? Can he view the actions and costs of other players? All of these (and other related) questions are important as they can help determine how the player should learn and act. Theoretically, the more information an player has about the game, the better he should be able to do. In short, the information an player has about the game can vary along the following dimensions: knowledge of the player's own actions; knowledge of the player's own costs; knowledge of the existence of other players; knowledge of the other players' actions; knowledge of the other players' costs and in case learning is used, knowledge of the other players' learning algorithms.

In a game with *complete information* each player has knowledge of the payoffs and possible strategies of other players. Thus, *incomplete information* refers to situations in which the payoffs and strategies of other players are not completely known. The term *perfect information* refers to situations in which the actual actions taken by associates are fully observable. Thus, *imperfect information* implies that the exact actions taken by associates are not fully known.

2.6 Solution Concepts

Given a game's specification $\mathcal{G}(\mathcal{M}, \Omega_i, J_i)$ an important issue is to predict how the game will be played, i.e., to determine its outcome. These predictions are called solutions, and describe which strategies will be adopted by players, therefore predicting the result of the game. A solution concept briefly describes how to use a certain set of mathematical rules to decide how to play the game. Various solution concepts have been developed, in trying to indicate/predict how players will behave when they play a generic game. Herein we only introduce these solution concepts in short.

2.6.1 *Minimax Solution*

One of the most basic properties of every game is the *minimax solution* (or minimax strategy), also called security strategy. The minimax solution is the strategy that minimizes a player's maximum expected loss (cost). There is an alternate set of terminology we can use (often used in the literature as we mentioned before). Rather than speak of minimizing our maximum expected loss, we can talk of maximizing our minimum expected payoff. This is known as the maximin solution. Thus, the terms minimax and maximin can be used interchangeably. The minimax solution is an essential concept for zero-sum games.

Let us look at the Prisoner's Dilemma matrix game above. In the prisoner's dilemma, both players are faced with the choice of cooperating or defecting. If both players cooperate, they both receive a relatively low cost (which is -3 in this case). However, if one of the players cooperates and the other defects, the defector receives a very low cost (-8 in this case) (called the temptation cost), and the cooperator receives a relatively high cost (0 in this case). If both players defect, then both receive a higher cost (which is -1 in this case). So what should you do in this game? Well, there are a lot of ways to look at it, but if you want to play conservatively, you might want to invoke the minimax solution concept, which follows from the following reasoning. If you play cooperate, the worst you can do is get a cost of 0 (thus, we say that the security of cooperating is 0). Likewise, if you play defect, the worst you can do is get a cost of -1 (security of defecting is -1). Alternately, we can form a mixed strategy over the two actions. However, it just so happens in this game that no mixed strategy has higher security than defecting, so the minimax strategy in this game is to defect. This means that the minimax value (which is the maximum cost one can incur when plays the minimax strategy) is -1.

However, even though the minimax value is the lowest cost you can guarantee yourself without the cooperation of your associates, you might be able to do much better on *average* than the minimax strategy if you can either outsmart your associates or get them to cooperate or compromise with you (in a game that is not fully competitive). So we need other solution concepts as well.

2.6.2 Best Response

Another basic solution concept in multi-player games is to play the strategy that gives you the lowest cost given your opponents' strategies. That is exactly what the notion of the *best response* suggests. Suppose that you are player i, and your opponents' play $\mathbf{u}_{-i}$. Then the your best response in terms of pure strategies is u_i^* such that

$$J_i\left(\mathbf{u}_{-i}, u_i^*\right) \le J_i(\mathbf{u}_{-i}, u_i), \quad \forall u_i \in \Omega_i$$

In the case of mixed strategies, assuming your opponents' play the strategy $\mathbf{x}_{-i}$, your best response is the strategy x_i^* such that

$$J_i\left(\mathbf{x}_{-i}, x_i^*\right) \le J_i(\mathbf{x}_{-i}, x_i), \quad \forall x_i \in \Delta_i$$

where Δ_i is the probability simplex. The best-response idea has had a huge impact on learning algorithms. If you know what your other players are going to do, why not get the lowest cost (highest payoff) you can get (i.e., why not play a best response)? Taking this one step further, you might reason that if you think you know what other players are going to do, why not play a best response to that belief? While this obviously is not an unreasonable idea, it has two problems. The first problem is that your belief may be wrong, which might expose you to terrible risks. Secondly, this "best-response" approach can be quite unproductive in a repeated game when other players are also learning/adapting [48].

2.6.3 Nash Equilibrium Solution

We now introduce briefly a most celebrated solution concept for a N-player noncooperative game $\mathcal{G}$. John Nash's identification of the *Nash equilibrium* concept has had perhaps the single biggest impact on game theory. Simply put, in a Nash equilibrium, no player has an incentive to unilaterally deviate from its current strategy. Put another way, if each player plays a best response to the strategies of all other players, we have a Nash equilibrium.

We will discuss the extent to which this concept is satisfying by looking at a few examples later on.

Definition 2.4 Given a game $\mathcal{G}$ a strategy N-tuple (profile) $\mathbf{u}^* = (u_1^*, \dots, u_N^*)$ is said to be a *Nash equilibrium* (or in equilibrium) if and only if

$$J_i(u_1^*, \dots, u_N^*) \le J_i(u_1^*, \dots, u_{i-1}^*, u_i, u_{(i+1)}^*, \dots, u_N^*), \quad \forall u_i \in \Omega_i, \ \forall i \in \mathcal{M} \quad (2.2)$$

or, in compact notation,

$$J_i(\mathbf{u}_{-i}^*, u_i^*) \le J_i(\mathbf{u}_{-i}^*, u_i), \quad \forall u_i \in \Omega_i, \ \forall i \in \mathcal{M}$$

where $\mathbf{u}^* = (\mathbf{u}_{-i}^*, u_i^*)$ and $\mathbf{u}_{-i}^*$ denotes $\mathbf{u}^*$ of all strategies except the ith one.

Thus $\mathbf{u}^*$ is an equilibrium if no player has a positive incentive for *unilateral* chance of his strategy, i.e., assuming the others keep their same strategies. In particular this means that once all choices of pure strategies have been revealed no player has any cause for regret (hence the point of no regret concept). A similar definition holds for $\mathbf{x}$ mixed strategies (as seen above in the best response).

Example 2.5 Consider the game with normal form

		player 2	
		$u_{2,1}$	$u_{2,2}$
player 1	$u_{1,1}$	$(3,1)$	$(0,0)$
	$u_{1,2}$	$(0,0)$	$(1,3)$

and note that both $(u_{1,1}, u_{2,1})$ and $(u_{1,2}, u_{2,2})$ are equilibrium pairs. For matrix games we shall use the matrix notation and for the above we will say that $(3, 1)$ and $(1, 3)$ are equilibria.

If we look at another game (a coordination game),

$$M = \begin{bmatrix} (-3,-3) & (0,0) \\ (0,0) & (-1,-1) \end{bmatrix}$$

By inspection as in the above, we can conclude that both the joint actions $(j,k) = (1,1)$ and $(j,k) = (2,2)$ are Nash equilibria since in both cases, neither player can benefit by unilaterally changing its strategy. Note, however, that this illustrates that not all Nash equilibria are created equally. Some give better costs than others (and some players might have different preference orderings over Nash equilibrium).

While all the Nash equilibria we have identified so far for these two games are *pure strategy* Nash equilibrium, they need not be so. In fact, there is also a third Nash equilibrium in the above coordination game in which both players play mixed-strategies. The next chapter we shall formally review this extension.

Here are a couple more observations about the Nash equilibrium as a solution concept:

- In constant-sum games, the minimax solution is a Nash equilibrium of the game. In fact, it is the unique Nash equilibrium of constant-sum games as long is there is not more than one minimax solution (which occurs only when two strategies have the same security level).
- Since a game can have multiple Nash equilibrium, this concept does not tell us how to play a game (or how we would guess others would play the game). This poses another question: Given multiple Nash equilibria, which one should (or will) be played? This leads to considering refinements of Nash equilibria.

Strategic dominance is another solution concept that can be used in many games. Loosely, an action is strategically dominated if it never produces lower costs (higher payoffs) and (at least) sometimes gives higher costs (lower payoffs) than some other action. An action is strategically dominant if it strategically dominates all other actions. We shall formally define this later on. For example, in the Prisoner's Dilemma (PD) game, the action defect strategically dominates cooperate in the one-shot game. This concept of strategic dominance (or just dominance, as we will sometimes call it) can be used in some games (called iterative dominance solvable games) to compute a Nash equilibrium.

2.6.4 Pareto Optimality

One of the features of a Nash equilibrium (NE) is that in general it does not correspond to a socially optimal outcome. That is, for a given game it is possible for all the players to improve their costs (payoffs) by collectively agreeing to choose a strategy different from the NE. The reason for this is that a posteriori some players may choose to deviate from such a cooperatively agreed-upon strategy in order to improve their payoffs further at the group's expense. A Pareto optimal equilibrium describes a social optimum in the sense that no individual player can improve his payoff (or lower his cost) without making at least one other player worse off. Pareto optimality is not a solution concept, but it can be an important attribute in determining what solution the players should play (or learn to play). Loosely, a Pareto optimal (also called *Pareto efficient*) solution is a solution for which there exists no other solution that gives *every* player in the game a higher payoff (lower cost). A PE solution is formally defined as follows.

Definition 2.6 A solution $\mathbf{u}^*$ is *strictly Pareto dominated* if there exists a joint action $\mathbf{u} \in \Omega$ for which $J_i(\mathbf{u}) < J_i(\mathbf{u}^*)$ for all i, and *weakly Pareto dominated* if there exists a joint action $\mathbf{u} \neq \mathbf{u}^* \in \Omega$ for which $J_i(\mathbf{u}) \leq J_i(\mathbf{u}^*)$ for all i.

Definition 2.7 A solution $\mathbf{u}^*$ is *weakly Pareto efficient (PE)* if it is not strictly Pareto dominated and *strictly Pareto efficient (PE)* if it is not weakly Pareto dominated.

Often, a Nash equilibrium (NE) is not Pareto efficient (optimal). Then one speaks of a loss of efficiency, which is also referred to as the Price of Anarchy. An interesting problem is how to design games with improved Nash efficiency, and pricing or mechanism design is concerned with such issues.

In addition to these solution concepts other important ones include the Stackelberg equilibrium [20], which is relevant in games where the information structure plays an important role, and correlated equilibria [48, 98], which case is relevant in games where the randomization used to translate players' mixed strategies into actions are correlated.

2.7 The Rationality Assumption

Given the number of available solution concepts, NE refinements and the apparent arbitrariness with which they may be applied, why would one expect that in an actual noncooperative game players would choose any particular refined NE? This question turns out to be a valid objection, namely the perfect rationality of all the participating agents. In the literature rationality is often discussed, without being precisely defined. One possible formulation is as follows: a player is rational if it consistently acts to improve its payoff without the possibility of making mistakes, has full knowledge of other players' intentions and the actions available to them, and has an infinite capacity to calculate a priori all possible refinements to NE($\mathcal{G}$) in an attempt to find the "best one". If a game involves only rational agents, each of whom believe all other agents to be rational, then theoretical results offer accurate predictions of the game outcomes.

2.8 Learning in Classical Games

Yet another game classification is related to this rationality assumption or the lack of it. In this monograph we will be concerned with rational players and one-shot games. A more realistic modeling scenario involves players that are less than rational and a repeated game play. We review here very briefly the conceptual differences for completeness. The reader is referred to extensive references on this topic such as [48]. We will use the term bounded rationality to describe players that do not necessarily have access to full or accurate information about the game, and who have a limited capacity to perform calculations on the information that is available to them. Instead of immediately playing a perfect move, boundedly rational players adapt their strategy based on the outcomes of previous matches [48, 49, 137]. We can refer to this modeling scenario as a classical game with learning. All solution concepts studied in classical game theory remain important in games with learning.

The important problem to study is not only to classify games for which equilibria exist and have favorable properties such as uniqueness, but also, in conjunction, to classify the strategy update laws that yield convergence to these equilibria under repeated play. In the terminology of [158], a game with learning can be said to have an "inner game" (i.e. the underlying classical game $\mathcal{G}$) and an "outer game" (i.e. the dynamics of the strategy update laws).

One of earliest strategy update laws to be studied is fictitious play (FP) [31]. In FP, each player keeps a running average, known as an empirical frequency of his opponent's actions, and chooses his next move as a best response to this average. The term "fictitious play" comes from the consideration of playing under the unrealistic assumption that the opponent is playing a constant strategy, hence he is a fictitious player. It has been proved that under a number of different sufficient conditions FP is guaranteed to converge to one. One of the earlier cases studied was the class of finite, two-player, zero sum games, for which convergence of FP was proved in [127], followed by other results [75].

A continuous time version of fictitious play can be derived by considering the infinitesimal motions of the empirical frequencies [58]. It can be shown that if both players are updating their strategies so that these are the best response to the other one, we may write

$$\dot{u}_i = -u_i + r_i(\mathbf{u}_{-i})$$

This equation known as the best-response (BR) dynamic, and clearly displays the very convenient feature that $\dot{\mathbf{u}} = 0$ if and only if $\mathbf{u}$ has reached a fixed point of the reaction function $r(\mathbf{u})$. By the characterization of NE given above (see (4.5)), we conclude that the set of equilibria of the best-response dynamic coincides precisely with NE($\mathcal{G}$). Most of these studies are done within the framework of evolutionary game theory [132, 142], while this monograph is concerned only with the setup of classical games.

2.9 Notes

As a preface to the remaining chapters, this chapter provided a brief overview of basic game concepts.

Chapter 3
Matrix Games

Abstract This chapter considers normal-form games with finite action sets, hence matrix games. Two-player bimatrix cases are treated first, followed by m-player matrix games, both introducing pure- and mixed-strategy Nash equilibrium concepts. The concepts of dominance and best replies are reviewed, as well as Nash equilibria theorem and Nash equilibria refinements. Results are mostly adapted from (Basar and Olsder in Dynamic Noncooperative Game Theory. SIAM Series Classics in Applied Mathematics, 1999; Owen in Game Theory, Academic Press, San Diego, 1995).

3.1 Introduction

In this chapter we consider the class of m-player matrix games, where each player i has a finite number of discrete options to choose from, i.e., there are n_i possible actions, so that the set of its actions is simply identified with a set of indices $\{1, \dots, n_i\}$ corresponding to these possible actions. On the other hand, in a continuous game each player can choose its action from a continuum of (possibly vector-valued) alternatives, that is, $\Omega_i \subset \mathbf{R}^{n_i}$. This will be the focus of the next chapter.

We consider first the two-player matrix game case, or bimatrix games, and generalize then to m-players matrix games. We discuss pure and mixed- strategy game formulations, review concepts of dominance and best replies, and then prove the important Nash equilibria theorem, followed by a brief review of Nash equilibria refinements.

3.2 Bimatrix Games

Consider a two-player matrix game, where player 1 and 2 each have a finite number n_1 and n_2 of discrete options or pure strategies to choose from. Then the set of their actions Ω_1 and Ω_2 can be simply identified with the set of indices $M_1 := \{1, \dots, n_1\}$ and $M_2 := \{1, \dots, n_2\}$ corresponding to these possible actions.

Let the action of player 1 be denoted by $u_1 \in \Omega_1$; the jth action can be identified with the index $j \in M_1$. Similarly for player 2, let its action be denoted by $u_2 \in \Omega_2$;

L. Pavel, *Game Theory for Control of Optical Networks*,
Static & Dynamic Game Theory: Foundations & Applications,
DOI 10.1007/978-0-8176-8322-1_3,

the kth action can be identified with the index $k \in M_2$. Each of the two players have cost matrix A and B, respectively. Then we can write for the game outcome denoted by J_1 when strategy pair (j, k) is used

$$a_{jk} = \left(\mathbf{e}_j^1\right)^T A\mathbf{e}_k^2 := J_1\left(\mathbf{e}_j^1, \mathbf{e}_k^2\right)$$

where $\mathbf{e}_j^1 \in R^{n_1}$ and $\mathbf{e}_k^2 \in R^{n_2}$ are the jth unit vector in R^{n_1} and the kth unit vector in R^{n_2}, respectively. Thus player 1's cost when strategy pair (u^1, u^2) is set to be the (j, k) pair is

$$J_1(u_1, u_2) = J_1\left(\mathbf{e}_j^1, \mathbf{e}_k^2\right) = \left(\mathbf{e}_j^1\right)^T A\mathbf{e}_k^2$$

Then for player 1 we can write $u_1 \in \Omega_1 := \{\mathbf{e}_1^1, \dots, \mathbf{e}_j^1, \dots, \mathbf{e}_{n_1}^1\}$ and correspondingly for player 2, $u_2 \in \Omega_2 := \{\mathbf{e}_1^2, \dots, \mathbf{e}_k^2, \dots, \mathbf{e}_{n_2}^2\}$, and its cost when (u_1, u_2) is set to be the (j, k) pair is

$$J_2(u_1, u_2) = J_2\left(\mathbf{e}_j^1, \mathbf{e}_k^2\right) = \left(\mathbf{e}_j^1\right)^T B\mathbf{e}_k^2$$

Each of the players aims to minimize its own cost, J_1 and J_2, respectively, described by its own cost matrix A and B, respectively. We call this a (A, B) bimatrix game. In the case when $B = -A$ the game is called a zero-sum game.

The definition of a Nash equilibrium in pure strategies for a bimatrix game is given next.

Definition 3.1 (Pure-strategy Nash equilibrium (NE)) For a given (A, B) bimatrix game let (j^*, k^*) be pair of pure strategies chosen by the two players. Then if for any other pure strategies (j, j) both

$$a_{j^*k^*} \le a_{jk^*}, \quad \forall j = 1, \dots, n_1$$
$$b_{j^*k^*} \le b_{j^*k}, \quad \forall k = 1, \dots, n_2$$

hold, then (j^*, k^*) is a *pure-strategy Nash equilibrium* (NE) solution for the bimatrix game, and $(a_{j^*k^*}, b_{j^*k^*})$ is a NE equilibrium outcome of the game.

In the case when more than a single NE solution exist, we shall denote this set by $NE(\mathcal{G})$. The game outcome can be different for different NE solutions and the question of ordering the elements in $NE(\mathcal{G})$ arises. Since we cannot have total ordering between pairs of numbers one can resort to partial ordering in order to arrive at some preferential choice.

Definition 3.2 A pair of pure strategies (j^*, k^*) is said to be *better* than another pair $(\tilde{j}, \tilde{k})$ if

$$a_{j^*k^*} \le a_{\tilde{j}\tilde{k}}, \quad \text{and} \quad b_{j^*k^*} \le b_{\tilde{j}\tilde{k}}$$

and if at least one of these inequalities is strict.

It can be shown that the set of NE equilibrium points is invariant to positive affine transformations on the cost functions.

Definition 3.3 Two bimatrix games (A, B) and $(\tilde{A}, \tilde{B})$ are said to be *strategically equivalent games* if there exist positive constants $\alpha_1, \alpha_2 > 0$ and scalar β_1, β_2 such that

$$\tilde{a}_{jk^*} = \alpha_1 a_{jk^*} + \beta_1, \quad \alpha_1 > 0, \ \beta_1 \in \mathbf{R}, \ \forall j$$

$$\tilde{b}_{j^*k} = \alpha_2 b_{j^*k} + \beta_2, \quad \alpha_2 > 0, \ \beta_2 \in \mathbf{R}, \ \forall k$$

(operation on columns in A and on rows in B).

Proposition 3.4 *All strategically equivalent bimatrix games have the same NE equilibria.*

We consider below a couple of examples and their pure NE equilibria.

Example 3.5 (Prisoner's Dilemma) Recall the Prisoner's Dilemma (PD) game reviewed in Chap. 2, where cost matrices (A, B) for the two players are

$$A = \begin{bmatrix} -3 & 0 \\ -8 & -1 \end{bmatrix}, \qquad B = \begin{bmatrix} -3 & -8 \\ 0 & -1 \end{bmatrix}$$

for the choices of to "Confess" (first strategy) or to "Not Confess" (second strategy) available for the two prisoners. This is a game in which there are gains from cooperation between players; the best outcome is for the players to not confess (second strategy), but each player has an incentive to be a "free rider". Whatever one player does, the other prefers "Confess" so that game has a unique pure Nash equilibrium (Confess, Confess) or pair $(1, 1)$. Indeed player 1's first pure strategy ("Confess") gives a smaller loss (higher payoff) than its second pure strategy, irrespective of what strategy player 2 uses. Similarly, player 2's first pure strategy (defect) gives always a smaller loss (higher payoff) than its second pure strategy: each entry in first column of B matrix is less than the corresponding entry in the second column. Hence individual rationality of minimizing its cost or loss leads to each player to select the first strategy, hence both "Confess". The dilemma arises because both players would have lower loss if they were to select their second strategy ("Not Confess"), but this would require some coordination.

Unfortunately not every game has a pure equilibrium. For example the game of Matching Pennies in Example 2.3 in Chap. 2 does not have a (pure) Nash equilibrium. This leads to the mixed-strategy extension of a game.

3.2.1 Mixed Strategies

A player is said to use a *mixed strategy* whenever he/she chooses to randomize over the set of available pure actions. Formally, a mixed strategy is a probability distribution that assigns to each available action a likelihood of being selected. If only one action has a positive probability of being selected, the player is said to use a pure strategy. Thus when an equilibrium cannot be found in pure strategies the space of strategies can be enlarged, and the players are allowed to base their decisions on random events, hence mixed strategies.

In the Matching Pennies game if both players choose each of their pure strategies with probability $\frac{1}{2}$ then each will have an expected cost (loss) of 0 which seems an acceptable solution. This is the case when the game is played repeatedly and each player aims to minimize the expected (average) outcome of individual plays.

Because randomization is more complex and cognitively demanding than is the deterministic selection of a single action, this raises the question of how mixed strategies should be interpreted. An interpretation is given in [57] by John Harsanyi who showed that a mixed-strategy equilibrium of a game with perfect information can be viewed as the limit point of a sequence of pure-strategy equilibria of games with imperfect information. Specifically, starting from a game with perfect information, one can obtain a family of games with imperfect information by allowing for the possibility that there are small random variations in payoffs and that each player is not fully informed of the payoff functions of the other players. Harsanyi showed that the frequency with which the various pure strategies are chosen in these perturbed games approaches the frequency with which they are chosen in the mixed-strategy equilibrium of the original game as the magnitude of the perturbation becomes vanishingly small. The introduction of mixed strategies was successful as a new solution concept because von Neuman [159] was able to show (even from 1928) that for any matrix game the minimax value is equal to the maximin value in mixed strategies, hence any such game has a solution in mixed strategies (von Neumann's Minimax Theorem). Nash extended this to any m-player finite matrix game.

Now for a two-player game $N = 2$, with pure strategies $\mathbf{u} = (u_1, u_2)$ we denote by x and y the mixed strategy x_1 of player 1 and x_2 player 2, respectively, and $\mathbf{x} = (x_1, x_2) = (x, y)$. For player 1 a *mixed strategy* denoted x is a probability distribution defined over the set $\Omega_1 := \{\mathbf{e}_1^1, \dots, \mathbf{e}_j^1, \dots, \mathbf{e}_{n_1}^1\}$ of pure strategies, $x : \Omega_1 \to [0, 1]$. Let x_j denote the probability that player 1 will choose action j from his n_1 available (pure) alternatives in Ω_1, i.e.,

$$\Pr[\chi = \mathbf{e}_j^1] = x(\mathbf{e}_j^1) := x_j$$

where χ denotes random variable of choosing one of the pure strategies in set Ω_1. Since Ω_1 is finite this probability distribution is a vector of probabilities associated with the pure action. A mixed strategy is identified as the vector composed of the probabilities associated with available actions, i.e., $x = [x_j]$, $j \in \{1, \dots, n_1\}$,

$x \in \Delta_1$, where

$$\Delta_1 := \left\{ x \in \mathbb{R}^{n_1} \;\middle|\; \sum_{j=1}^{n_1} x_j = 1,\ x_j \geq 0,\ \forall j = 1, \ldots, n_1 \right\} \tag{3.1}$$

is the unit simplex of $(n_1 - 1)$ dimension. Note that pure strategies are just extreme case of mixed strategies (vertices of the simplex Δ_1). Similarly, for player 2 we shall identify his mixed strategy with $y = [y_k]$, $k \in \{1, \ldots, n\}$, $y \in \Delta_2$, where y_k denotes the probability that player 2 will choose action k from his n_2 available (pure) alternatives in Ω_2 and $y \in \Delta_2$,

$$\Delta_2 := \left\{ y \in \mathbb{R}^{n_2} \;\middle|\; \sum_{k=1}^{n_2} y_k = 1,\ y_k \geq 0,\ \forall k = 1, \ldots, n_2 \right\}$$

Assuming the two players' strategies to be jointly independent, the probability of selecting pure-strategy pair $(u_1, u_2) = (\mathbf{e}_j^1, \mathbf{e}_k^2) \in \Omega_1 \times \Omega_2$ is given by $x(u_1) \cdot y(u_2) = x(\mathbf{e}_j^1) \cdot y(\mathbf{e}_k^2) = x_j \cdot y_k$. In terms of the cost for player 1, the *expected* (averaged) cost when a (x, y) mixed-strategy pair is used is given as

$$\begin{aligned} E\big[J_1(\mathbf{u})\big] = E\big[J_1(u_1, u_2)\big] &= \sum_{u_1 \in \Omega_1} \sum_{u_2 \in \Omega_2} J_1(u_1, u_2) x(u_1) y(u_2) \\ &= \sum_{j=1}^{n_1} \sum_{k=1}^{n_2} J_1\big(\mathbf{e}_j^1, \mathbf{e}_k^2\big) x_j y_k \end{aligned} \tag{3.2}$$

and we denote this by $\overline{J}_1(x, y)$ or $\overline{J}_1(\mathbf{x})$, as the *mixed-strategy cost function*, when $x \in \Delta_1$, $y \in \Delta_2$. Using $J_1(\mathbf{e}_j^1, \mathbf{e}_k^2) = a_{jk}$ into (3.2) this gives

$$\overline{J}_1(\mathbf{x}) = \overline{J}_1(x, y) = \sum_{j=1}^{n_1} \sum_{k=1}^{n_2} a_{jk} x_j y_k = x^T A y$$

Similarly for player 2

$$\overline{J}_2(\mathbf{x}) = \overline{J}_2(x, y) = \sum_{j=1}^{n_1} \sum_{k=1}^{n_2} b_{jk} x_j y_k = x^T B y$$

where x, y are n_1, n_2-dimensional vectors.

Remark 3.6 In the case when $B = -A$, the game is a *two-player zero-sum matrix game*

$$\overline{J}_1(x, y) + \overline{J}_2(y, x) = 0, \quad \forall x,\ \forall y$$

So in this game each player minimizes its cost $\overline{J}_1$ and $\overline{J}_2$, but with respect to $\overline{J}$ player 1 is the minimizer, while player 2 is regarded as the maximizer of $\overline{J}$.

Definition 3.7 (Mixed-strategy Nash equilibrium (NE)) For a given (A, B) bimatrix game let $\mathbf{x}^* = (x^*, y^*)$ be pair of mixed strategies chosen by the two players. Then if for any other mixed strategies (x, y) both

$$x^{*T} A y^* \le x^T A y^*, \quad \forall x \in \Delta_1$$

$$x^{*T} B y^* \le x^{*T} B y, \quad \forall y \in \Delta_2$$

hold, then $\mathbf{x}^* = (x^*, y^*) \in \Delta_X$ is a Nash equilibrium in *in mixed strategies* for the bimatrix game.

Computational methods for finding a mixed NE solution strategies for bimatrix games include the Lemke–Howson algorithm or are based on conversion to a nonlinear programming problem [20].

3.3 m-Player Games

The game formulation can easily be extended from two-player (bimatrix) games to m-player finite (matrix) games. Let the set of players be denoted by $\mathcal{M} = \{1, \dots, m\}$, and assume that $u_i \in \Omega_i$ where Ω_i is the finite action set.

3.3.1 Pure and Mixed Strategies

Let the strategy of player i, $i \in \mathcal{M} = \{1, \dots, m\}$ be denoted by u_i (pure strategy) or by x_i (mixed strategy). In a matrix case, each player i has a finite number n_i of actions at its disposal. Player i, $i \in \mathcal{M}$, uses *pure strategy* u_i by simply choosing an action j from his n_i possible ones in $M_i = \{1, \dots, n_i\}$, or by u_i taking a value $\mathbf{e}^i_j$ from his n_i available alternatives in $\Omega_i = \{\mathbf{e}^i_1, \dots, \mathbf{e}^i_j, \dots, \mathbf{e}^i_{n_i}\}$. Let $\mathbf{u} = (u_1, \dots, u_m) \in \Omega$ denote the pure-strategy (action) profile used by all m players, where $\Omega = \Omega_1 \times \dots \times \Omega_m$. The cost function is denoted by $J_i = J_i(u_1, \dots, u_m)$ corresponding to a cost matrix A_i.

In the case of mixed strategies, the actual action selected is the outcome of some randomization process. A *mixed strategy* for player i, $i \in \mathcal{M}$, denoted by x_i is a probability distribution on the set Ω_i of its pure u_i strategies, hence $x_i(u_i)$. In this matrix game case let $x_{i,j}$ denote the probability that player i will choose action j, $u_i = \mathbf{e}^i_j$, from his n_i available (pure) alternatives in Ω_i. Then a *mixed strategy* x_i is defined as the vector composed of the probabilities associated with available actions, i.e., $x_i(u_i) = [x_{i,j}(\mathbf{e}^i_j)]$, $j \in M_i$, $i \in \mathcal{M}$. We sometimes drop the argument so that $x_i = [x_{i,j}]$, $j \in M_i$, where $x_i \in \Delta_i$, and

$$\Delta_i := \left\{ x_i \in \mathbf{R}^{n_i} \,\middle|\, \sum_{j=1}^{n_i} x_{i,j} = 1,\ x_{i,j} \ge 0,\ \forall j \in M_i \right\} \tag{3.3}$$

denotes the unit simplex in R^{n_i} space, which has dimension $n_i - 1$. The vertices of Δ_i are the unit vectors $\mathbf{e}^i_j$. Every mixed strategy $x_i \in \Delta_i$ is some convex combination of the unit vectors (pure strategies) $\mathbf{e}^i_j \in \Omega_i$,

$$x_i = \sum_{j=1}^{n_i} x_{i,j} \mathbf{e}^i_j$$

and the mixed-strategy simplex Δ_i is the *convex hull* of its vertices.

Definition 3.8 For a mixed strategy $x_i \in \Delta_i$, we define its *support* or *carrier* as the set of *pure* strategies that have assigned positive probabilities,

$$\text{supp}(x_i) = \{j \in M_i \mid x_{i,j} > 0\}$$

The subset

$$\text{int}(\Delta_i) = \{x_i \in \Delta_i \mid x_{i,j} > 0, \ \forall j \in M_i\}$$

is called the *interior* of Δ_i. Hence, mixed strategies in the interior $x_i \in \text{int}(\Delta_i)$ are called *completely mixed* or *interior*, in that they assign positive probabilities to all player's pure strategies, hence have full support, i.e., $\text{supp}(x_i) = M_i$.

3.3.2 Mixed-Strategy Cost Functions

A game could be denoted by $\mathcal{G}(\mathcal{M}, \Omega_i, J_i)$ in its pure-strategy representation or when we refer to its mixed-strategy extension $\mathcal{G}(\mathcal{M}, \Delta_i, \overline{J}_i)$. The overall strategy space is $\Omega = \bigtimes_{i \in \mathcal{M}} \Omega_i$ indicating a pure-strategy game, or $\Delta_X = \bigtimes_{i \in \mathcal{M}} \Delta_i$, indicating a mixed-strategy game.

Since the game is non-cooperative, the mixed strategies viewed as probability distributions are assumed jointly independent, that is the probability of arriving at a pure-strategy profile (m-tuple) $\mathbf{u} = (u_1, \dots, u_m) \in \Omega$, denoted by $\mathbf{x}(\mathbf{u})$ is given as

$$\mathbf{x}(\mathbf{u}) = \prod_{i=1}^{m} x_i(u_i)$$

where $\mathbf{x}$ is the m-tuple $\mathbf{x} = (x_1, \dots, x_m) \in \Delta_X$. Thus the probability of using pure-strategy profile (m-tuple) $(k_1, \dots, k_m) \in M_1 \times M_m$, or $\mathbf{u} = (\mathbf{e}^1_{k_1}, \dots, \mathbf{e}^m_{k_m})$, is

$$\mathbf{x}(\mathbf{e}^1_{k_1}, \dots, \mathbf{e}^m_{k_m}) = \prod_{i=1}^{m} x_i(\mathbf{e}^i_{k_i}) = \prod_{i=1}^{m} x_{i,k_i}, \quad k_i \in M_i$$

In terms of pure strategies the cost for player i when the pure-strategy m-tuple $\mathbf{u}$ is used is denoted by $J_i(\mathbf{u}) \in R$. If mixed strategies x_i, $i \in \mathcal{M}$ are used according to the distribution $\mathbf{x}(\mathbf{u})$ then the cost will be the statistical expectation of J_i with respect to the probability distribution $\mathbf{x}(\mathbf{u})$, namely

$$E\big[J_i(\mathbf{u})\big] = \sum_{\mathbf{u}\in\Omega} J_i(\mathbf{u})\mathbf{x}(\mathbf{u}) := \overline{J}_i(\mathbf{x})$$

Expanding Ω, $\mathbf{u}$, and $\mathbf{x}(\mathbf{u})$ this is written as

$$\overline{J}_i(\mathbf{x}) = \overline{J}_i(x_1,\dots,x_m) = \sum_{u_1\in\Omega_1}\cdots\sum_{u_m\in\Omega_m} J_i(u_1,\dots,u_m)\prod_{l=1}^{m} x_l(u_l) \tag{3.4}$$

We often use the following notation: a strategy profile $\mathbf{x} = (x_1,\dots,x_i,\dots,x_m) \in \Delta_X$ is written as $\mathbf{x} = (\mathbf{x}_{-i}, x_i)$, where $x_i \in \Delta_i$ and $\mathbf{x}_{-i} = (x_1,\dots,x_{i-1},x_{i+1},\dots,x_m)$ is the $(m-1)$-tuple obtained from $\mathbf{x}$ without the ith player component $i \in \mathcal{M}$, $\mathbf{x}_{-i} \in \Delta_{-i}$ where $\Delta_{-i} = \times_{l\neq i} \Delta_l$. We denote by $(\mathbf{x}_{-i}, u_i)$ the strategy profile where player i has replaced $x_i \in \Delta_i$ by its pure strategy $u_i \in \Omega_i$, while all other players use strategies according to $\mathbf{x} \in \Delta_X$, i.e.,

$$(\mathbf{x}_{-i}, u_i) = (x_1,\dots,x_{i-1},u_i,x_{i+1},\dots,x_m)$$

Using this notation we can, based on (3.4), write

$$\overline{J}_i(\mathbf{x}_{-i}, u_i) = \sum_{u_1\in\Omega_1}\cdots\sum_{u_{i-1}\in\Omega_{i-1}}\sum_{u_{i+1}\in\Omega_{i+1}}\cdots\sum_{u_m\in\Omega_m} J_i(\mathbf{u})\prod_{l=1,l\neq i}^{m} x_l(u_l) \tag{3.5}$$

We also write $(\mathbf{x}_{-i}, z_i) \in \Delta_X$ for the strategy profile in which player i plays mixed strategy $z_i \in \Delta_i$ while all other players use strategies according to $\mathbf{x} \in \Delta_X$. This notation is particularly useful when a single player considers "deviations" $z_i \in \Delta_i$ from a given profile $\mathbf{x} \in \Delta_X$.

The definition of an NE equilibrium point in mixed strategies (MSNE) is given next.

Definition 3.9 (Mixed-strategy NE) Given a non-cooperative m-player finite game $\mathcal{G}$ a mixed-strategy m-tuple $\mathbf{x}^* = (x_1^*,\dots,x_m^*) \in \Delta_X$, $x_i^* \in \Delta_i$, $i \in \mathcal{M}$ is an *equilibrium point* (or a *mixed-strategy Nash equilibrium point* if

$$\overline{J}_i\big(x_1^*,\dots,x_m^*\big) \le \overline{J}_i\big(x_1^*,\dots,x_{i-1}^*, w_i, x_{(i+1)*},\dots,x_m^*\big), \quad \forall w_i \in \Delta_i,\ \forall i \in \mathcal{M}$$

or, in compact notation,

$$\overline{J}_i\big(\mathbf{x}^*\big) \le \overline{J}_i\big(\mathbf{x}_{-i}^*, w_i\big), \quad \forall w_i \in \Delta_i,\ \forall i \in \mathcal{M} \tag{3.6}$$

where $\mathbf{x}^* = (\mathbf{x}_{-i}^*, x_i^*)$ and $\mathbf{x}_{-i}^*$ denotes $\mathbf{x}^*$ of all mixed strategies except the ith one, $i \in \mathcal{M}$.

Since there can be more than a single NE we denote this set as $NE(\mathcal{G})$, and $\mathbf{x}^*$ is an element of this set, $\mathbf{x}^* \in NE(\mathcal{G})$. Then, based on Definition 3.9,

$$NE(\mathcal{G}) = \left\{\mathbf{x}^* \in \Delta_X \mid \overline{J}_i(\mathbf{x}^*) \le \overline{J}_i(\mathbf{x}^*_{-i}, w_i),\ \forall i \in \mathcal{M},\ \forall w_i \in \Delta_i\right\} \tag{3.7}$$

Now, based on the above identification of Ω_1 with $\{\mathbf{e}^1_1, \dots, \mathbf{e}^1_{k_1}, \dots, \mathbf{e}^1_{M_1}\}$, we can replace $u_1 \in \Omega_1$ by $\mathbf{e}^1_{k_1}$ for $k_1 \in M_1$ and we can write

$$\begin{aligned}
\overline{J}_i(x^*_1, \dots, x^*_m) &= \sum_{u_1 \in \Omega_1} \cdots \sum_{u_m \in \Omega_m} J_i(u_1, \dots, u_m) \prod_{l=1}^{m} x^*_{l,k_l} \\
&= \sum_{k_1 \in M_1} \cdots \sum_{k_m \in M_m} J_i(\mathbf{e}^1_{k_1}, \dots, \mathbf{e}^m_{k_m}) \prod_{l=1}^{m} x^*_{l,k_l}
\end{aligned}$$

with the second form explicitly showing the finite matrix form. Using this form and identifying $J_i(\mathbf{e}^1_{k_1}, \dots, \mathbf{e}^m_{k_m}) = A^i_{k_1,\dots,k_m}$ (element in the cost matrix of player i), we see that (3.6) can be re-written as a set of m inequalities, and the ith one, $i \in \mathcal{M}$, $\forall w_i = [x_{i,j}] \in \Delta_i$, $j \in M_i$

$$\sum_{k_1 \in M_1} \cdots \sum_{k_m \in M_m} A^i_{k_1,\dots,k_m} \prod_{l=1}^{m} x^*_{l,k_l} \le \sum_{k_1 \in M_1} \cdots \sum_{k_m \in M_m} A^i_{k_1,\dots,k_m} \prod_{l=1, l \ne i}^{m} x^*_{l,k_l} x_{i,j} \tag{3.8}$$

We prefer (3.6) as the more compact notation.

3.4 Dominance and Best Replies

One could define partial ordering of a player's (pure or mixed) strategy set as given by the consequences on the outcome for that player. We shall define everything herein in terms of mixed strategies since pure strategies are special cases of these, so we work on the mixed-strategy simplex Δ_i of each player i. A strategy *weakly* dominates another strategy if it never results in a worse outcome (higher loss) than the second and sometimes results in lower loss. A strategy is *undominated* if there is no strategy that weakly dominates it. A strategy *strictly* dominates another strategy if it always results in lower loss.

Definition 3.10 A strategy $z_i \in \Delta_i$ *weakly dominates* strategy $w_i \in \Delta_i$ if $\forall \mathbf{x}_{-i} \in \Delta_{-i}$ the following holds: $\overline{J}_i(\mathbf{x}_{-i}, z_i) \le \overline{J}_i(\mathbf{x}_{-i}, w_i)$, with strict inequality for some $\mathbf{x}_{-i}$. A strategy $w_i \in \Delta_i$ is *undominated* if no such strategy w_i exists.

Definition 3.11 A strategy $z_i \in \Delta_i$ *strictly dominates* strategy $w_i \in \Delta_i$ if $\forall \mathbf{x}_{-i} \in \Delta_{-i}$ the following holds: $\overline{J}_i(\mathbf{x}_{-i}, z_i) < \overline{J}_i(\mathbf{x}_{-i}, w_i)$.

One of the basic assumptions in non-cooperative game theory is that players are rational and do not ever use strictly dominated strategies. This is the reason why *pure* dominated pure strategies can be deleted without affecting the game outcome.

3.4.1 Best-Response Correspondence

We review next the concept of best-reply or best-response (BR) correspondence in mixed strategies. Consider a game $\mathcal{G}(I, \Delta_i, \overline{J}_i)$. Given a mixed-strategy profile $\mathbf{x}_{-i} \in \Delta_{-i}$, a **mixed**-*strategy best-reply (-response) correspondence* for player i is a mixed strategy $z_i \in \Delta_i$ such that no other mixed strategy gives him lower cost (loss) against $\mathbf{x}_{-i}$. We denote this set by $\Phi_i : \Delta_{-i} \to \Delta_i$,

$$\Phi_i(\mathbf{x}_{-i}) := \{z_i \in \Delta_i \mid \overline{J}_i(\mathbf{x}_{-i}, z_i) \leq \overline{J}_i(\mathbf{x}_{-i}, w_i),\ \forall w_i \in \Delta_i\} \tag{3.9}$$

for any given $\mathbf{x}_{-i} \in \Delta_{-i}$, also called the *optimal response set (or rational reaction set)* [20]. The overall *mixed-strategy best response* of the m-players to strategy profile $\mathbf{x} \in \Delta_X$ is $\Phi : \Delta_X \to \Delta_X$

$$\Phi(\mathbf{x}) = (\Phi_1(\mathbf{x}_{-1}), \ldots, \Phi_i(\mathbf{x}_{-i}), \ldots, \Phi_m(\mathbf{x}_{-m})) \subset \Delta_X$$

which can be multi-valued, hence a correspondence. Now since every mixed strategy is a convex combination of pure strategies and $\overline{J}_i(\mathbf{x}_{-i}, z_i)$ is linear in z_i, it follows that no mixed strategy $z_i \in \Delta_i$ can give a lower cost against $\mathbf{x}_{-i}$ than one of the pure best replies to $\mathbf{x}_{-i}$ of player i. The best-response set $\Phi_i(\mathbf{x}_{-i})$ is a face of Δ_i, i.e., a convex hull of some pure strategies (vertices) in Δ_i) and is always non-empty, closed, and convex.

3.5 Nash Equilibria Theorem

In this section we give a proof for existence of a *mixed-strategy Nash equilibria* (NE) for any m-player finite matrix game [95]. Consider an m-player finite game $\mathcal{G}(\mathcal{M}, \Delta_i, \overline{J}_i)$, and let $NE(\mathcal{G}) \subset \Delta_X$ denote the set of its NE points, as in (3.7), i.e.,

$$NE(\mathcal{G}) = \{\mathbf{x}^* \in \Delta_X \mid \overline{J}_i(\mathbf{x}^*_{-i}, x_i^*) \leq \overline{J}_i(\mathbf{x}^*_{-i}, w_i),\ \forall w_i \in \Delta_i,\ \forall i \in \mathcal{M}\}$$

Based on the best-reply correspondence Φ, (3.9) for any mixed-strategy Nash profile (equilibrium point) $\mathbf{x}^* = (\mathbf{x}^*_{-i}, x_i^*) \in NE(\mathcal{G})$, we can write

$$x_i^* \in \Phi_i(\mathbf{x}^*_{-i}), \quad \forall i \in \mathcal{M}$$

In fact, $\mathbf{x}^* \in \Delta_X$ is an NE solution if x_i^* is a solution to its optimization problem, given all other players taking equilibrium actions, $\mathbf{x}^*_{-i}$, i.e., x_i^* is player i's best response to all his opponents' actions, and the same is true for all the players $i \in \mathcal{M}$. Hence

$$NE(\mathcal{G}) = \{\mathbf{x}^* \in \Delta_X \mid x_i^* \in \Phi_i(\mathbf{x}_{-i}^*),\ \forall i \in \mathcal{M}\} = \{\mathbf{x}^* \in \Delta_X \mid \mathbf{x}^* \in \Phi(\mathbf{x}^*)\} \quad (3.10)$$

which says that in terms of best replies, a mixed-strategy profile $\mathbf{x}^* \in \Delta_X$ is an NE if it is a best reply to itself, or if it is a *fixed point* of the mixed-strategy best-reply correspondence Φ. Note that this is not the fixed point of a function but of a multi-valued (set-valued) mapping. This last interpretation is what leads to one of the proofs for existence of an NE in any m-player finite game, namely the proof based on Kakutani fixed-point theorem (and best-response mapping even if not continuous) given later on.

Definition 3.12 An NE $\mathbf{x}^* \in \Delta_X$ is called a *strict NE* if each player's component x_i^* is the *unique* best reply to $\mathbf{x}^*$ (singleton) hence if

$$\Phi(\mathbf{x}^*) = \{\mathbf{x}^*\}$$

This means that while an NE requires that no unilateral deviation be profitable (lower cost), a strict NE requires that *all* such deviations are costly (higher cost). Thus a strict NE cannot involve any randomization at all, since the there would exist two pure strategies that give the same lowest cost to some player (over which he can randomize). As a consequence every strict NE is a pure-strategy profile, i.e., a vertex in Δ_X.

Another property is that an NE strategy x_i^* cannot be *strictly* dominated, but can be *weakly dominated.* As seen from Definition 3.10, there can be another best reply z_i to the NE profile $\mathbf{x}^*$ that is as good as x_i^* against all strategy profiles and that does better (has lower cost) against some profile.

We now give a proof of Nash equilibrium existence based on Brouwer's fixed-point Theorem A.18 (see Appendix A). The proof uses the following preliminary results.

Lemma 3.13 *Let* $\mathbf{x}^* = (\mathbf{x}_{-i}^*, x_i^*) \in \Delta_X$. *Then* $\mathbf{x}^* \in NE(\mathcal{G})$, *i.e., it is a mixed-strategy Nash equilibrium (NE) if and only if for every player* i,

$$\overline{J}_i(\mathbf{x}_{-i}^*, x_i^*) \le \overline{J}_i(\mathbf{x}_{-i}^*, \mathbf{e}_j^i), \quad \forall j \in M_i,\ \forall i \in \mathcal{M} \quad (3.11)$$

Proof (*Necessity*) If $\mathbf{x}^*$ is an mixed NE point, (3.6) holds for any w_i. Since any pure strategy is a particular case of a mixed strategy, we can take $w_i = \mathbf{e}_j^i$ which shows that (3.11) holds.

(*Sufficiency*) Choose an arbitrary mixed strategy $w_i \in \Delta_i$, hence

$$w_i = \sum_{j \in M_i} \mathbf{e}_j^i \alpha_j^i, \quad \text{for some } \alpha_j^i \ge 0,\ \sum_{j \in M_i} \alpha_j^i = 1$$

Multiply (3.11) by α^i_j, followed by summing over $j \in M_i$, so that

$$\sum_{j\in M_i} \alpha^i_j \overline{J}_i(\mathbf{x}^*_{-i}, x^*_i) \le \sum_{j\in M_i} \overline{J}_i(\mathbf{x}^*_{-i}, \mathbf{e}^i_j)\alpha^i_j$$

This yields

$$\overline{J}_i(\mathbf{x}^*_{-i}, x^*_i) \sum_{j\in M_i} \alpha^i_j \le \overline{J}_i(\mathbf{x}^*_{-i}, w_i)$$

Since $\sum_{j\in M_i} \alpha^i_j = 1$, by (3.9) this implies that

$$x^*_i \in \Phi_i(\mathbf{x}^*_{-i})$$

and this holds for $\forall i \in \mathcal{M}$ hence by (3.10) $\mathbf{x}^*$ is an NE. □

Lemma 3.14 *Consider any mixed strategy* $\mathbf{x} = (\mathbf{x}_{-i}, x_i) \in \Delta_X$, $x_i \in \Delta_i$. *For every player* $i \in \mathcal{M}$ *there exists a* $k \in M_i$ *such that*

$$x_i(\mathbf{e}^i_k) = x_{i,k} > 0 \quad \textit{and} \quad \overline{J}_i(\mathbf{x}_{-i}, x_i) \le \overline{J}_i(\mathbf{x}_{-i}, \mathbf{e}^i_k)$$

Proof We prove the statement by contradiction. Consider any $x_i \in \Delta_i$, $x_i = \sum_{j\in M_i} x_{i,j}\mathbf{e}^i_j = \sum_{k\in \mathrm{supp}(x_i)} x_{i,k}\mathbf{e}^i_k$ as a convex combination of pure strategies, where $\sum_{k\in \mathrm{supp}(x_i)} x_{i,k} = 1$, and $\mathrm{supp}(x_i) \neq 0$ since for at least one $\mathbf{e}^i_k$ we have $x_{i,k} > 0$. Assume by contradiction that for all pure strategies $k \in \mathrm{supp}(x_i)$, i.e., with $x_{i,k} > 0$, we have $\overline{J}_i(\mathbf{x}_{-i}, x_i) > \overline{J}_i(\mathbf{x}_{-i}, \mathbf{e}^i_k)$. Hence for all such strategies,

$$\overline{J}_i(\mathbf{x})x_{i,k} > \overline{J}_i(\mathbf{x}_{-i}, \mathbf{e}^i_k)x_{i,k}$$

Summing the foregoing over k yields

$$\sum_{k\in \mathrm{supp}(x_i)} \overline{J}_i(\mathbf{x})x_{i,k} > \sum_{k\in \mathrm{supp}(x_i)} \overline{J}_i(\mathbf{x}_{-i}, \mathbf{e}^i_k)x_{i,k}$$

Then factoring $\overline{J}_i(\mathbf{x})$ on the LHS and using the representation of x_i gives

$$\overline{J}_i(\mathbf{x}) > \overline{J}_i(\mathbf{x}_{-i}, x_i) = \overline{J}_i(\mathbf{x})$$

hence a contradiction. □

Theorem 3.15 (Existence of mixed-strategy NE [95]) *Every finite (static) m-player non-cooperative game* $\mathcal{G}$ *has at least one mixed-strategy equilibrium point, called* mixed-strategy Nash equilibrium (NE).

Proof If player i has n_i pure strategies, then the set of his mixed strategies can be represented geometrically by an $(n_i - 1)$-dimensional simplex, Δ_i. Each Δ_i is a convex, closed, and bounded (hence compact) subset of the R^{n_i} Euclidean space.

Hence the overall $\Delta = \Delta_1 \times \cdots \times \Delta_m$ is a convex, closed, and bounded (hence compact) subset of the R^n, where $n = \sum_{i=1}^m n_i$ Euclidean space, and any mixed-strategy m-tuple $\mathbf{x} = (x_1, \ldots, x_m)$ can be viewed as a point $\mathbf{x} \in \Delta_X$. For any such $\mathbf{x} \in \Delta_X$ and any *pure* strategy $u_i = \mathbf{e}_j^i \in \Omega_i$ for player i we define

$$C_j^i(\mathbf{x}) = \max\{\overline{J}_i(\mathbf{x}) - \overline{J}_i(\mathbf{x}_{-i}, \mathbf{e}_j^i), 0\}$$

and we see that C_j^i is a function that measures the decrease in the cost for player i when in the mixed-strategy m-tuple (profile) $\mathbf{x}$, player i replaces strategy x_i by its pure strategy $u_i = \mathbf{e}_j^i$. We also see that $C_j^i(\mathbf{x}) \geq 0$. Any possible increase in cost is not indicated since in this case the function vanishes.

Now for each $i \in \mathcal{M}$, $j \in M_i$ we consider the transformation

$$\eta_j^i := \frac{x_i(\mathbf{e}_j^i) + C_j^i(\mathbf{x})}{1 + \sum_{j \in M_i} C_j^i(\mathbf{x})}$$

where $x_i(\mathbf{e}_j^i) = x_{i,j}$. Each of these is non-negative also. Note that, since $x_i \in \Delta_i$,

$$\sum_{j \in M_i} x_i(\mathbf{e}_j^i) = \sum_{j \in M_i} x_{i,j} = 1$$

and summing over all $j \in M_i$ (or $u_i \in \Omega_i$) we have

$$\sum_{j \in M_i} \frac{x_i(\mathbf{e}_j^i) + C_j^i(\mathbf{x})}{1 + \sum_{j \in M_i} C_j^i(\mathbf{x})} = 1$$

Hence $\sum_{j \in M_i} \eta_j^i = 1$ and for fixed $\mathbf{x}$ and i this η_j^i can be taken a probability attached to the strategy $u_{i,j}$, and $\eta^i \in \Delta_i$ itself.

Moreover, since $\overline{J}_i(\mathbf{x})$ is polynomial in $x_{i,j}$, $C_j^i(\mathbf{x})$ is a continuous function of $\mathbf{x}$. Since $1 + \sum_{j \in M_i} C_j^i(\mathbf{x}) \neq 0$, it follows that η^i is itself a continuous function of $\mathbf{x}$. Thus gathering all in a m-tuple (vector), we see that $\eta = [\eta^1, \ldots, \eta^m]^T$ is a continuous vector-valued function of $\mathbf{x}$ that we denote by $\eta = T(\mathbf{x})$, that maps the convex closed bounded set Δ_X into itself. Then by Brouwer's fixed-point Theorem A.18 (see Appendix A) it follows that there exists a fixed point for T, hence

$$T(\tilde{\mathbf{x}}) = \tilde{\mathbf{x}}$$

The theorem is proved if we can show that every such fixed point is necessarily a Nash equilibrium and vice versa that every Nash equilibrium is a fixed point of this mapping. Since $\tilde{\mathbf{x}} = (\tilde{x}^1, \ldots, \tilde{x}^m)$ is a fixed point for T it follows that for all $i \in \mathcal{M}$ and for all $l \in M_i$

$$\tilde{x}^i(\mathbf{e}_l^i) := \frac{\tilde{x}^i(\mathbf{e}_l^i) + C_l^i(\tilde{\mathbf{x}})}{1 + \sum_{j \in M_i} C_j^i(\tilde{\mathbf{x}})} \tag{3.12}$$

For this $\tilde{\mathbf{x}} \in \Delta_X$, by Lemma 3.14 and the definition of C^i_j it follows that for any player i there exist a k and a *pure* strategy $\mathbf{e}^i_k$ such that $\tilde{x}^i(\mathbf{e}^i_k) = \tilde{x}_{i,k} > 0$

$$\tilde{x}^i(\mathbf{e}^i_k) > 0 \quad \text{and} \quad \overline{J}_i(\tilde{\mathbf{x}}) - \overline{J}_i(\tilde{\mathbf{x}}_{-i}, \mathbf{e}^i_k) \le 0$$

hence

$$\tilde{x}^i(\mathbf{e}^i_k) > 0 \quad \text{and} \quad C^i_k(\tilde{\mathbf{x}}) = 0$$

For this particular strategy (3.12) becomes

$$\tilde{x}^i(\mathbf{e}^i_k) := \frac{\tilde{x}^i(\mathbf{e}^i_k)}{1 + \sum_{j \in M_i} C^i_j(\tilde{\mathbf{x}})}$$

so that

$$\tilde{x}^i(\mathbf{e}^i_k) \sum_{j \in M_i} C^i_j(\tilde{\mathbf{x}}) = 0$$

which since $\tilde{x}^i(\mathbf{e}^i_k) > 0$ implies $\sum_{j \in M_i} C^i_j(\tilde{\mathbf{x}}) = 0$. Since all $C^i_j \ge 0$ the foregoing implies

$$C^i_j(\tilde{\mathbf{x}}) = 0, \quad \forall i \in \mathcal{M}, \ \forall j \in M_i$$

and from the definition of C^i_j this means that for each player i we have

$$\overline{J}_i(\tilde{\mathbf{x}}) \le \overline{J}_i(\tilde{\mathbf{x}}_{-i}, \mathbf{e}^i_j), \quad \forall i \in \mathcal{M}, \ \forall j \in M_i$$

By Lemma 3.14, $\tilde{x}$ is a NE equilibrium point and proof is complete. □

An alternative proof based on Kakutani's fixed-point theorem is the variant that John Nash used in his description of Nash equilibrium for m-player games, work that would later earn him a Nobel Prize in Economics, i.e., that every finite game with mixed strategies for any number of players has a Nash equilibrium. Kakutani's fixed-point theorem is a generalization of Brouwer's fixed-point theorem for *set-valued maps or correspondences* that requires it to be *hemi-continuous*, or that it has a *closed graph*. Unlike the previous proof where we constructed a single-valued function C^i, here we can simply use the best-response map (correspondence) Φ_i and Φ. We give this proof for completeness.

Proof (based on Kakutani's fixed-point theorem) Recall that $\Phi_i(\mathbf{x}_{-i})$ is the set of tuples of mixed strategies chosen by each player in a game, and $\Phi(\mathbf{x})$ is the overall best response. Since there may be a number of responses which are equally good, $\Phi_i(\mathbf{x}_{-i})$ is set-valued rather than single-valued. Then the Nash equilibrium of the game is defined as a fixed point of Φ.

Note that Δ_X is a non-empty, compact, and convex set. The best-reply correspondence (mapping) $\Phi : \Delta_X \to \Delta_X$ is upper hemi-continuous by the continuity of the payoff function $\overline{J}_i$, which in itself follows form Berge's maximum Theorem A.17

(see Appendix A). For every $\mathbf{x} \in \Delta_X$, the image $\Phi(x) \subset \Delta_X$ is a non-empty, closed, and convex set. Hence, by Kakutani's fixed-point Theorem A.19, Appendix A, Φ has at least one fixed point in Δ_X, i.e., there exists $\mathbf{x}^* \in \Delta_X$ such that $\mathbf{x}^* \in \Phi(\mathbf{x}^*)$. □

3.6 Nash Equilibria Refinements

In this section we review some NE refinements. In order to go beyond certain weakness of the Nash equilibrium when several of these exist, a large range of refinements have been proposed since about the late 1970. We shall review only a few of these for normal-form game only.

Let $NE(\mathcal{G}) \subset \Delta_X$ denote the set of NE points of a m-player finite game $\mathcal{G}(I, \Delta_i, \overline{J}_i)$.

3.6.1 *Perfect Equilibrium*

This refinement is due to Selten [136], best known as the "*trembling hand perfection*". This intuitively means that NEs that are not robust to "trembles" in players' strategies are discarded.

Consider a game $\mathcal{G}(\mathcal{M}, \Delta_i, \overline{J}_i)$ and let μ be an *error function* such that a number $\mu_{i,k} \in (0, 1)$ is assigned for each player $i \in \mathcal{M}$, and pure strategy $k \in M_i$ which defines the probability that this k pure strategy will be played by mistake (trembling hand), with $\sum k \in M_i \mu_{i,k} < 1$. Note that such a small probability > 0 is assigned to *all* pure strategies.

This means that for each player $i \in \mathcal{M}$ the error function μ defines a subset of mixed strategies

$$\Delta_i(\mu) := \{x_i \in \Delta_i \mid x_{i,k} \geq \mu_{i,k}\} \subset \text{int}(\Delta_i)$$

that player i can use. Define a *perturbed game* $\mathcal{G}(\mu) = \mathcal{G}(\mathcal{M}, \Delta_i(\mu), \overline{J}_i)$, where $\Delta_X(\mu) = \times_{i \in \mathcal{M}} \Delta_i(\mu) \subset \text{int}(\Delta_X)$. By a similar argument based on Kakutani's frixed-point theorem, every such perturbed game has an non-empty set $NE(\mathcal{G}(\mu))$ of NEs. The smaller all error probabilities are, the larger $\Delta_X(\mu)$ is and as $\mu \to 0$ the perturbed game approaches the original game.

Definition 3.16 An NE, $\mathbf{x}^* \in NE(\mathcal{G})$ is a *perfect equilibrium* if, for some sequence $\{\mathcal{G}(\mu_t)\}_{\mu_t \to 0}$ of perturbed games, there exist profiles $\mathbf{x}_t \in NE(\mathcal{G}(\mu_t))$ such that $\mathbf{x}_t \to \mathbf{x}^*$.

Let us call $PE(\mathcal{G})$ the set of perfect equilibria. Note that every interior (completely mixed) or inner NE is perfect. This is seen because if $\mathbf{x}^* \in \text{int}(\Delta_X)$, then for sufficiently small $\mu_{i,k}$, $\mathbf{x}^* \in \text{int}(\Delta_X(\mu))$. If in addition $\mathbf{x}^* \in NE(\mathcal{G})$ then

$\mathbf{x}^* \in NE(\mathcal{G}(\mu))$. The definition above requires only that the NE be robust with respect to some trembles (error function), so existence can be established even if there are no interior NE.

Proposition 3.17 *For any finite game, $PE(\mathcal{G}) \neq \varnothing$.*

Proof For any sequence $\{\mathcal{G}(\mu_t)\}_{\mu_t \to 0}$ let $\mathbf{x}_t \in NE(\mathcal{G}(\mu_t))$ for each t. Since $\{\mathbf{x}_t\}$, $t = 1, \dots, \infty$ is a sequence in a compact set Δ_X it has a convergent subsequence, $\{\mathbf{x}_{t_s}\}$, $s = 1, \dots, \infty$ with limit $\mathbf{x}^* \in \Delta_X$. For each s, $\mathcal{G}(\mu_{t_s})$ is the associated perturbed game. Then since $\mathbf{x}_{t_s} \in NE(\mathcal{G}(\mu_{t_s}))$ it follows that $\mathbf{x}^* \in NE(\mathcal{G})$ by continuity arguments, and $\mathbf{x}^*$ is perfect since $\mathbf{x}_{t_s} \to \mathbf{x}^*$ and $\mathbf{x}_{t_s} \in NE(\mathcal{G}(\mu_{t_s}))$ for all s. □

We give without proof the following useful result (for a proof see [157]).

Proposition 3.18 *Every perfect equilibrium $\mathbf{x} \in PE(\mathcal{G})$ is undominated, i.e., every player's strategy component x_i is undominated, $i \in \mathcal{M}$. In a two-player game $\mathcal{M} = \{1, 2\}$ if an NE $\mathbf{x} \in NE(\mathcal{G})$ is undominated, then it is perfect, i.e., $\mathbf{x} \in PE(\mathcal{G})$.*

As a most stringent condition on the "trembles", the concept of strictly perfect equilibrium requires robustness with respect to *any* low probability tremble.

Definition 3.19 An NE, $\mathbf{x}^* \in NE(\mathcal{G})$ is a *strictly perfect equilibrium* if, for every sequence $\{\mathcal{G}(\mu_t)\}_{\mu_t \to 0}$ of perturbed games, there exist profiles $\mathbf{x}_t \in NE(\mathcal{G}(\mu_t))$ such that $\mathbf{x}_t \to \mathbf{x}^*$.

Note that any interior (inner) NE equilibrium $\mathbf{x} \in NE(\mathcal{G})$ is strictly proper, since one could take $\mathbf{x}_t = \mathbf{x}^*$ for all t sufficiently large such that $\mathbf{x}^* \in NE(\mathcal{G}(\mu_t))$. In fact a perfect NE equilibrium that is not strictly perfect is *vulnerable* against some sequence of trembles.

3.6.2 Proper Equilibrium

An intermediary refinement is one introduced by Myerson [93] which imposes some conditions on the "trembles" to which the NE should be robust (not only to some unqualified trembles). Specifically, an NE equilibrium should be robust with respect to those trembles that are such that more costly "errors" are less probable. This is like requesting the perturbed ("trembled") strategy to be "proper" is some sense, formalized as follows. Given some $\epsilon > 0$, a strategy profile $\mathbf{x} \in \text{int}(\Delta_X)$ is called ϵ-proper if

$$\overline{J}_i(\mathbf{x}_{-i}, \mathbf{e}_k^i) > \overline{J}_i(\mathbf{x}_{-i}, \mathbf{e}_j^i) \quad \Longrightarrow \quad x_{i,k} \le \epsilon x_{i,j}$$

3.6.3 Strategically Stable Equilibrium

The concept of a strictly perfect NE can be extended to robustness of NE sets, called *strategically stable* sets.

Definition 3.20 A set of NEs, $X^* \subset NE(\mathcal{G})$ is *strategically stable* if it is the smallest non-empty, closed set such that for every $\epsilon > 0$ there exists some $\delta > 0$ such that every strategy perturbed game $\mathcal{G}(\mu) = (\mathcal{M}, \Delta_X(\mu), \overline{J}_i)$ with errors $\mu_{i,k} < \delta$ has some NE within distance ϵ from the set X^*.

3.7 Notes

In this chapter we reviewed basic concepts and results for matrix games, i.e., games with finite action sets. In the next chapter we consider games with infinite (continuous) action sets.

Chapter 4
Games with Continuous Action Spaces

Abstract This chapter focuses on Nash games with continuous kernel, i.e., games with continuous action spaces and cost functions. Basic concepts and results are reviewed focused on game formulations, Nash equilibria, reaction curves, and existence results. These are mostly adapted from (Basar and Olsder in Dynamic Noncooperative Game Theory. SIAM Series Classics in Applied Mathematics, 2nd edn., 1999). For more in-depth treatment the reader is referred to this and other more extensive references on the subject.

4.1 Introduction

In this chapter we consider games with infinite action spaces (hence continuous) and with players' payoff (cost) functions as continuous in the actions. We mostly treat pure strategies (actions). We introduce basic NE concepts and existence results for the case in which no coupling exists between those action spaces, results mostly adapted from [20, 98]. For more in-depth treatment the reader is referred to those and other more extensive references on the subject. A review of the associated optimization results is presented in one of the appendices.

The chapter is organized as follows. We present basic concepts and game formulation for games with continuous action spaces and uncoupled constraints, and then we present fundamental existence and uniqueness results for the Nash equilibrium.

4.2 Game Formulations

A game consists of a set of players, an action set action set (also referred to as a set of strategies) available to those players and an individual objective function for each player. In a game, each player individually takes an optimal action which optimizes its own objective function and each player's success in making decisions depends on the decisions of the others.

Let us consider games with continuous action spaces and *pure* strategies. Specifically, as the focus of this chapter we consider games with continuous action spaces and uncoupled constraints. The following chapter will treat such games, but with

L. Pavel, *Game Theory for Control of Optical Networks*,
Static & Dynamic Game Theory: Foundations & Applications,
DOI 10.1007/978-0-8176-8322-1_4,

coupled constraints. Consider $\Omega_i = [m_i, M_i]$, where $m_i > 0$ and $M_i > m_i$ are scalars. Each player $i \in \mathcal{M}$ has an *action variable* $u_i \in \Omega_i$ and such a game is called a game with orthogonal (uncoupled constraints), as before the overall *action space* Ω being defined by the Cartesian product of Ω_i

$$\Omega = \Omega_1 \times \cdots \times \Omega_m \tag{4.1}$$

We also let $\Omega_{-i} := \Omega_1 \times \cdots \times \Omega_{i-1} \times \Omega_{i+1} \times \cdots \times \Omega_m$. It can be seen that Ω is compact, convex, and has a nonempty interior set. Moreover, the action space Ω is separable by construction since there is no coupling among Ω_i. Any two players i and j can take their actions independently from separate action sets Ω_i and Ω_j, $j \neq i$, respectively. An *action vector* $\mathbf{u} \in \Omega$ can be written as $\mathbf{u} = [u_1, \ldots, u_m]^T$, or $\mathbf{u} = (\mathbf{u}_{-i}, u_i)$ with the vector $\mathbf{u}_{-i} \in \Omega_{-i}$ obtained by deleting the ith element from $\mathbf{u}$, i.e., $\mathbf{u}_{-i} = [u_1, \ldots, u_{i-1}, u_{i+1}, \ldots, u_m]^T$.

Given a game's specification $\mathcal{G}(I, \Omega_i, J_i)$ an important issue is to predict how the game will be played, i.e., to determine its outcome. These predictions are called solutions, and describe which strategies will be adopted by players, therefore predicting the result of the game. The outcome of a game is referred to as its equilibrium, or solution, and it is given by a point $\mathbf{u}^* \in \Omega$. This solution corresponds to a set of strategies (actions) that the players will adopt according to their knowledge of the game, and of other players' actions and intentions. We shall discuss next the Nash equilibrium [96] in the context of such games. Recall that in brief Nash equilibrium (NE) is a solution concept of a game in which each player is assumed to know the equilibrium strategies of the other players, and no player has anything to gain by changing its own strategy unilaterally.

Before we discuss the NE concept we shall briefly remark on the mixed-strategy extension for game with continuous action spaces.

4.3 Extension to Mixed Strategies

Similarly to the concept of a mixed strategy for matrix games, this concept can be defined for continuous (or infinite) games by employing the appropriate distribution function. We only consider for simplicity the $N = 2$ case. Let the two players' pure strategies be denoted $u_1 \in \Omega_1$, $u_2 \in \Omega_2$, where this time Ω_i are infinite set such as for example the interval [0, 1].

A *mixed strategy* for player 1 is a *cumulative distribution function* σ_1 defined on Ω_1: for every $u_1 \in \Omega_1$, $\sigma_1(u_1)$ is a random process for choosing a number not greater than u_1. That is,

$$\sigma_1(u_1) = \Pr(\xi \leq u_1)$$

where *Pr* denote the probability and ξ the random variable of the process. (i) $\sigma_1(u_1) \geq 0$, $\forall u_1 \in \Omega_1$, (ii) $\sigma_1(0) = 0$, $\sigma(1) = 0$, (iii) $\sigma_1(u_1)$ is non-decreasing, (iv) $\sigma_1(u_1)$ is right-continuous in the open $(0, 1)$. Similarly $\sigma_2(u_2)$ is the mixed strategy of player 2 having similar properties. If player 1 chooses *pure* strategy u_1 and player 2 chooses mixed strategy $\sigma_2(u_2)$ the expected cost for player 1 is

$$\overline{J}_1(u_1, \sigma_2) = E\big(J_1(u_1, \sigma_2)\big) = \int_{\Omega_2} J_1(u_1, u_2)\, d\sigma_2(u_2)$$

where the integral is Lebesgue–Stieljes integral. Similarly if player 2 chooses *pure* strategy u_2 and player 1 chooses mixed strategy $\sigma_1(u_1)$ the expected cost for player 1 is

$$\overline{J}_1(\sigma_1, u_2) = E\big(J_1(\sigma_1, u_2)\big) = \int_{\Omega_1} J_1(u_1, u_2)\, d\sigma_1(u_1)$$

Now if player 1 and 2 use mixed strategies $\sigma_1(u_1)$ and $\sigma_2(u_2)$, then joint probability distributions have to be considered, and assuming them independent, the expected cost for player 1 is

$$\overline{J}_1(\sigma_1, \sigma_2) = \int_{\Omega_1} \int_{\Omega_2} J(u_1, u_2)\, d\sigma_1(u_1)\, d\sigma_2(u_2)$$

and similarly the expected cost for player 2. Then definition for a Nash equilibrium can be given in the mixed-strategy case using the set of distribution functions in terms of these expected costs.

4.4 Nash Equilibria and Reaction Curves

Return to the basic case of games with continuous action spaces in the pure-strategy case, $\mathcal{G}(I, \Omega_i, J_i)$. Let us define an NE solution for this case. In such a game an *individual cost function* $J_i : \Omega \to \mathbb{R}$ is assigned to each player i, $J_i(\mathbf{u})$, and each player aims to minimize its own cost function $J_i(u^i, \mathbf{u}_{-i})$, in the presence of all other players.

Then a pure-strategy Nash equilibrium (NE), or simply a *Nash equilibrium (NE)* of this noncooperative game is the defined as a point $\mathbf{u}^* \in \Omega$ that satisfies

$$J_i\big(\mathbf{u}^*\big) \leq J_i\big(\mathbf{u}^*_{-i}, u_i\big), \quad \forall u_i \in \Omega_i, \ \forall i \in \mathcal{M}$$

or, alternatively given the actions of all other players, $\mathbf{u}_{-i} \in \Omega_{-i}$, each player i independently minimizes its own cost:

$$\begin{array}{ll} \min & J_i(\mathbf{u}_{-i}, u_i) \\ \text{subject to} & u_i \in \Omega_i \end{array} \tag{4.2}$$

The formal definition of a Nash equilibrium (NE) is given next.

Definition 4.1 (Nash equilibrium (NE)) A vector $\mathbf{u}^* \in \Omega$ is called a *Nash equilibrium (NE) solution* of $\mathcal{G}(\mathcal{M}, \Omega_i, J_i)$ if for all $i \in \mathcal{M}$ and for every $\mathbf{u}^*_{-i} \in \Omega_{-i}$,

$$J_i(\mathbf{u}^*_{-i}, u^*_i) \leq J_i(\mathbf{u}^*_{-i}, u_i), \quad \forall u_i \in \Omega_i \tag{4.3}$$

If in addition $\mathbf{u}^*$ is not on the boundary of the action space Ω, it is called an *inner NE solution.*

As we remarked before, at an NE no player can benefit by altering its action unilaterally. In other words, $\mathbf{u}^*$ is an NE if given the action $\mathbf{u}^*_{-i}$ taken by his opponents, no player has an incentive to choose a different action on his own, i.e. an NE is a so-called no-regret strategy.

For a given $\mathbf{u}_{-i} \in \Omega_{-i}$, player i takes an action $\xi \in \Omega_i$ such that

$$J_i(\mathbf{u}_{-i}, \xi) \leq J_i(\mathbf{u}_{-i}, u_i), \quad \forall u_i \in \Omega_i$$

Then for a game $\mathcal{G}(\mathcal{M}, \Omega_i, J_i)$, we may define a best response correspondence

$$R_i(\mathbf{u}_{-i}) = \{\xi \in \Omega_i \mid J_i(\mathbf{u}_{-i}, \xi) \leq J_i(\mathbf{u}_{-i}, u_i), \ \forall u_i \in \Omega_i\} \tag{4.4}$$

which is called the *optimal reaction set* of player $i \in \mathcal{M}$, i.e., the set of all optimal responses by player i to any fixed $\mathbf{u}_{-i}$. The concepts of best response correspondence and the NE concept, defined here for pure-strategy game $\mathcal{G}(\mathcal{M}, \Omega_i, J_i)$, correspond to those reviewed in Chap. 3 for a mixed-strategy matrix game $\mathcal{G}(\mathcal{M}, \Delta_i, \overline{J}_i)$.

Note that $\mathbf{u}^* \in \Omega$ is an NE solution if u^*_i is a solution to the optimization problem (4.2), given all other players taking equilibrium actions, $\mathbf{u}^*_{-i}$, i.e., u^*_i is player i's best response to all his opponents' actions, and the same is true for all the players $i \in \mathcal{M}$. Based on this we define next the reaction function of a player.

Definition 4.2 (Reaction function) If the optimal reaction set $R_i(\mathbf{u}_{-i})$ (4.4) is a singleton (i.e., a set with exactly one element) for every given $\mathbf{u}_{-i} \in \Omega_{-i}$, then $R_i(\mathbf{u}_{-i})$ is called the *reaction function* of player i, and is specifically denoted by $r_i : \Omega_{-i} \to \Omega_i$.

We present next properties of reaction functions of $\mathcal{G}(\mathcal{M}, \Omega_i, J_i)$ under a typical assumption.

Assumption 4.3 *The cost function $J_i(\mathbf{u}_{-i}, u_i)$ is twice continuously differentiable in all its arguments and strictly convex in u_i for every $\mathbf{u}_{-i} \in \Omega_{-i}$.*

Proposition 4.4 *Consider $\mathcal{G}(\mathcal{M}, \Omega_i, J_i)$ under Assumption* 4.3. *For each $i \in \mathcal{M}$, a unique continuous reaction function $r_i : \Omega_{-i} \to \Omega_i$ exists.*

Proof Under Assumption 4.3, the cost function $J_i(\mathbf{u}_{-i}, u_i)$ is strictly convex in u_i. By Proposition A.2, there exists a unique minimum u_i, for any given $\mathbf{u}_{-i}$, such that

$$J_i(\mathbf{u}_{-i}, u_i) < J_i(\mathbf{u}_{-i}, \xi), \quad \forall \xi \in \Omega_i, \ \xi \neq u_i$$

By Definition 4.2, this implies that a unique mapping, i.e., a reaction function $r_i : \Omega_{-i} \rightarrow \Omega_i$ exists for each player i. The continuity can be obtained directly from Berge's Maximum Theorem (Theorem A.17). □

4.5 Existence and Uniqueness Results

An important question is whether a game admits a (possible unique) Nash equilibrium (NE), and conversely to characterize classes of games for which there exists a unique NE. When equilibria do exist, they may not be unique and in such cases the NE is said to be weak as a solution concept.

From the definition of an NE (Definition 4.1) and the definition of the best response (reaction function), (4.4), it follows that for each player an NE (if it exists) must lie on his own reaction curve, i.e. $\mathbf{u}_i^* = r_i(\mathbf{u}_{-i}^*)$. Therefore in graphical terms, the existence of a Nash equilibrium (NE) is equivalent to requiring that a plot of the graph of each $r_i(\mathbf{u}_{-i}) \in \Omega$ results in at least one common intersection point. Let us denote the set of all NE by $NE(\mathcal{G})$. Then the set of all such intersections in Ω forms $NE(\mathcal{G})$. We are thus led to another important characterization of $NE(\mathcal{G})$ as the set of fixed points of the vector function $r : R^m \rightarrow R^m$, i.e.

$$\mathbf{u}^* = r(\mathbf{u}^*) \tag{4.5}$$

where $r(\mathbf{u}) = [r_1(u_{-1}), \dots, r_i(u_{-i}), \dots r_m(u_{-m})]^T$.

Intuitively, this implies that an NE is a strategy profile that constitutes the best reply to itself. When $\mathbf{u}^*$ is the unique best reply to itself we say that $\mathbf{u}^*$ is a *strict NE*. In that case it will be true that

$$J_i(\mathbf{u}^*) < J_i(\mathbf{u}_{-i}^*, u_i), \quad \forall u_i \in \Omega_i, \ \forall i \in \mathcal{M}$$

and the set $NE(\mathcal{G})$ is a singleton.

Provided that reaction functions are well defined and a common intersection point of the reaction functions exists, the existence of an NE solution can be established. This reasoning leads to the following proposition (adapted from Theorem 4.3 in [20]).

Proposition 4.5 *Under Assumption* 4.3, $\mathcal{G}(\mathcal{M}, \Omega_i, J_i)$ *admits an NE solution.*

Proof By strict convexity of the cost function and Proposition 4.4, for every given $\mathbf{u}_{-i} \in \Omega_{-i}$, there exists a unique reaction function r_i of player $i \in \mathcal{M}$ such that

$$u_i = r_i(\mathbf{u}_{-i}) \in \Omega_i, \quad \forall i \in \mathcal{M}.$$

The corresponding vector form is $\mathbf{u} = r(\mathbf{u}) \in \Omega$ with $r(\mathbf{u}) = [r_1(u_{-1}), \dots, r_m(u_{-m})]^T$. Furthermore, by Proposition 4.4, r_i is continuous, and so is $r(\mathbf{u})$. On the other hand, $r(\mathbf{u})$ maps a compact and convex set Ω into itself. From (Theorem A.18), there exists a $\mathbf{u}^*$ such that $u^* = r(\mathbf{u}^*)$. By Definition 4.1, it follows that $\mathbf{u}^*$ is an NE solution of $\mathcal{G}(\mathcal{M}, \Omega_i, J_i)$. □

As to the property of an NE solution of being inner, the following result is immediate from Definition 4.1 and Proposition A.1.

Proposition 4.6 *Under Assumption* 4.3, *if a vector* $\mathbf{u}^*$ *is an inner NE solution of* $\mathcal{G}(\mathcal{M}, \Omega_i, J_i)$, *then* $\mathbf{u} = \mathbf{u}^*$ *satisfies the following set of necessary conditions*:

$$\frac{\partial J_i}{\partial u_i}(\mathbf{u}_{-i}, u_i) = 0, \quad \forall i \in \mathcal{M} \tag{4.6}$$

Now let us define an associated optimization problem from the cost functions J_i, i.e.,

$$\begin{aligned} \min \quad & \sum_{j=1}^{m} J_j(\mathbf{u}) \\ \text{subject to} \quad & \mathbf{u} \in \Omega \end{aligned}$$

Then in general an NE solution is not an optimal solution to this associated optimization problem. It is well-known that in most cases Nash equilibria do not optimize the system performance with the Prisoner's Dilemma [98] being the best-known example. Let us take a look at a two-player Nash game for illustration.

Example 4.7 Consider a two-player Nash game with the cost functions

$$J_1(u_2, u_1) = 2u_1^2 - 2u_1 - u_1 u_2$$

$$J_2(u_1, u_2) = u_2^2 - \frac{1}{2}u_2 - u_1 u_2$$

and where $u_i \in \Omega_i = [0, 8]$, $i = 1, 2$.

The constraint set $\Omega = \Omega_1 \times \Omega_2$ is convex and compact and has a nonempty interior set. From the cost functions, it follows that $\frac{\partial^2 J_1}{\partial u_1^2} = 4$ and $\frac{\partial^2 J_2}{\partial u_2^2} = 2$. Thus Assumption 4.3 is satisfied. From Proposition 4.4, the reaction functions r_i, $i = 1, 2$, exist and are continuous. By Definition 4.2, the reaction function $r_1(u_2)$ can be obtained by optimizing $J_1(u_2, u_1)$ with respect to u_1 for every given u_2. It follows that $r_1(u_2) = \frac{1}{4}(u_2 + 2)$. Similarly, the reaction function $r_2(u_1)$ is obtained as $r_2(u_1) = \frac{1}{2}(u_1 + \frac{1}{2})$. The reaction curves are shown in Fig. 4.1. By Definition 4.1, an NE solution lies on both reaction curves. Therefore the intersection point of $r_1(u_2)$ and $r_2(u_1)$, $(u_1^*, u_2^*) = (\frac{9}{14}, \frac{4}{7})$, is an NE solution. The corresponding optimal cost values are $J_1^* = -\frac{81}{98}$ and $J_2^* = -\frac{16}{49}$. Figure 4.1 indicates that this game admits a unique NE solution.

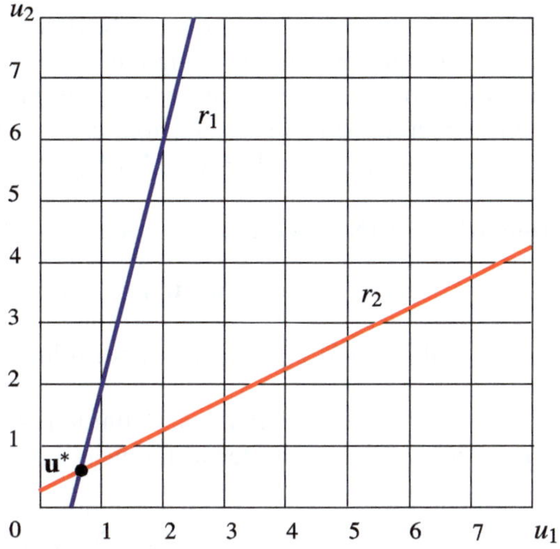

Fig. 4.1 Reaction curves, $r_1(u_2)$ and $r_2(u_1)$

Next consider the associated optimization problem

$$\begin{array}{ll} \min & J_0(u_1, u_2) \\ \text{subject to} & 0 \leq u_i \leq 8, \quad i = 1, 2, \end{array}$$

where $J_0(u_1, u_2) = J_1(u_2, u_1) + J_2(u_1, u_2) = 2u_1^2 + u_2^2 - 2u_1 - \frac{1}{2}u_2 - 2u_1u_2$. The constraint set Ω is still convex and compact. The Hessian of J_0 is positive definite for every $\mathbf{u} \in \Omega$ with

$$\nabla^2 J_0 = \begin{bmatrix} 4 & -2 \\ -2 & 2 \end{bmatrix}$$

such that the cost function J_0 is strictly convex over Ω. From Proposition A.2 the associated optimization problem admits a unique global minimum. By using Proposition A.3, the optimal solution is $\mathbf{u}^{\text{opt}} = (\frac{5}{4}, \frac{3}{2})$ and the optimal value is $J_0^{\text{opt}} = -\frac{13}{8}$. It is clear that $u^* \neq \mathbf{u}^{\text{opt}}$. The following is a further discussion. In the game framework, given $\mathbf{u}^{\text{opt}}$, the cost values are $J_1(\mathbf{u}^{\text{opt}}) = -\frac{5}{4}$, which is less than $J_1(\mathbf{u}^*)$ and $J_2(\mathbf{u}^{\text{opt}}) = -\frac{3}{8}$ which is less than $J_2(\mathbf{u}^*)$. It follows that in the presence of selfishness, the NE solution is not necessarily the optimal solution from a social point of view.

From the individual cost functions $J_i(\mathbf{u}_{-i}, u_i)$ let us define now an augmented *system-like* two-argument cost function $\tilde{J} : \Omega \times \Omega \to \mathbb{R}$ as in [20]

$$\tilde{J}(\mathbf{u}; \mathbf{x}) := \sum_{i=1}^{m} J_i(\mathbf{u}_{-i}, x_i), \tag{4.7}$$

where each $J_i(\mathbf{u}_{-i}, u_i)$ satisfies Assumption 4.3. We shall call this two-argument function $\tilde{J}$ the *Nash game (NG) cost function* of the game. Based on it another useful characterization of an NE can be derived as follows. We provide an alternate proof of Proposition 4.5 by using the NG cost function. In addition we show how necessary conditions for an inner NE solution of $\mathcal{G}(\mathcal{M}, \Omega_i, J_i)$ can be reformulated with respect to the NG cost function. As we will show in the next chapter this NG cost function can be used to solve Nash games with coupled constraints.

Definition 4.8 (NE-NG sense) A vector $\mathbf{u}^* \in \Omega$ satisfying

$$\tilde{J}(\mathbf{u}^*; \mathbf{u}^*) \leq \tilde{J}(\mathbf{u}^*; \mathbf{x}), \quad \forall \mathbf{x} \in \Omega \tag{4.8}$$

with $\tilde{J}$ defined in (4.7), is called an NE solution of $\mathcal{G}(\mathcal{M}, \Omega_i, J_i)$.

Remark 4.9 The two definitions, Definitions 4.1 and 4.8 are equivalent. Indeed (4.8) can be equivalently rewritten as for every given $\mathbf{u}^*_{-i}$,

$$\sum_{i=1}^{m} J_i(\mathbf{u}^*_{-i}, u^*_i) \leq \sum_{i=1}^{m} J_i(\mathbf{u}^*_{-i}, x_i), \quad \forall \mathbf{x} \in \Omega$$

Thus it is immediately seen that $\mathbf{u}^*$ satisfies Definition 4.8 if $\mathbf{u}^*$ is an NE solution in the sense of Definition 4.1. Conversely, if $\mathbf{u}^*$ satisfies Definition 4.8, then it constitutes an NE solution in the sense of Definition 4.1 as shown next by contradiction. Assume to the contrary that such a $\mathbf{u}^*$ is not an NE solution in the sense of Definition 4.1. This implies that for some $i \in \mathcal{M}$, there exists a $\tilde{u}_i \neq u^*_i$ such that $J_i(\mathbf{u}^*_{-i}, \tilde{u}_i) < J_i(\mathbf{u}^*_{-i}, u^*_i)$. By adding $\sum_{j \in \mathcal{M}, j \neq i} J_j(\mathbf{u}^*_{-j}, u^*_j)$ to both sides, the following inequality holds:

$$\tilde{J}(\mathbf{u}^*; \tilde{\mathbf{u}}) < \tilde{J}(\mathbf{u}^*; \mathbf{u}^*),$$

where $\tilde{\mathbf{u}} := (\mathbf{u}^*_{-i}, \tilde{u}_i)$ and $\tilde{u}_i \neq u^*_i$. This contradicts the hypothesis (4.8).

Based on the concept of NG cost function, (4.7), we give next a proof of Proposition 4.5. The proof essentially follows arguments similar to those used in Theorem 4.4 in [20]. We present it here to give readers a better interpretation of the two-argument NG cost function, because this will prove a very useful analytical vehicle in later chapters for games with coupled constraints (Chaps. 5 and 7). The proof helps reformulate the necessary conditions (4.6) with respect to the NG cost function.

Proof of Proposition 4.5 From (4.7), the two-argument NG cost function $\tilde{J}(\mathbf{u}; \mathbf{x})$ is separable in the second argument $\mathbf{x}$ for every given $\mathbf{u}$, i.e., each component cost function in $\tilde{J}(\mathbf{u}; \mathbf{x})$ is decoupled in $\mathbf{x}$ for every given $\mathbf{u}$. Therefore by using (4.7), for every given $\mathbf{u}$, the gradient of $\tilde{J}(\mathbf{u}; \mathbf{x})$ with respect to $\mathbf{x}$ is written as

$$\nabla_x \tilde{J}(\mathbf{u}; \mathbf{x}) := \begin{bmatrix} \frac{\partial J_1}{\partial x_1}(u_{-1}, x_1) \\ \vdots \\ \frac{\partial J_m}{\partial x_m}(u_{-m}, x_m) \end{bmatrix} \tag{4.9}$$

The Hessian of $\tilde{J}(\mathbf{u};\mathbf{x})$ with respect to $\mathbf{x}$, $\nabla^2_{xx}\tilde{J}(\mathbf{u};\mathbf{x})$, is a diagonal matrix with the diagonal elements being $\frac{\partial^2 J_i}{\partial x_i^2}(\mathbf{u}_{-i}, x_i)$, $i = 1,\dots,m$. Under Assumption 4.3, from the strict convexity of $J_i(\mathbf{u}_{-i}, x_i)$ with respect to x_i (for every given $\mathbf{u}_{-i}$), it follows that $\tilde{J}(\mathbf{u};\mathbf{x})$ is strictly convex with respect to its second argument $\mathbf{x}$, for every given $\mathbf{u}$. Moreover, $\tilde{J}(\mathbf{u};\mathbf{x})$ is continuous in its arguments. Let us define a reaction set:

$$R(\mathbf{u}) = \{\mathbf{x} \in \Omega \mid \tilde{J}(\mathbf{u};\mathbf{x}) \le \tilde{J}(\mathbf{u},\mathbf{w}),\ \forall \mathbf{w} \in \Omega\},$$

or, equivalently,

$$R(\mathbf{u}) = \arg\min_{\mathbf{x}\in\Omega} \tilde{J}(\mathbf{u};\mathbf{x}),$$

where minimization on the right-hand side is done with respect to the second argument $\mathbf{x}$ in $\tilde{J}(\mathbf{u};\mathbf{x})$. Recall that Ω is compact. By the continuity and convexity property of $\tilde{J}$, it follows from Theorem A.17 (Berge's Maximum Theorem) that R is an upper-semi-continuous mapping that maps each point $\mathbf{u}$ in Ω into a compact and convex subset of Ω. Then by the Kakutani Fixed-Point Theorem (Theorem A.19), there exists a point $\mathbf{u}^*$ such that $\mathbf{u}^* \in R(\mathbf{u}^*)$, i.e., $\mathbf{u}^*$ satisfies (4.8). By Definition 4.8, $\mathbf{u}^*$ is an NE solution of $\mathcal{G}(\mathcal{M}, \Omega_i, J_i)$. □

Thus as seen in the proof, an NE solution $\mathbf{u}^*$ is a fixed-point solution that satisfies $\mathbf{u}^* \in R(\mathbf{u}^*)$, i.e., $\mathbf{u} = \mathbf{u}^*$ is a solution to the implicit equation

$$\mathbf{u} = \arg\min_{\mathbf{x}\in\Omega} \tilde{J}(\mathbf{u};\mathbf{x}) \tag{4.10}$$

Moreover, if $\mathbf{u}^*$ is inner, then $\mathbf{u} = \mathbf{u}^*$ equivalently satisfies

$$\nabla_x \tilde{J}(\mathbf{u};\mathbf{x})|_{\mathbf{x}=\mathbf{u}} = 0, \tag{4.11}$$

where the notation "$|_{\mathbf{x}=\mathbf{u}}$" denotes finding a fixed-point solution. By using (4.9), we get the component-wise form of (4.11):

$$\left.\frac{\partial J_i}{\partial x_i}(\mathbf{u}_{-i}, x_i)\right|_{x_i=u_i} = 0, \quad \forall i \in \mathcal{M}, \tag{4.12}$$

which are equivalent to the necessary conditions (4.6) with respect to J_i presented in Proposition 4.6.

Based on this one can summarize the procedure to find an inner NE solution with respect to $\tilde{J}$ as follows. As a first step solve $\nabla_x \tilde{J}(\mathbf{u};\mathbf{x}) = 0$ for every given $\mathbf{u}$ (i.e., $\mathbf{x}$ is the only variable), which gives $\mathbf{x}$ as a function of $\mathbf{u}$. Then look for a fixed-point solution for these equations, this fixed-point solution being denoted by $\mathbf{x} = \mathbf{u}$. Solving the resulting set of m equations, (4.12), leads to an inner NE solution. We give next an illustration of this procedure.

Example 4.10 (Consider the two-player Nash game in Example 4.7 and its NG cost function)

$$\begin{aligned}\tilde{J}(\mathbf{u};\mathbf{x}) &= J_1(u_2,x_1) + J_2(u_1,x_2)\\ &= \left(2x_1^2 - 2x_1 - x_1u_2\right) + \left(x_2^2 - \frac{1}{2}x_2 - u_1x_2\right)\end{aligned} \tag{4.13}$$

Assume there exists an inner NE solution. Then given every $\mathbf{u}$, in the first step we solve $\nabla_x \tilde{J}(\mathbf{u};\mathbf{x}) = 0$, i.e.,

$$\begin{aligned}\frac{\partial J_1}{\partial x_1}(u_2,x_1) &= 4x_1 - 2 - u_2 = 0\\ \frac{\partial J_2}{\partial x_2}(u_1,x_2) &= 2x_2 - \frac{1}{2} - u_1 = 0\end{aligned}$$

This leads to $\mathbf{x}$ as a function of $\mathbf{u}$, i.e.,

$$x = \begin{bmatrix} 0 & \frac{1}{4}\\ \frac{1}{2} & 0\end{bmatrix} u + \begin{bmatrix}\frac{1}{2}\\ \frac{1}{4}\end{bmatrix}$$

In the second step we solve for a fixed-point solution, by setting $\mathbf{x} = \mathbf{u}$ in the above, which leads to $(u_1^*, u_2^*) = (\frac{9}{14}, \frac{4}{7})$.

While existence can be guaranteed under relatively standard conditions, uniqueness of an NE depends on the particular situation and general results are hard to arrive at. In games where an NE is not unique and $NE(\mathcal{G})$ is not a singleton, one may seek various refinements of the NE concept as described in the previous chapter. These refinements could include testing an equilibrium for additional attributes such as stability or robustness, among others. The reader can consult 4.3 in [20] for such definitions. Besides refinements such as stability and robustness, sometimes it is possible to partially order the set $NE(\mathcal{G})$ by Pareto dominance. However, since the best one can do is only partially order the set $NE(\mathcal{G})$, we cannot expect in general that even Pareto dominance will produce a unique solution. Work such as [54] attempts to address the problem of equilibrium selection.

4.6 Notes

This chapter provided an overview of game with continuous action spaces and cost functions (continuous kernels) and uncoupled constraints. A review of optimization results is presented in the appendix.

Chapter 5
Computational Results for Games with Coupled Constraints

Abstract This chapter provides some results for Nash games with coupled constraints, i.e., coupled action sets. Work on games with coupled action spaces has been going on for more than 50 years. These are also called generalized Nash games, games with coupled constraints, or social equilibria. Game theoretical formulations of problems and computational approaches towards solving coupled or generalized Nash games have been areas of much recent interest. We present some new results mainly based on the Lagrangian approach extension proposed in Pavel (Automatica 43(2):226–237, 2007). We review a relaxation via an augmented optimization, the Lagrangian extension in a game setup, followed by duality and hierarchical decomposition in a game setup.

5.1 Introduction

As seen in the previous chapter, in Nash games with uncoupled constraints, the action space is the Cartesian product of the individual action sets and players can affect only the cost functions of the other players but not their feasible action sets.

On the other hand, in Nash games with coupled constraints, each player's action affects the feasible action sets of the other players. In Example 4.7, the action sets are $\Omega_i = [0, 8]$, $i = 1, 2$, such that the action space Ω is rectangular in $\mathbb{R}^2$.

Now consider the following example with a modified action space.

Example 5.1 Consider the two-player Nash game in Example 4.7 with an action space, $\Omega := \Omega_1 \times \Omega_2$ and $u_i \in \Omega_i = [0, 8]$, $i = 1, 2$. An additional constraint is considered: $u_1 + u_2 \leq 8$. Now the action space is modified to be

$$\overline{\Omega} = \{\mathbf{u} \in \Omega \mid u_1 + u_2 - 8 \leq 0\}.$$

Figure 5.1 shows that Ω is rectangular (constraints have no coupling), while $\overline{\Omega}$ is triangular (constraints are coupled). In this latter case $\overline{\Omega}$, it is not possible to obtain separate action sets from which the players can take actions independently, i.e.,

L. Pavel, *Game Theory for Control of Optical Networks*,
Static & Dynamic Game Theory: Foundations & Applications,
DOI 10.1007/978-0-8176-8322-1_5, © Springer Science+Business Media New York 2012

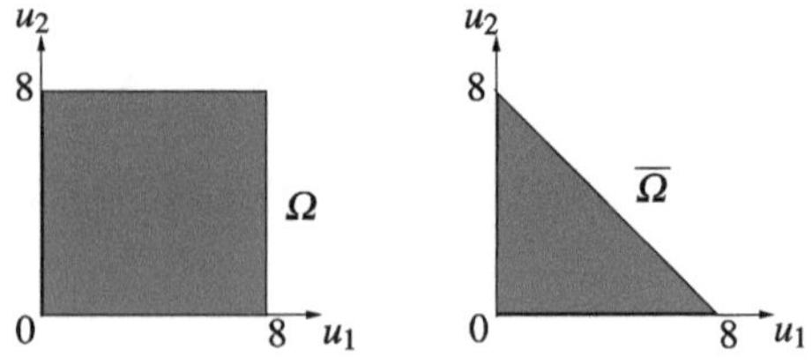

Fig. 5.1 Constraints: rectangular and triangular

$u_1 \in [0, 8 - u_2]$ and $u_2 \in [0, 8 - u_1]$. That is the action space $\overline{\Omega}$ is *coupled* and this game is called a two-player Nash game with coupled constraints.

Starting from this example, we present in this chapter some theoretically results for Nash games with coupled constraints, i.e., coupled action sets. We call such a game a coupled Nash game. Coupled Nash games are also called generalized Nash games (games with non-disjoint strategy sets) [56], games with coupled constraints [128], social equilibria games or pseudo-Nash equilibria games [16, 37]. Game theoretical formulations of problems and computational approaches towards solving coupled or generalized Nash games have been areas of much recent interest [15, 40, 90, 91, 120, 135, 138]. The work in [40] shows that a generalized Nash equilibrium can be calculated by solving a variational inequality and the results express conditions in terms of variational inequality problem and Karush–Kuhn–Tucker (KKT) conditions for the pseudo-gradient. Another related work is [90] where the authors present a scheme that associates to a generalized variational inequality, a dual problem and KKT conditions, thus allowing to solve primal and dual problems in the spirit of classical Lagrangian duality for constrained optimization problems, using set theoretic concepts and set-valued operators. Extensions based on this approach have been developed for non-compact and non-smooth settings in [138].

The treatment is this chapter mostly follows the Lagrangian approach extension proposed in [120]. This is a procedural method for computing a Nash equilibrium, based on an extension of duality to a game theoretical framework [32]. The setting of the construction uses the two-argument NG cost function (see Chap. 4) and relaxes also the constraints into a two-argument form. Thus the problem is enlarged into a constrained optimization problem in a space of twice the dimension followed by projection back into a one dimension (with a fixed-point solution). Moreover, for convex constraints, duality leads to hierarchical decomposition into a lower-level game with no coupled constraints and an optimization problem for Lagrangian prices.

The chapter is organized as follows. In Sect. 5.2 some results on Nash equilibria existence are reviewed and relaxation via an augmented optimization problem is considered. This is followed by results for Lagrangian extension in a game setup in Sect. 5.3. Sections 5.4 and 5.5 present results for duality extension and hierarchical decomposition in a game setup.

5.2 Nash Equilibria and Relaxation via an Augmented Optimization

In order to formally define a coupled Nash game, let us consider the following coupled inequality constraints:

$$g_r(\mathbf{u}) \le 0, \quad r = 1, \dots, R \tag{5.1}$$

or, in vector form, $\mathbf{g}(\mathbf{u}) \le 0$ where

$$\mathbf{g}(\mathbf{u}) = \begin{bmatrix} g_1(\mathbf{u}) \\ \vdots \\ g_R(\mathbf{u}) \end{bmatrix} \tag{5.2}$$

for $\mathbf{u} \in \Omega$. The coupled constraint set associated with each $g_r(\mathbf{u})$ is denoted by

$$\overline{\Omega}_r = \{\mathbf{u} \in \Omega \mid g_r(\mathbf{u}) \le 0\} \tag{5.3}$$

where Ω is defined (4.1), i.e., $\Omega = \Omega_1 \times \cdots \times \Omega_m$ with $\Omega_i = [m_i, M_i]$.

Thus the resulting action space $\overline{\Omega} \subset \mathbb{R}^m$ is coupled, defined as

$$\overline{\Omega} = \bigcap_{r=1}^{R} \overline{\Omega}_r = \{\mathbf{u} \in \Omega \mid \mathbf{g}(\mathbf{u}) \le 0\} \tag{5.4}$$

As before, for every player $i \in \mathcal{M}$ an individual cost function $J_i : \overline{\Omega} \to \mathbb{R}$ is defined that satisfies Assumption 4.3. For every given $\mathbf{u}_{-i} \in \Omega_{-i}$, a *projection action set* is also defined for each $i \in \mathcal{M}$,

$$\widehat{\Omega}_i(\mathbf{u}_{-i}) = \{\xi \in \Omega_i \mid \mathbf{g}(\mathbf{u}_{-i}, \xi) \le 0\} \tag{5.5}$$

For each $i \in \mathcal{M}$, this projection action set $\widehat{\Omega}_i(\mathbf{u}_{-i})$ is the *feasible* action set under the given $\mathbf{u}_{-i}$. A vector $\mathbf{u} = (\mathbf{u}_{-i}, u_i)$ is called *feasible* if $\mathbf{u} \in \overline{\Omega}$. The resulting coupled Nash game is denoted by $\mathcal{G}(\mathcal{M}, \widehat{\Omega}_i, J_i)$, and an NE solution is defined as follows.

Definition 5.2 (NE for coupled Nash games) A vector $\mathbf{u}^* \in \overline{\Omega}$ is called an *NE solution* NE solution of $\mathcal{G}(\mathcal{M}, \widehat{\Omega}_i, J_i)$ if for all $i \in \mathcal{M}$ and for every given $\mathbf{u}^*_{-i}$,

$$J_i(\mathbf{u}^*_{-i}, u_i^*) \le J_i(\mathbf{u}^*_{-i}, u_i), \quad \forall u_i \in \widehat{\Omega}_i(\mathbf{u}^*_{-i}), \tag{5.6}$$

where $\widehat{\Omega}_i(\mathbf{u}^*_{-i})$ is defined in (5.5).

Let us look again at Example 5.1. Note that $(u_1^*, u_2^*) = (\frac{9}{14}, \frac{4}{7}) \in \overline{\Omega}$ (see Fig. 5.2), where (u_1^*, u_2^*) is the unique NE solution obtained in Example 4.7. Then obviously (u_1^*, u_2^*) is the NE solution of the associated Nash game with coupled constraints.

The following proposition (adapted from Theorem 4.4 in [20]) gives sufficient conditions for existence of an NE solution.

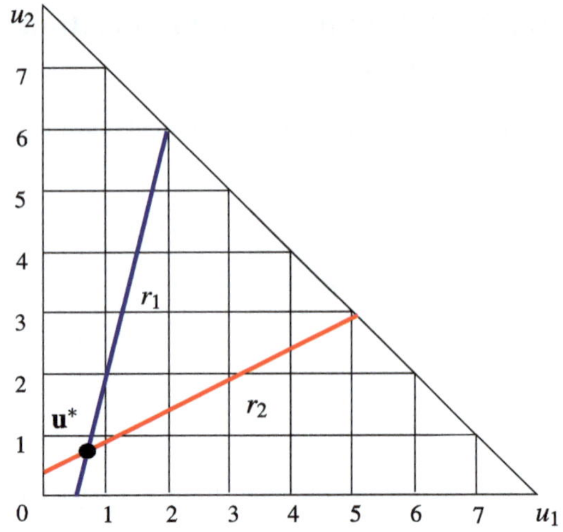

Fig. 5.2 Reaction curves in coupled Nash game: $r_1(u_2)$ and $r_2(u_1)$

Proposition 5.3 (Theorem 4.4, [20]) *Let the action space* $\overline{\Omega}$ *be a compact and convex subset of* $\mathbb{R}^m$. *Under Assumption* 4.3, $\mathcal{G}(\mathcal{M}, \widehat{\Omega}_i, J_i)$ *admits an NE solution.*

Next let us use the concept of NG cost function presented in Chap. 4 to characterize an NE solution of $\mathcal{G}(\mathcal{M}, \widehat{\Omega}_i, J_i)$.

Recall the two-argument NG cost function $\tilde{J}(\mathbf{u}; \mathbf{x})$ defined in (4.7),

$$\tilde{J}(\mathbf{u}; \mathbf{x}) := \sum_{i=1}^{m} J_i(\mathbf{u}_{-i}, x_i)$$

and note that $\tilde{J}(\mathbf{u}; \mathbf{x})$ is in its second argument $\mathbf{x}$. Then with respect to $\tilde{J}(\mathbf{u}; \mathbf{x})$, an NE solution $\mathbf{u}^*$ of $\mathcal{G}(\mathcal{M}, \widehat{\Omega}_i, J_i)$ satisfies

$$\tilde{J}(\mathbf{u}^*; \mathbf{u}^*) \leq \tilde{J}(\mathbf{u}^*; \mathbf{x}), \quad \forall \mathbf{x} \in \Omega, \text{ with } \mathbf{g}(\mathbf{u}^*_{-i}, x_i) \leq 0, \ \forall i \in \mathcal{M} \tag{5.7}$$

This result can be obtained by using Definition 5.2, the definition of $\tilde{J}(\mathbf{u}; \mathbf{x})$, (4.7), and the projection action set $\widehat{\Omega}_i(\mathbf{u}_{-i})$, (5.5).

Now, let us also augment the coupled constraints $\mathbf{g}(\mathbf{u})$ in (5.2) into an equivalent two-argument form, $\tilde{\mathbf{g}}$,

$$\tilde{\mathbf{g}}(\mathbf{u}; \mathbf{x}) = \sum_{i=1}^{m} \mathbf{g}(\mathbf{u}_{-i}, x_i), \tag{5.8}$$

where $\tilde{\mathbf{g}} = [\tilde{g}_1, \ldots, \tilde{g}_R]^T$ and

$$\tilde{g}_r(\mathbf{u}; \mathbf{x}) = \sum_{i=1}^{m} g_r(\mathbf{u}_{-i}, x_i), \quad \forall r = 1, \ldots, R$$

If $\mathbf{x} = \mathbf{u}$, then it follows that

$$\tilde{\mathbf{g}}(\mathbf{u}; \mathbf{u}) = \sum_{i=1}^{m} \mathbf{g}(\mathbf{u}) = m \times \mathbf{g}(\mathbf{u}) \tag{5.9}$$

Not only $\tilde{J}$ but $\tilde{\mathbf{g}}(\mathbf{u}; \mathbf{x})$ also is in its second argument $\mathbf{x}$.

We can remark that the use of the $\tilde{J}$ [20], defined on a space of twice the dimension of the original game is instrumental in what follows. This is because it allows to find a solution of the original Nash game with coupled constraints by solving a constrained optimization problem for $\tilde{J}$ and searching for a fixed-point solution. Two main features allow this. Firstly, the is in the second argument $\mathbf{x}$ for every given $\mathbf{u}$, i.e., each component cost function J_i in $\tilde{J}(\mathbf{u}; \mathbf{x})$ is decoupled in $\mathbf{x}$ and depends only on x_i. Secondly, the constraints $\mathbf{g}$ have been augmented into a separable two-argument form $\tilde{\mathbf{g}}$ thus enlarging the search set. *NG-feasibility* is equivalent to $\tilde{\mathbf{g}}(\mathbf{u}; \mathbf{u}) \le 0$. Intuitively, by introducing the NG cost function $\tilde{J}(\mathbf{u}; \mathbf{x})$, the coupled Nash game is related to a *constrained* optimization problem for $\tilde{J}(\mathbf{u}; \mathbf{x})$ that has a fixed-point solution. Recall that this methodology has been used for uncoupled Nash games in Chap. 4 where the associated optimization problem is unconstrained.

Herein, the optimization problem for $\tilde{J}(\mathbf{u}; \mathbf{x})$ is a constrained minimization of $\tilde{J}(\mathbf{u}; \mathbf{x})$ with respect to $\mathbf{x}$, with constraints $\tilde{\mathbf{g}}(\mathbf{u}; \mathbf{x})$, (5.8). A solution $\mathbf{u}^*$ of the constrained minimization of $\tilde{J}(\mathbf{u}; \mathbf{x})$ satisfies

$$\tilde{J}(\mathbf{u}^*; \mathbf{u}^*) \le \tilde{J}(\mathbf{u}^*; \mathbf{x}), \quad \forall \mathbf{x} \in \Omega, \text{ with } \tilde{\mathbf{g}}(\mathbf{u}^*; \mathbf{x}) \le 0, \tag{5.10}$$

with $\tilde{\mathbf{g}}(\mathbf{u}^*; \mathbf{u}^*) \le 0$.

Specifically one arrives at the following result.

Proposition 5.4 *A solution $\mathbf{u}^*$ of the constrained minimization of $\tilde{J}(\mathbf{u}; \mathbf{x})$ satisfying* (5.10) *with $\tilde{\mathbf{g}}(\mathbf{u}^*; \mathbf{u}^*) \le 0$ is an NE solution of $\mathcal{G}(\mathcal{M}, \widehat{\Omega}_i, J_i)$.*

This result has been proved by contradiction in [120]. We briefly review the argument here. By contradiction assume that $\mathbf{u}^*$ is not an NE solution of $\mathcal{G}(\mathcal{M}, \widehat{\Omega}_i, J_i)$. It follows that for some $i \in \mathcal{M}$, there exists an $\tilde{x}_i \in \Omega_i$ with $\mathbf{g}(\mathbf{u}^*_{-i}, \tilde{x}_i) \le 0$, such that

$$J_i(\mathbf{u}^*_{-i}, \tilde{x}_i) < J_i(\mathbf{u}^*_{-i}, u^*_i)$$

By adding a term $\sum_{j \in \mathcal{M}, j \ne i} J_j(\mathbf{u}^*_{-j}, u^*_j)$ to both sides, the following inequality holds:

$$\tilde{J}(\mathbf{u}^*; \tilde{\mathbf{x}}) < \tilde{J}(\mathbf{u}^*; \mathbf{u}^*),$$

with $\tilde{x} := (\mathbf{u}^*_{-i}, \tilde{x}_i) \in \Omega$. From $\tilde{\mathbf{g}}(\mathbf{u}^*; \mathbf{u}^*) \le 0$ and (5.9), it follows that $\mathbf{g}(\mathbf{u}^*_{-j}, u^*_j) \le 0, \forall j \in \mathcal{M}$, such that

$$\tilde{\mathbf{g}}(\mathbf{u}^*; \tilde{\mathbf{x}}) := \sum_{j \in \mathcal{M}, j \ne i} \mathbf{g}(\mathbf{u}^*_{-j}, u^*_j) + \mathbf{g}(\mathbf{u}^*_{-i}, \tilde{x}_i) \le 0$$

It follows that there exists an $\tilde{x}$ such that

$$\tilde{J}(\mathbf{u}^*; \tilde{\mathbf{x}}) < \tilde{J}(\mathbf{u}^*; \mathbf{u}^*), \quad \tilde{x} \in \Omega, \text{ with } \tilde{\mathbf{g}}(\mathbf{u}^*; \tilde{\mathbf{x}}) \leq 0,$$

which contradicts the hypothesis in (5.10). Thus $\mathbf{u}^*$ is an NE solution of $\mathcal{G}(\mathcal{M}, \widehat{\Omega}_i, J_i)$ as needed.

As seen in Proposition 5.4, an NE solution of $\mathcal{G}(\mathcal{M}, \widehat{\Omega}_i, J_i)$ can be found by solving the problem for $\tilde{J}(\mathbf{u}; \mathbf{x})$ and searching for a fixed-point solution.

5.3 Lagrangian Extension in a Game Setup

As seen in the above section, for every given $\mathbf{u} \in \Omega$, the constrained minimization (5.10) is a standard constrained optimization problem (see the appendix). In a game context one can use a Lagrangian extension for a two-argument constrained optimization as proposed in [116, 120].

Next we give a brief introduction of this methodology as it leads to an elegant hierarchical decomposition. Some results reviewed here have further use in the following chapters. Readers are referred to [120] for some of the detailed proofs.

As in standard optimization, associated to (5.10) a two-argument Lagrangian function $\tilde{L}$ is defined for $\tilde{J}$ and $\tilde{\mathbf{g}}$, where

$$\tilde{L}(\mathbf{u}; \mathbf{x}; \boldsymbol{\mu}) = \tilde{J}(\mathbf{u}; \mathbf{x}) + \boldsymbol{\mu}^T \tilde{g}(\mathbf{u}; \mathbf{x}), \tag{5.11}$$

with $\boldsymbol{\mu} = [\mu_1, \dots, \mu_R]^T$ being the Lagrange multiplier vector.

The next result (extension to Propositions A.3 and A.4) gives optimality conditions for an NE solution in terms of Lagrange multipliers.

Proposition 5.5 *Consider $\mathcal{G}(\mathcal{M}, \widehat{\Omega}_i, J_i)$. Let the action space $\overline{\Omega}$ be a compact and convex subset of $\mathbb{R}^m$ and Assumption 4.3 is satisfied.*

(a) (Necessity): *If u is an NE solution of $\mathcal{G}(\mathcal{M}, \widehat{\Omega}_i, J_i)$, then there exists a unique vector $\boldsymbol{\mu}^* \geq 0$ such that*

$$\nabla_x \tilde{L}(\mathbf{u}; \mathbf{x}; \boldsymbol{\mu}^*)\big|_{x=u} = 0 \tag{5.12}$$

$$\boldsymbol{\mu}^{*T} \mathbf{g}(\mathbf{u}) = 0, \tag{5.13}$$

where the notation "$|_{x=u}$" is defined in (4.11) and denotes finding a fixed-point solution.

(b) (Sufficiency): *Let $\mathbf{u}^*$ be a feasible vector together with a vector $\boldsymbol{\mu} = [\boldsymbol{\mu}_1, \dots, \boldsymbol{\mu}_R]^T$, such that $\boldsymbol{\mu} \geq 0$ and $\boldsymbol{\mu}^T g(\mathbf{u}^*) = 0$. Assume that $\mathbf{u}^*$ minimizes the Lagrangian function, $\tilde{L}$ (5.11), over $\mathbf{x} \in \Omega$, as a fixed-point solution, i.e., $\mathbf{u} = \mathbf{u}^*$ satisfies*

$$\mathbf{u} = \arg\min_{\mathbf{x} \in \Omega} \tilde{L}(\mathbf{u}; \mathbf{x}; \boldsymbol{\mu}), \tag{5.14}$$

where minimization on RHS of (5.14) *is done with respect to* $\mathbf{x}$ *in* $\tilde{L}$. *Then* $\mathbf{u}^*$ *is an NE solution of* $\mathcal{G}(\mathcal{M}, \widehat{\Omega}_i, J_i)$.

Note that if u is an inner NE solution, then $\boldsymbol{\mu}^* = 0$ such that (5.12) is equivalent to (4.11), the necessary condition for an NE solution in an uncoupled Nash game.

Remark 5.6 The Lagrangian optimality condition in Proposition 5.5 shows that $\mathbf{u}^*$ is obtained by first minimizing the augmented Lagrangian function $\tilde{L}(\mathbf{u}; \mathbf{x}; \boldsymbol{\mu})$ with respect to the second argument $\mathbf{x}$, which gives $x = \phi(\mathbf{u})$ for every given u. The next step involves finding a fixed-point solution $\mathbf{u}^*$ of ϕ by setting $\mathbf{x} = \mathbf{u}$, i.e., solving $\mathbf{u} = \phi(\mathbf{u})$. This $\mathbf{u}^*$, a fixed-point solution to the minimization of $\tilde{L}$ over $\mathbf{x} \in \Omega$, the following holds: $\tilde{L}(\mathbf{u}^*; \mathbf{u}^*; \mu) \le \tilde{L}(\mathbf{u}^*; \mathbf{x}; \mu)$, $\forall \mathbf{x} \in \Omega$. Note that $\mathbf{u}^*$ thus obtained depends on $\boldsymbol{\mu}$, $\mathbf{u}^*(\boldsymbol{\mu})$. An optimal $\boldsymbol{\mu}^*$ is achieved by solving

$$\boldsymbol{\mu}^{*T}\mathbf{g}\big(\mathbf{u}^*(\boldsymbol{\mu}^*)\big) = 0 \quad \text{and} \quad \boldsymbol{\mu}^* \ge 0 \tag{5.15}$$

That is, in compact notation we write

$$\tilde{L}\big(\mathbf{u}^*; \mathbf{u}^*; \mu\big) = \Big[\min_{\mathbf{x}\in\Omega} \tilde{L}(\mathbf{u}; \mathbf{x}; \mu)\Big]\Big|_{\mathbf{x}=\mathbf{u}} \tag{5.16}$$

The obtained $\mathbf{u}^*(\boldsymbol{\mu}^*)$ is a solution of (5.10) and hence an NE solution of $\mathcal{G}(\mathcal{M}, \widehat{\Omega}_i, J_i)$. We shall call $(\mathbf{u}^*(\boldsymbol{\mu}^*), \boldsymbol{\mu}^*)$ an *NE solution–Lagrange multiplier pair*.

Example 5.7 Consider the two-player Nash game presented in Example 5.1 with a coupled constraint. The corresponding $\tilde{J}(\mathbf{u}; \mathbf{x})$ is

$$\tilde{J}(\mathbf{u}; \mathbf{x}) = \big(2x_1^2 - 2x_1 - x_1u_2\big) + \left(x_2^2 - \frac{1}{2}x_2 - u_1x_2\right)$$

The augmented constraint is

$$\tilde{\mathbf{g}}(\mathbf{u}; \mathbf{x}) = g(u_2, x_1) + g(u_1, x_2) = (x_1 + u_2 - 8) + (u_1 + x_2 - 8) \le 0$$

Thus the Lagrangian function is obtained as

$$\begin{aligned}\tilde{L}(\mathbf{u}; \mathbf{x}; \boldsymbol{\mu}) &= \tilde{J}(\mathbf{u}; \mathbf{x}) + \boldsymbol{\mu}\tilde{\mathbf{g}}(\mathbf{u}; \mathbf{x}) \\ &= \big(2x_1^2 - 2x_1 - x_1u_2\big) + \left(x_2^2 - \frac{1}{2}x_2 - u_1x_2\right) \\ &\quad + \boldsymbol{\mu}(x_1 + u_2 - 8 + u_1 + x_2 - 8)\end{aligned}$$

To find an NE solution, u, and the corresponding Lagrange multiplier vector, $\boldsymbol{\mu}$, one needs to solve the necessary conditions, (5.12), with $x_i = u_i$, $i = 1, 2$. Then it follows that

$$4u_1 = u_2 + 2 - \boldsymbol{\mu} \quad \text{and} \quad 2u_2 = u_1 + \frac{1}{2} - \boldsymbol{\mu}$$

with u and $\boldsymbol{\mu}$ satisfying $\boldsymbol{\mu}(u_1 + u_2 - 8) = 0$ and $\boldsymbol{\mu} \geq 0$. Thus if $\boldsymbol{\mu} = 0$, then $(u_1, u_2) = (\frac{9}{14}, \frac{4}{7})$, which is an NE solution. If $\boldsymbol{\mu} \neq 0$, $(u_1, u_2) = (\frac{51}{16}, \frac{77}{16})$ with $\boldsymbol{\mu} = -\frac{95}{16} < 0$ such that $(\frac{51}{16}, \frac{77}{16})$ is not an NE solution.

Note that as J_i and g_r are differentiable convex functions and $\Omega = \mathbb{R}^m$, the Lagrangian function $\tilde{L}(\mathbf{u}; \mathbf{x}; \boldsymbol{\mu})$ is convex with respect to $\mathbf{x}$, so the Lagrangian minimization is equivalent to the first order necessary condition. Thus in the presence of convexity the first order optimality conditions are also sufficient.

5.4 Duality Extension

In this section we review duality results for Nash games with coupled constraints, based on the Lagrangian extension. A dual cost function related to the minimization of the associated Lagrangian function (cf. Proposition 5.5) is introduced, similar to standard optimization [24]. For Nash games with convex coupled constraints, one can show that duality enables decomposition into a lower-level Nash game with no coupled constraints, and a higher-level optimization problem.

Consider a Nash game with coupled constraints and recall the associated Lagrangian function $\tilde{L}$ and its minimization in a fixed-point sense, as in (5.16). with the resulting fixed-point solution as a function of μ, $\mathbf{u}^* = \mathbf{u}^*(\mu)$.

We define a function called the dual cost function $D(\boldsymbol{\mu})$ as

$$D(\boldsymbol{\mu}) := \tilde{L}\big(\mathbf{u}^*; \mathbf{u}^*; \boldsymbol{\mu}\big), \tag{5.17}$$

where $\mathbf{u}^*$ minimizes $\tilde{L}$ defined in (5.11) over $\mathbf{x} \in \Omega$ as a fixed-point solution, i.e., $\mathbf{u} = \mathbf{u}^*$ satisfies

$$\mathbf{u} = \arg\min_{\mathbf{x} \in \Omega} \tilde{L}(\mathbf{u}; \mathbf{x}; \boldsymbol{\mu})$$

In a fixed-point notation, $D(\boldsymbol{\mu})$ can be written as

$$D(\boldsymbol{\mu}) := \Big[\min_{\mathbf{x} \in \Omega} \tilde{L}(\mathbf{u}; \mathbf{x}; \boldsymbol{\mu})\Big]\Big|_{\arg\min_{\mathbf{x} \in \Omega} \tilde{L} = \mathbf{u}},$$

where $\tilde{\mathbf{g}}(\mathbf{u}; \mathbf{u}) \leq 0$, and the dual optimal value is defined as

$$D^* = \max_{\mu \geq 0} D(\boldsymbol{\mu}) \tag{5.18}$$

The primal and dual optimal solution pairs are characterized by the following result (Theorem 2 in [120]).

Theorem 5.8 $(\mathbf{u}^*; \boldsymbol{\mu}^*)$ *is an optimal NE solution–Lagrange multiplier pair in the sense of* (5.10) *and* (5.18), *if and only if*:

(1) $\mathbf{u}^* \in \Omega$ *and* $\tilde{\mathbf{g}}_l(\mathbf{u}^*; \mathbf{u}^*) \leq 0$ *(NG-feasibility)*;

(2) $\boldsymbol{\mu}^* \geq 0$ (*dual feasibility*);
(3) $\mathbf{u}^* = \arg\{\min_{\mathbf{x}\in\Omega} \tilde{L}(\mathbf{u}; \mathbf{x}; \mu^*)|_{\mathbf{x}=\mathbf{u}}\}$ (*Lagrangian optimality*);
(4) $\mu^{*T} \tilde{g}(\mathbf{u}^*; \mathbf{u}^*) = 0$ (*complementary slackness*).

Proof If $(\mathbf{u}^*; \mu^*)$ is an optimal NE solution–Lagrange multiplier pair, then $\mathbf{u}^*$ is feasible and μ^* is dual feasible and the first two relations follow directly. The last two relations follow from Proposition 5.5.

For sufficiency, using Lagrangian optimality one obtains

$$\tilde{L}(\mathbf{u}^*; \mathbf{u}^*, \mu^*) = \Big[\min_{\mathbf{x}\in\Omega} \tilde{L}(\mathbf{u}; \mathbf{x}; \boldsymbol{\mu}^*)\Big]\Big|_{\mathbf{x}=\mathbf{u}}$$

so that

$$\tilde{L}(\mathbf{u}^*; \mathbf{u}^*, \mu^*) \leq \tilde{L}(\mathbf{u}^*; \mathbf{x}; \boldsymbol{\mu}^*), \quad \forall \mathbf{x} \in \Omega$$

Using (5.11) and complementary slackness yields

$$\tilde{L}(\mathbf{u}^*; \mathbf{u}^*, \mu^*) = \tilde{J}(\mathbf{u}^*; \mathbf{u}^*)$$

Then from the foregoing inequality one can write

$$\tilde{J}(\mathbf{u}^*; \mathbf{u}^*) \leq \tilde{J}(\mathbf{u}^*; \mathbf{x}), \quad \forall x \in \Omega, \qquad \tilde{\mathbf{g}}(\mathbf{u}^*; \mathbf{x}) \leq 0$$

Therefore (5.10) holds and $\mathbf{u}^*$ is an optimal NE game solution with $\tilde{J}^* = \tilde{J}(\mathbf{u}^*; \mathbf{u}^*)$. Using (5.17), evaluated at μ^*, and the foregoing relations yields

$$D(\mu^*) = \min_{\mathbf{x}\in\Omega} \tilde{L}(\mathbf{u}; \mathbf{x}; \boldsymbol{\mu}^*)\big|_{\mathbf{x}=\mathbf{u}} = \tilde{J}(\mathbf{u}^*; \mathbf{u}^*)$$

and for the optimal dual cost D^*, (5.18),

$$D^* \geq D(\mu^*) = \tilde{J}(\mathbf{u}^*; \mathbf{u}^*) = \tilde{J}^*$$ □

If a Lagrange multiplier μ is known then all optimal NE solutions can be found by minimizing the Lagrangian $\tilde{L}$ over $\mathbf{x} \in \Omega$, in the fixed-point sense as in (5.16). However among those solutions $\mathbf{u}^*(\mu)$, there may be vectors that do not satisfy the coupled NG-feasibility condition $\mathbf{g}(\mathbf{u}^*) \leq 0$, so this has to be checked. Such an $\mathbf{u}^*$ is a solution to (5.10) and hence an NE solution if the complementary slackness condition in Theorem 5.8 holds. Note that $\mathbf{u}^*$ thus obtained depends on $\boldsymbol{\mu}$, $u^*(\boldsymbol{\mu})$.

To summarize, we saw that a Nash equilibrium for games with coupled constraints can be found by solving a constrained optimization problem for the two-argument NG cost function. This can be used as a convenient vehicle for computation of a Nash equilibrium. The problem is relaxed into an optimization problem in a space of twice the dimension and a fixed-point solution is used to project back into the original space and game problem. The formalism of standard optimization enables one to set up a Lagrangian extension and duality results. Conditions for the existence of an NE solution are given in Proposition 5.3, the same as the ones for the existence of a normalized equilibrium in [128]. In fact, $\mathbf{u}^*(\boldsymbol{\mu}^*)$ is a normalized equilibrium point with all r_i in [128] equal to 1. Note that uniqueness of the Nash equilibrium is not presumed. Rather this technique offers a procedural (computational) method to find one such solution.

5.5 Hierarchical Decomposition in a Game Setup

The separability in the second argument of both NG cost function and constraints ensures that $D(\boldsymbol{\mu})$ in (5.17) can be decomposed. Such a decomposition result for the minimization of $\tilde{L}(\mathbf{u}; \mathbf{x}; \boldsymbol{\mu})$ (Theorem 3, [120]) is presented next. The result shows that the minimum of $\tilde{L}(\mathbf{u}; \mathbf{x}; \boldsymbol{\mu})$ with respect to $\mathbf{x} \in \Omega$ can be obtained by minimizing a set of one-argument Lagrangian functions. Thus the fact that both the NG-game cost and the constraints are separable in the second argument is exploited to show that the dual NG cost function $D(\mu)$ can be decomposed and, equivalently, found by solving a modified Nash game with no coupled constraints.

The decomposition result is given next.

Proposition 5.9 (Theorem 3, [120]) *Consider $\mathcal{G}(\mathcal{M}, \widehat{\Omega}_i, J_i)$. Let the action space $\overline{\Omega}$ be a compact and convex subset of $\mathbb{R}^m$ and Assumption 4.3 is satisfied. The associated dual cost function $D(\boldsymbol{\mu})$, (5.17), can be decomposed as*

$$D(\boldsymbol{\mu}) = \sum_{i=1}^{m} L_i\big(\mathbf{u}_{-i}^*(\boldsymbol{\mu}), u_i^*(\boldsymbol{\mu}), \boldsymbol{\mu}\big), \tag{5.19}$$

where

$$L_i(\mathbf{u}_{-i}, x_i, \boldsymbol{\mu}) = J_i(\mathbf{u}_{-i}, x_i) + \boldsymbol{\mu}^T \mathbf{g}(\mathbf{u}_{-i}, x_i) \tag{5.20}$$

and $\mathbf{u}^*(\boldsymbol{\mu}) = [u_i^*(\boldsymbol{\mu})] \in \Omega$ minimizes a set of L_i defined in (5.20) over $x_i \in \Omega_i$ as a fixed-point solution, $\forall i \in \mathcal{M}$. In other words, $u_i = u_i^*(\boldsymbol{\mu})$ satisfies

$$u_i = \arg\min_{x_i \in \Omega_i} L_i(\mathbf{u}_{-i}, x_i, \boldsymbol{\mu}), \quad \forall i \in \mathcal{M}$$

Proof By Proposition 5.5, the necessary conditions for NE optimality with respect to the Lagrangian $\tilde{L}$, (5.11), require one to solve

$$\nabla_x \tilde{L}(\mathbf{u}; \mathbf{x}; \boldsymbol{\mu})|_{\mathbf{x}=\mathbf{u}} = 0 \tag{5.21}$$

or, equivalently component-wise, $\frac{\partial \tilde{L}(\mathbf{u};\mathbf{x};\boldsymbol{\mu})}{\partial x_i}|_{x_i=u_i} = 0$, $i = 1, m$. Using the definitions for $\tilde{J}$ and $\tilde{g}$ one can write $\tilde{L}$, (5.11), as

$$\tilde{L}(\mathbf{u}; \mathbf{x}; \boldsymbol{\mu}) = \sum_{i=1}^{m} L_i(\mathbf{u}_{-i}, x_i, \mu) \tag{5.22}$$

with L_i as in (5.20). Using (5.22) into the foregoing yields

$$\sum_{j=1}^{m} \frac{\partial}{\partial x_i} L_j(\mathbf{u}_{-j}, x_j; \boldsymbol{\mu}) = 0, \quad i = 1, m$$

Due to separability with respect to $\mathbf{x}$, this yields

$$\frac{\partial}{\partial x_i} L_i(\mathbf{u}_{-i}, x_i; \boldsymbol{\mu}) = 0, \quad i = 1, m \tag{5.23}$$

Therefore, component-wise (5.21) is the same as (5.23). To either (5.21) or (5.23) one needs to find a fixed-point solution. Now (5.23) are the first order necessary conditions for minimizing L_i (5.20) with respect to x_i. Since J_i and g_r are convex, they are also sufficient. For each given $\mathbf{u}_{-i}$, from (5.23) one obtains $x_i^* = x_i(\mathbf{u}_{-i})$, so that

$$x_i(\mathbf{u}_{-i}) = \arg \min_{x_i \in \Omega_i} \left[L_i(\mathbf{u}_{-i}, x_i, \boldsymbol{\mu})\right], \quad i = 1, m \tag{5.24}$$

Moreover, because a fixed-point solution is sought one sets $\mathbf{x} = \mathbf{u}$, i.e., component-wise one needs to solve

$$x_i(\mathbf{u}_{-i}) = u_i, \quad \forall i = 1, \ldots, m$$

for a fixed-point vector denoted $\mathbf{u}^* = [u_i^*]$ and $\mathbf{x} = [u_i^*]$, which depends on μ. With this $\mathbf{u}^*$ let us return now to the value functional in (5.22). The first step taken in order to obtain $\mathbf{u}^*$ was minimization with respect to $\mathbf{x}$, so that from (5.22) one has

$$\min_{\mathbf{x} \in \Omega} \tilde{L}(\mathbf{u}; \mathbf{x}; \boldsymbol{\mu}) = \min_{\mathbf{x} \in \Omega} \sum_{i=1}^{m} L_i(\mathbf{u}_{-i}, x_i, \boldsymbol{\mu}), \quad \mathbf{x} \in \Omega$$

for any given $\mathbf{u}$, with L_i as in (5.20). Since $\Omega = \Omega_1 \times \cdots \times \Omega_m$ and the right-hand side is separable with respect to $\mathbf{x} = [x_i]$, $x_i \in \Omega_i$, it follows that

$$\min_{\mathbf{x} \in \Omega} \tilde{L}(\mathbf{u}; \mathbf{x}; \boldsymbol{\mu}) = \sum_{i=1}^{m} \min_{x_i \in \Omega_i} L_i(\mathbf{u}_{-i}, x_i, \mu) \tag{5.25}$$

for any given $\mathbf{u}$. Now evaluating (5.25) at the fixed-point $\mathbf{u}^* = [u_i^*]$, $\mathbf{x} = [u_i^*]$ obtained as above, one can write

$$\begin{aligned} \left[\min_{\mathbf{x} \in \Omega} \tilde{L}(\mathbf{u}; \mathbf{x}; \boldsymbol{\mu})\right]\Big|_{\mathbf{u}=\mathbf{u}^*, \mathbf{x}=\mathbf{u}^*} &= \sum_{i=1}^{m} \left[\min_{x_i \in \Omega_i} L_i(\mathbf{u}_{-i}, x_i, \mu)\right]\Big|_{u_i=u_i^*, x_i=u_i^*} \\ &= \sum_{i=1}^{m} L_i\big(\mathbf{u}_{-i}^*(\mu), u_i^*(\mu), \mu\big) \end{aligned}$$

We write this in a compact fixed-point notation

$$\Big[\min_{\mathbf{x}\in\Omega} \tilde{L}(\mathbf{u};\mathbf{x};\boldsymbol{\mu})\Big]\Big|_{\mathbf{x}=\mathbf{u}} = \sum_{i=1}^{m}\Big[\min_{x_i\in\Omega_i} L_i(\mathbf{u}_{-i},x_i,\mu)\Big]\Big|_{x_i=u_i}$$

The proof is completed by using (5.17) and recalling that $\mathbf{u}^*$ is a fixed-point solution to the set of m optimizations (5.24), i.e., equivalently $\mathbf{u}^*$ is an NE solution to the Nash game with cost functions L_i, (5.20). □

Proposition 5.9 yields a decomposition into a lower-level modified Nash game with cost functions L_i, (5.20), with no coupled constraints, and a higher-level optimization problem. In general, $\mathbf{u}^*(\mu)$ may not be NE optimal for the given μ, in the sense of attaining the minimum NG cost such that $L_i^* = J_i^*$. However , by Theorem 5.8 there exists a dual optimal price $\mu^* \geq 0$ such that $\mathbf{u}(\mu^*) = [u_i(\mu^*)]$ is NE optimal. Hence μ^* can be found as the maximizer in (5.18). A sufficient condition is that the dual cost $D(\mu)$ is strictly concave in μ, for $\mathbf{u}^*(\mu)$ as obtained from the lower-level game, (5.20). Alternatively, the price μ can be adjusted until the slackness conditions in Theorem 5.8 are satisfied indicating that the dual optimal price μ^* is found.

This decomposition result has a hierarchical game interpretation [20]. At the upper-level is a Stackelberg game ([20], p. 179): the system is the leader that sets "prices" (Lagrange multipliers) and the m players are the followers. Given prices as set by the leader, a Nash game is played at the lower-level between m players, with cost functions L_i, (5.20). Each player reacts to given "prices" and the price acts as a coordination signal.

As in standard optimization, the hierarchical game decomposition may offer computational advantages. For example, the lower-level game may admit a closed-form explicit solution, or the higher-level problem may have a reduced dimension. One such application of these results is presented in the following chapters.

Provided that the action space $\overline{\Omega}$ is compact and convex and Assumption 4.3 is satisfied, the NE solution–Lagrange multiplier pair $(\mathbf{u}^*(\boldsymbol{\mu}^*), \boldsymbol{\mu}^*)$ can be obtained by solving a lower-level uncoupled Nash game with individual cost functions L_i, (5.20), so that $\mathbf{u}^*(\boldsymbol{\mu})$ is obtained and a higher-level problem with respect to the Lagrange multiplier, i.e., solving (5.15).

The duality approach offers a natural way to hierarchically decompose a Nash game with coupled convex constraints into a lower-level Nash game with individual cost functions $L_i(u_{-i}, x_i, \mu)$ and with no coupled constraints, and a higher-level system optimization problem. In effect, the interpretation is that a procedural method for finding a solution to a Nash game with coupled constraints can be based on solving a modified game with no coupled constraints and an optimization problem.

5.6 Notes

Game theoretical formulations of problems and computational approaches towards solving generalized or coupled Nash games have been areas of much recent interest. The study of conditions for existence and uniqueness of Nash equilibrium [95] in pure strategies continues to be a fundamental issue. Only sufficient conditions for existence are available, while for the study of uniqueness by now only partial results are available. Uniqueness results exist only for special classes of games. Recent work on this topic has focused on S-modular games [10], potential games [135], routing games in parallel links [17]. Uniqueness of a normalized equilibrium point is studied in [128]. From a computation point of view, the study of generalized Nash equilibrium presents severe analytical difficulties [40]. Insightful theoretical results have been obtained for computation of equilibria in classes of games with structures, such as two-player polynomial games [113], separable games [151] or potential games [89].

Duality has also received interest from the perspective of games. Duality and dual games are studied for repeated two-player zero-sum games in [35]. The results in [120] and their extension herein have provided a way of finding one such Nash equilibrium, assuming that at least one exists but not presuming uniqueness of the Nash equilibrium. The related similar results in [40] that have appeared almost at the same time as [120] in parallel and independently indicate continued interest in this area after 40 years. In the following chapters we shall mostly follow this approach.

Part II
Game Theory in Optical Networks

Part II
[illegible]

Chapter 6
Optical Networks: Background and Modeling

Abstract This chapter provides an overview of basic background on transmission in optical networks and on general topologies to be studied. Most of the material is adapted from Agrawal (Fiber-optic Communication Systems, 3rd edn. Wiley, New York, 2002) and Ramaswami and Sivarajan (Optical Networks: A Practical Perspective, 2nd edn., Academic Press, San Diego, 2002), and the reader is referred to these references for more in-depth material. The concepts of OSNR and link power capacity constraint in optical networks are introduced as a preface to the remaining chapters.

6.1 Introduction

This book is focused on methodologies and game theory for optical networks from a control perspective. The general setup is that given an optical communication network, some performance measures to be optimized among many networks units/players/channels, one must design an algorithm that achieves as good a performance as possible for each channel. In the following two chapters we consider the problem of developing such algorithms from a game theoretical perspective. Channels seek to re-adjust parameters such that they maintain an optimal quality of service (QoS) (OSNR). Because of optical networks' characteristics, game theoretic formulations and results cannot be ported directly from other application domains (e.g. wireless networks or congestion control). These characteristics are either physical constraints or network specific topologies that translate into various mathematical constraints in the game formulation: coupled as opposed to disjoint action sets, constraints on player's interaction (global versus localized interaction, i.e., one shot end-to-end games versus stage or partitioned games), constraints on players' actions (global constraints versus propagated/modified constraints). In this chapter we present the background and mathematical modeling on optical networks needed to develop these game theoretically inspired algorithms.

Optical wavelength-division multiplexed (WDM) communication networks are evolving beyond statically designed point-to-point links. That is, one goal is how to realize reconfigurable networks with arbitrary topologies, while at the same time maintaining network stability, optimal channel transmission performance and quality of service (QoS), [92].

L. Pavel, *Game Theory for Control of Optical Networks*,
Static & Dynamic Game Theory: Foundations & Applications,
DOI 10.1007/978-0-8176-8322-1_6,

At the physical transmission level, one parameter that directly determines channel performance and QoS is the bit-error rate (BER). BER in turn, depends on optical signal-to-noise ratio (OSNR), dispersion, and nonlinear effects, [2]. Typically in link optimization, OSNR is considered as the dominant performance parameter, with dispersion and nonlinearity effects being kept low by proper link design. This is called the *noise-limited regime* versus the *dispersion-limited regime* as these two regimes can be treated separately. Detailed OSNR computations are needed to determine whether the channel OSNR on a given optical device or route is acceptable [152]. Therefore, OSNR optimization can be directly translated to QoS optimization.

The dominant impairment in OSNR is given by noise accumulation in chains of optical amplifiers and its effect on OSNR, [45]. A traditional approach uses a static budget of fiber and device impairments along a fiber link with sufficient tolerance margins added [126]. Therefore at least a desired OSNR is achieved for each channel. The process is repeated manually after reconfiguration. An optimum map of the amplifier gain distribution has been proposed in [85]. The map not only minimized the effects of the nonlinearity, but also gave rise to the minimum OSNR for a given average power and the minimum average power for a given OSNR. In fact, *power control* has been used to solve OSNR equalization problems since the 1990s. The primary objective of power control is to obtain a satisfactory channel OSNR by regulating channel input power. In [34], a heuristic algorithm was proposed for on-line OSNR equalization in single point-to-point links, while an analysis of the performance limits associated with end-to-end equalization was performed in [154, 155]. The algorithm in [34] is centralized, in that each channel needs to know the OSNR of all other channels. OSNR equalization has been formulated as a static optimization problem in a point-to-point link in [45]. Optimal channel optical power vector at the transmitter is found by solving an eigenvalue problem for the system transmission matrix, composed of link gains of all channels. The proposed algorithm, developed only for a single point-to-point link, requires centralized, global link information.

Current approaches that are developed for single links are not appropriate for reconfigurable optical networks, where the length of the links and the number of devices in a link are changing. Moreover, in these networks arbitrary topologies can be formed, different channels can travel via different optical paths, and also different channels can have different levels of QoS (OSNR) requirements. Therefore one seeks to adjust network parameters (optical power, gains) in an optimal way, based on on-line network feedback (from receiver to transmitter and various nodes). This adjustment can result in increased network flexibility and capacity, so that the needed quality of service (QoS) is ensured whenever a new channel is added/dropped.

These observations justify the need for on-line network control algorithms towards OSNR optimization that have provable convergence properties for general network configurations. Moreover, particularly useful are decentralized algorithms, such that channel power at the transmitter (Tx), can be adjusted based on feedback from the corresponding receiver (Rx) only, plus other channel measurements. Figure 6.1 depicts one such feedback system (encapsulating the mesh network above)

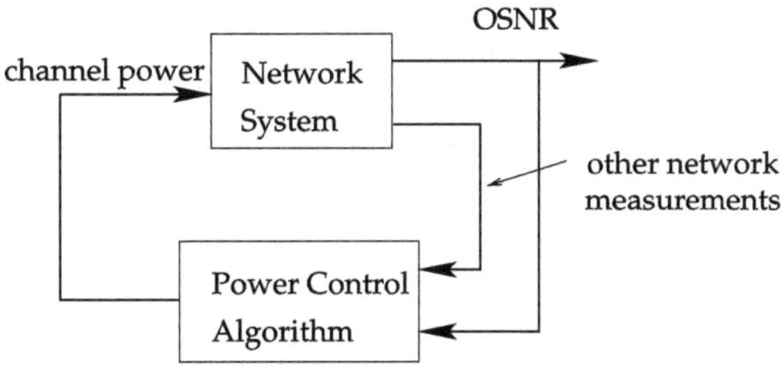

Fig. 6.1 Network control block

with a power control algorithm. We take channel signal power at the sources (channel power) as input to the optical network system and channel OSNR output and other network measurements as feedback signals.

This problem on how to design such algorithms can be approached from an optimization perspective or from a game theoretical perspective. A game theoretical one is suited particularly because it naturally enables one to capture user preferences and player incentives and leads to distributed algorithm. This is the approach we take in this book, although from time to time we refer to and compare it to an optimization approach.

Such power control algorithms can be implemented in an iterative feedback manner. Each channel OSNR is measured by the optical spectrum analyzer (OSA) at the optical switches and the destination. Feedback network measurements are transmitted via optical service channels (OSCs) in optical networks. Each source has a computation unit which performs an algorithmic calculation and determines the corresponding channel input power.

This problem is similar to power control in wireless communication systems, a topic which has been explored extensively, via either centralized approaches [164], or decentralized, noncooperative game approaches [8, 41, 61]. There are several differences that make this problem more challenging in optical networks. In wireless networks, channels are characterized by static loss/gain only, with single Tx to Rx links, or multiple links with no gains (ATM networks). Optical networks bring a new combination of challenges: amplified spans, multiple links, accumulation, and self-generation of optical (ASE) noise, as well as cross-talk generated at the routing elements.

In the following chapters we consider this problem of developing algorithms from a game theoretical perspective for re-adjusting network parameters such that channels maintain an optimal QoS (OSNR), while at the same time taking into account constraints imposed by dispersion and nonlinearity. The material is based on mathematical modeling, presented in this chapter, organized as follows. In the next two sections (Sects. 6.2 and 6.3) we provide basic background on transmission basics in optical networks and on general topologies to be studied. In Sect. 6.4 the analytical mathematical model for OSNR is presented for point-to-point link (Sect. 6.4.1) and for network topology (Sect. 6.4.2), both as recursive relations and end-to-end relations. Link power capacity constraint is introduced in Sect. 6.4.3.

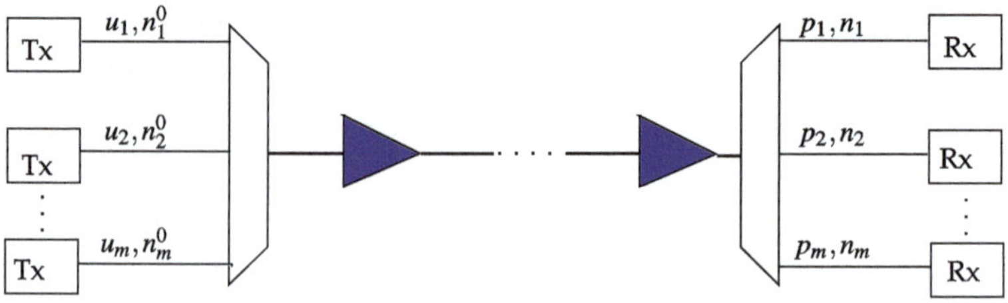

Fig. 6.2 A point-to-point WDM fiber link

6.2 Transmission Basics

In this preliminary section we present some basic principles of optical communications. The textbook [126] gives a broad introductory background on optical communication networks, as well as optical components. On the other hand [2] provides a much more detailed study of the physics involved in optical communication systems, and the reader is referred to these references for more in-depth treatment.

Optical networks are high-capacity communication networks where optical fiber is used as the transmission medium. A set of channels corresponding to a set of wavelengths, are transmitted across the network by intensity modulation of the corresponding wavelength of light. The signals (channels) are multiplexed into a single beam of light using wavelength-division multiplexing (WDM) before being transmitted over the fiber. Each channel is assigned a unique wavelength in the beam of light.

The network medium may be a single point-to-point WDM fiber link (see Fig. 6.2), or a network of optical switches and WDM fiber links. In turn the transmitter block consists of one or more transmitters. Each transmitter consists of a laser and a laser modulator to carry out the E/O conversion, one for each wavelength.

In first-generation optical networks, electronic switches were used; this means that a transmitted signal might go through electrical-to-optical (E/O) and optical-to-electrical (O/E) conversion many times before reaching its destination. Optical networks are called all-optical if a transmitted signal remains in the optical form from its source (transmitter, Tx) to its destination (receiver, Rx) without undergoing any O/E conversions.

Signals at the different wavelengths are combined onto a single fiber by using an optical multiplexer. Individual channel powers can be adjusted at the signal sources (Tx). Signals are demultiplexed at their destination points (Rx) to be processed outside the optical domain.

A receiver block may consist of a demultiplexer and a set of photodetectors. Each photodetector receives a signal at a wavelength and carries out the O/E conversion.

As an optical signal propagates through the fiber, its optical power (which is proportional to the magnitude of the light intensity) is *attenuated* due to absorption and scattering. Optical amplifiers (OAs) are deployed along fiber links throughout the network, about every 80 to 100 kilometers in an optical link, in order to amplify channel powers. Due to the broad spectrum of an optical amplifier gain as well as to there being virtually no cross-talk (slow response-time), a single OA can provide simultaneous amplification of many wavelengths (channels). However the amplifier does not have individual control of the gain or the power of each wavelength (channel); the spectral shape of the gain dictates the amount of gain that each channel will experience. While providing such benefits as simultaneous channel amplification, optical amplifiers have the disadvantage of introducing *amplified-spontaneous emission* (ASE) noise that gets itself amplified and cascaded, leading to optical signal-to-noise (OSNR) degradation [2, 126, 161].

Another important physical impairment is dispersion, which characterizes the propagation of different components of a optical pulse (wavelengths/frequencies, polarization states, modes, etc.) with different speeds in the fiber and this leads to pulse spreading. Effective means of dispersion control have to be imposed for high-performance transmission by using static or tunable dispersion control.

Nonlinear effects appear at transmission of high-intensity light (high power). Due to the fact that the refractive index depends nonlinearly on the light intensity, negative effects such as four-wave mixing (generation of additional frequencies), or coupling of light from one frequency to another (e.g. by stimulated Raman scattering) or self- or cross-phase modulation of the optical signal can develop. Nonlinear effects are negligible if the fiber link is designed properly with optimized total input power of all signals. Such a condition is regarded as a link capacity constraint. The effects of nonlinearity can also be minimized by a map of optical amplifier gain distribution as proposed in [85].

Thus network system performance can be improved by dispersion management techniques and optimization of the total input power. For further details on fiber nonlinearities, dispersion, and their effects, readers are referred to [19, 29, 33, 62, 88, 124, 152, 167], while an in-depth treatment of these effects can be found in [2].

Typically in link optimization, OSNR is considered as the dominant performance parameter, with dispersion and nonlinearity effects being kept low by proper link design. This is called the *noise-limited regime* versus the *dispersion-limited regime*, as these two regimes can be treated separately. ASE noise generated and accumulated in chains of OAs degrades associated channel OSNR.

Detailed OSNR computations are needed to determine whether the channel OSNR on a given optical device or route is acceptable [152]. This OSNR model will be the focus of Sect. 6.4. The next section, Sect. 6.3 gives some background on the type of topologies used in optical networks.

6.3 Topologies and Setup

We consider typical topologies for wavelength-routed optical networks.

A *lightpath* (also referred to as *channel*) is established for each transmitted signal from its associated transmitter to the corresponding receiver as per the set of transmission requests. A light path is routed and switched by optical switches on each intermediate fiber link.

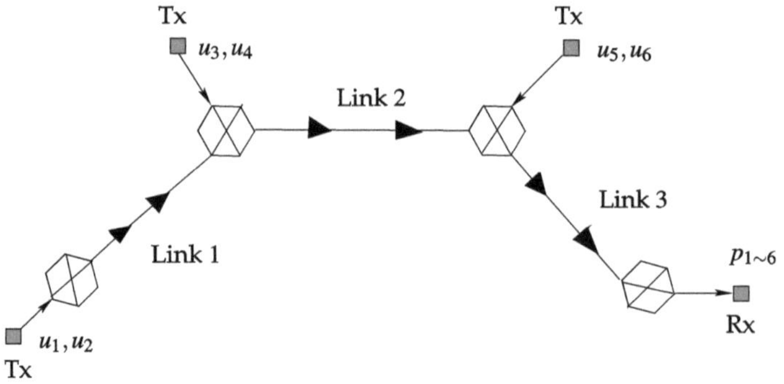

Fig. 6.3 A single-sink multi-link topology

The collection of lightpaths is called the *virtual topology* in the optical layer. The optical switch may be equipped with a wavelength converter. Then some channels may be converted from one wavelength to another as well along their route by *wavelength conversion*.

Consider an optical network that is defined by a set of optical fiber links $\mathcal{L} = \{1, \ldots, L\}$ connecting optical switch nodes. A link $l \in \mathcal{L}$ is composed of N_l optical amplified spans. Each span includes an optical fiber followed by an optical amplifier.

These nodes allow channels in the network to be added, dropped or routed and can also provide the flexibility of channel power adjustments [2].

Optical switch nodes could be optical cross-connects (OXCs) or optical add/drop multiplexers (OADMs). An OXC deals with multiple wavelengths and selectively drops some of these wavelengths locally or adds selected wavelengths while letting others pass through. An OADM provides a similar function but at a much smaller size.

There are various topologies that an optical network can be configured into, the simplest one being the point-to-point link topology (Fig. 6.2).

Another simple topology is a single-sink topology where channels are added at different links, but dropped off at the output of the same link (e.g. Fig. 6.3), multi-link topology (Fig. 6.4), moving to the most complex, mesh topology (Fig. 6.5(a)).

A set of channels, $\mathcal{M} = \{1, \ldots, m\}$, corresponding to a set of wavelengths, are transmitted across the network by intensity modulation and wavelength-multiplexing. We denote by $\mathcal{M}_l$ the set of channels transmitted over link $l \in \mathcal{L}$. Also, we denote by $\mathcal{R}_i$, $i \in \mathcal{M}$, the set of links from the associated Tx to the corresponding Rx that channel i uses in its optical route. For a link $l \in \mathcal{R}_i$, we sometimes use the notation $\mathcal{R}_{i,l}$ to denote the set of links previous to link l on the path $\mathcal{R}_i$, i.e., for $\forall l \in \mathcal{R}_i$,

$$\mathcal{R}_{i,l} := \{l' \in \mathcal{L} \mid l' \in \mathcal{R}_i,\ l' \text{ precedent to } l\}$$

We denote by u_i, n_i^0, p_i and n_i the channel signal power at Tx, the channel noise power at Tx, the channel signal power at Rx and the channel power at Rx, respectively, for channel $i \in \mathcal{M}$ (illustrated in Fig. 6.5(a)). We let $\mathbf{u} =$

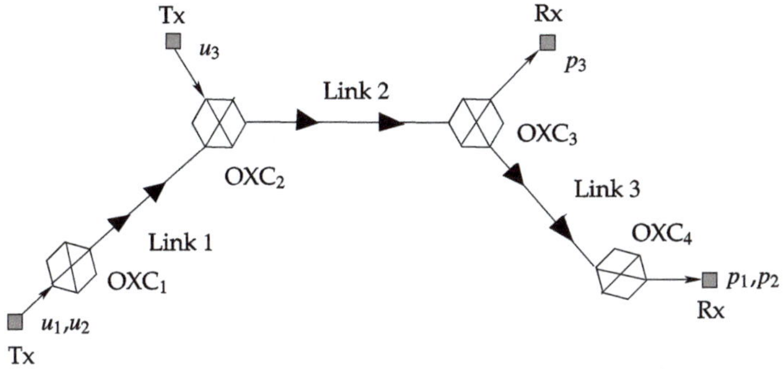

Fig. 6.4 Example: three-link topology with three channels

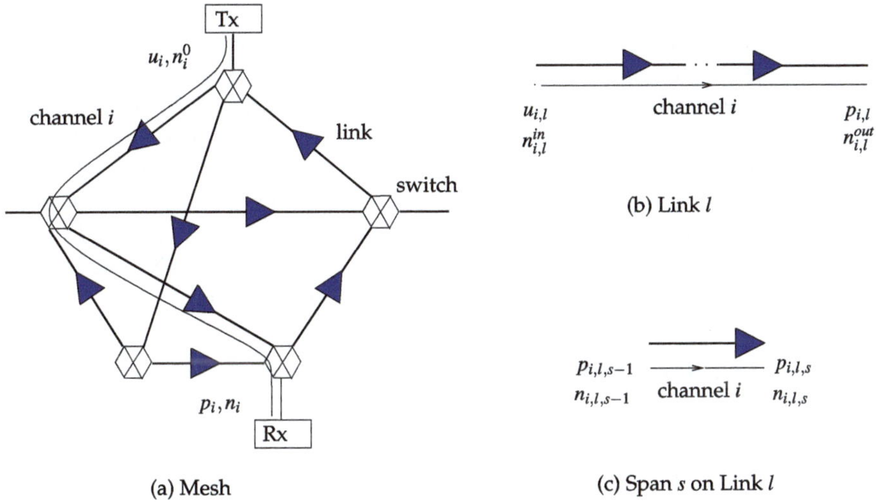

Fig. 6.5 A wavelength-routed optical network configuration: mesh, link, and span

$[u_1, \ldots, u_i, \ldots, u_m]^T$ or $\mathbf{u} = [u_i]$ denote the vector form. We also use $\mathbf{u} = (\mathbf{u}_{-i}, u_i)$ with $\mathbf{u}_{-i} = [u_1, \ldots, u_{i-1}, u_{i+1}, \ldots, u_m]^T$.

Sometimes we use $\mathbf{u} = [u_i]$ so that $[\mathbf{u}]_i = \mathbf{u}(i) = u_i$.

We sometimes use for convenience the following notation. For two vectors $\mathbf{p}$ and $\mathbf{n}$ of the same dimension, we denote by $\mathbf{p}./\mathbf{n}$ element-by-element division, such that

$$[\mathbf{p}./\mathbf{n}]_i = [\mathbf{p}]_i/[\mathbf{n}]_i = p_i/n_i$$

We sometimes also use $.*$ to denote element-by-element vector multiplication, such that

$$[\mathbf{y}. * \mathbf{n}]_i = [\mathbf{y}]_i * [\mathbf{n}]_i = y_i * n_i$$

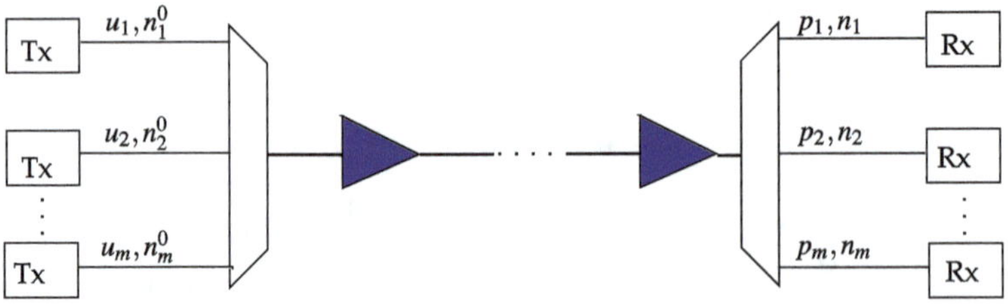

Fig. 6.6 A point-to-point WDM fiber link

6.4 Power Control and OSNR Model

As we mentioned in the previous sections, the bit-error rate of the transmission depends on the optical signal-to-noise ratio (OSNR) of the signal channels at reception, Rx. As the signal power in a channel increases, the OSNR value of the channel increases, and the bit-error rate decreases. That is, maximization of channel OSNR values is particularly important in optical networks. However, this channel OSNR optimization is not straightforward, in that a trade-off exists on the attainable OSNR values of each channel. As one channel increases its signal power, thereby increasing its OSNR, it may adversely affect the OSNR values of the other channels in the network. This will become more apparent after this section, where we present the modeling of channel OSNR in optical networks. The material in this section is based mostly on [2, 45, 119]. We start with the basic OSNR model for a point-to-point link, and then move on towards the network model.

6.4.1 Point-to-Point Link Model

Consider a single point-to-point WDM link as shown in Fig. 6.6, composed of N cascaded optical amplified spans, i.e., optical fiber and amplifiers.

A set $\mathcal{M} = \{1, \dots, m\}$ of channels are transmitted over this link. Each channel i corresponds to a wavelength λ_i, $i \in \mathcal{M}$. We denote by $p_{i,s}$ and $n_{i,s}$ the optical signal and noise power of channel i at the output of span s, respectively. For each $s = 1, \dots, N$, we denote these in vector form by $\mathbf{p}_s = [p_{i,s}]$, $\mathbf{n}_s = [n_{i,s}]$, where $i \in \mathcal{M}$. Let channel signal and noise power at the input of the link be denoted by u_i and n_i^0, respectively, in vector form $\mathbf{u} = [u_i]$, $\mathbf{n}^0 = [n_i^0]$, and by convention [alph]

$$u_i = p_{i,0}, \qquad \mathbf{u} = \mathbf{p}_0 \tag{6.1a}$$

$$n_i^0 = n_{i,0}, \qquad \mathbf{n}^0 = \mathbf{n}_0 \tag{6.1b}$$

Let channel signal and noise power at the output of the link (at Rx) be denoted by p_i, n_i^{out}, respectively, or in vector form $\mathbf{p}$ and $\mathbf{n}^{\text{out}}$. These are taken to be the same as the channel signal and noise power at the output of span N, i.e.,

$$p_i = p_{i,N}, \qquad \mathbf{p} = \mathbf{p}_N \tag{6.2a}$$

$$n_i^{\text{out}} = n_{i,N}, \qquad \mathbf{n}^{\text{out}} = \mathbf{n}_N \tag{6.2b}$$

The optical signal-to-noise ratio (OSNR) at Rx, for the ith channel, $i \in \mathcal{M}$, denoted by y_i is defined by

$$y_i := \frac{p_i}{n_i^{\text{out}}} \quad \text{or} \quad \mathbf{y} = \mathbf{p} \cdot / \mathbf{n}^{\text{out}} \tag{6.3}$$

in vector form.

The following basic propagation relations will be used. An optical amplified span s, $s = 1, \ldots, N$ on each link is composed of an optical fiber with attenuation (loss) coefficient, L_s, which is wavelength independent, and an optical amplifier. An optical amplifier simultaneously increases the optical power of all channels $i \in \mathcal{M}$ multiplexed (transmitted) over the link, with a wavelength-dependent gain (spectral gain shape).

Let us consider an amplifier on span s and let the gain it provides to channel i be denoted by $g_{i,s}$. Then the *span transmission* of span s for channel i is defined as

$$h_{i,s} = g_{i,s} L_s, \quad \forall s = 1, \ldots, N \tag{6.4}$$

denoted in vector form by $\mathbf{h}_s = [h_{1,s} \ldots h_{i,s} \ldots h_{m,s}]^T$. Similarly let *transmission* up to span s for channel i be denoted as $H_{i,s}$ and defined by

$$H_{i,s} = \prod_{q=1}^{s} h_{i,q} \tag{6.5}$$

and in vector form by $\mathbf{H}_s = [H_{1,s} \ldots H_{i,s} \ldots H_{m,s}]^T$.

An optical amplifier introduces broadband amplified-spontaneous emission noise (ASE noise). This is in fact the dominant impairment affecting channel performance and OSNR, i.e., accumulation of ASE noise in amplifiers. The reason is that typically dispersion and nonlinearity effects are considered to be limited (see [45, 85]). ASE noise generated by an optical amplifier is gain dependent which means that it is wavelength-dependent also, due to the spectral gain shape of the amplifier.

The following assumption is typically used with respect to ASE noise [45].

Assumption 6.1 *ASE noise does not contribute to the amplifier gain saturation.*

ASE noise power generated within the bandwidth of the ith channel by the sth amplifier is denoted by $ASE_{i,s}$ and is given as [2]

$$ASE_{i,s} = 2n_{\text{sp}}[g_{i,s} - 1]h\nu_i B_o \tag{6.6}$$

where $n_{\text{sp}} > 1$ is a constant factor called amplifier excess noise factor, h is the Planck constant, B_o is the optical bandwidth, and ν_i is the optical frequency corresponding to wavelength λ_i, where $\lambda_i = c/\nu_i$, with c the speed of light. Typical values are λ_i in the 1550 nm (infrared) range and ν_i in the THz range. In vector form we denote $\mathbf{ASE}_s = [ASE_{1,s} \ldots ASE_{i,s} \ldots ASE_{m,s}]^T$.

Noise is generated additively at each amplifier. Moreover in cascaded amplifiers, each OA amplifies simultaneously signal and noise power, since ASE is in-band noise (same wavelength or optical frequency as the signal channel). Thus, channel signal and noise power at the output of span s on the link, denoted by $p_{i,s}$, $n_{i,s}$ are given for $\forall s = 1, \dots, N$ as

$$p_{i,s} = h_{i,s} p_{i,s-1}, \tag{6.7a}$$

$$n_{i,s} = h_{i,s} n_{i,s-1} + ASE_{i,s}, \tag{6.7b}$$

where as before $p_{i,0} = u_i$, $n_{i,0} = n_i^0$, or in vector form

$$\mathbf{p}_s = \mathrm{Diag}(\mathbf{h}_s)\mathbf{p}_{s-1}, \tag{6.8a}$$

$$\mathbf{n}_s = \mathrm{Diag}(\mathbf{h}_s)\mathbf{n}_{s-1} + \mathbf{ASE}_s, \tag{6.8b}$$

where $\mathrm{Diag}(\mathbf{v})$ denotes the diagonal matrix with the elements of vector $\mathbf{v}$ as diagonal entries.

In the following we present the OSNR model, i.e., its relation to the input optical signal and noise power. We consider only forward propagation of signal and noise in steady-state, [45], i.e., we do not consider amplifier gain dynamics. This is justified for any OSNR game or optimization problem where power adjustments need to be made at steady-state, after updating network topology.

We can consider separately two cases depending on the amplifier operation mode: gain control mode (Lemma 6.1), and power control mode (Lemma 6.2) in [119]. In the gain control mode, the gain of an amplifier is kept constant, while in the power control mode the total output power is kept constant. The case of gain control mode leads to a simpler mathematical model (given in Lemma 6.1): OSNR on a particular channel does not depend on the power of the rest of the channels. However, the typical operation mode is the automatic power control mode because this ensures that a constant power is launched into the next span of a link, which leads to uniform total power distribution across a link and thus limits the nonlinear effects [2, 85]. The model for this case is given in Lemma 6.2.

The following lemma gives the model for gain control mode.

Lemma 6.1 *Let* $\mathbf{u}$, $\mathbf{n}^0$ *be the optical signal and noise power at the input* (*Tx*) *of a point-to-point optical link with* N *spans and amplifiers in gain control mode. Under Assumption* 6.1 *the optical signal power* $\mathbf{p}$, *noise power* $\mathbf{n}^{\text{out}}$ *and the OSNR at Rx* $\mathbf{y}$ (6.3) *are given in vector form as*

$$\mathbf{p} = \mathrm{Diag}(\mathbf{H}_N)\mathbf{u} \tag{6.9a}$$

$$\mathbf{n}^{\text{out}} = \mathrm{Diag}(\mathbf{H}_N)\left(\mathbf{n}^0 + \sum_{r=1}^{N} \left[\mathrm{Diag}(\mathbf{H}_r)\right]^{-1} \mathbf{ASE}_r\right) \tag{6.9b}$$

and

$$\mathbf{y} = \mathbf{u}./\left(\mathbf{n}^0 + \sum_{r=1}^{N} \left[\mathrm{Diag}(\mathbf{H}_r)\right]^{-1} \mathbf{ASE}_r\right) \tag{6.10}$$

where $\mathbf{H}_r$ *is transmission up to span* r, (6.5), $\mathbf{H}_N$ *is the link transmission, and* $\mathrm{Diag}(\mathbf{v})$ *denotes the diagonal matrix with the elements of vector* $\mathbf{v}$ *as diagonal entries.*

Proof The proof follows by developing propagation relations for both signal and noise. Using (6.8a), (6.8b) yields

$$\mathbf{p}_s = \mathrm{Diag}(\mathbf{h}_s)\,\mathrm{Diag}(\mathbf{h}_{s-1})\mathbf{p}_{s-2},$$
$$\mathbf{n}_s = \mathrm{Diag}(\mathbf{h}_s)\,\mathrm{Diag}(\mathbf{h}_{s-1})\mathbf{n}_{s-2} + \mathrm{Diag}(\mathbf{h}_s)\mathbf{ASE}_{s-1} + \mathbf{ASE}_s$$

Using these recursively after s together with (6.1a), (6.1b), yields for $\forall s = 1, N$

$$\mathbf{p}_s = \prod_{q=1}^{s} \mathrm{Diag}(\mathbf{h}_q)\mathbf{u},$$
$$\mathbf{n}_s = \prod_{q=1}^{s} \mathrm{Diag}(\mathbf{h}_q)\mathbf{n}^0 + \sum_{r=0}^{s-2}\left(\prod_{v=s-r}^{s} \mathrm{Diag}(\mathbf{h}_v)\right)\mathbf{ASE}_{s-r-1} + \mathbf{ASE}_s,$$

or

$$\mathbf{p}_s = \prod_{q=1}^{s} \mathrm{Diag}(\mathbf{h}_q)\mathbf{u},$$
$$\mathbf{n}_s = \prod_{q=1}^{s} \mathrm{Diag}(\mathbf{h}_q)\mathbf{n}^0$$
$$+ \sum_{r=0}^{s-2}\left(\prod_{v=1}^{s} \mathrm{Diag}(\mathbf{h}_v)\right)\left(\prod_{v=1}^{s-r-1} \mathrm{Diag}(\mathbf{h}_v)\right)^{-1} \mathbf{ASE}_{s-r-1} + \mathbf{ASE}_s$$

After some manipulation, factoring out $\prod_{v=1}^{s} \mathrm{Diag}(\mathbf{h}_v)$, denoting $s-r-1$ by new index r we obtain

$$\mathbf{p}_s = \prod_{q=1}^{s} \mathrm{Diag}(\mathbf{h}_q)\mathbf{u},$$
$$\mathbf{n}_s = \left(\prod_{q=1}^{s} \mathrm{Diag}(\mathbf{h}_q)\right)\left(\mathbf{n}^0 + \sum_{r=1}^{s}\left[\prod_{v=1}^{r} \mathrm{Diag}(\mathbf{h}_v)\right]^{-1} \mathbf{ASE}_r\right)$$

Now from (6.5) one can write

$$\mathrm{Diag}(\mathbf{H}_s) = \mathrm{Diag}(\mathbf{h}_s)\,\mathrm{Diag}(\mathbf{h}_{s-1})\cdots\mathrm{Diag}(\mathbf{h}_1) = \prod_{q=1}^{s} \mathrm{Diag}(\mathbf{h}_q)$$

so that, for $\forall s = 1, \ldots, N$, the foregoing is written as

$$\mathbf{p}_s = \mathrm{Diag}(\mathbf{H}_s)\mathbf{u},$$

$$\mathbf{n}_s = \mathrm{Diag}(\mathbf{H}_s)\left(\mathbf{n}^0 + \sum_{r=1}^{s} [\mathrm{Diag}(\mathbf{H}_r)]^{-1}\mathbf{ASE}_r\right)$$

Using these for $s = N$ with $\mathbf{p} = \mathbf{p}_N$, $\mathbf{n}^{\mathrm{out}} = \mathbf{n}_N$ (see (6.2a), (6.2b)) gives (6.9a), (6.9b). Then (6.10) follows immediately by using these into (6.3). □

Next one can use this result and develop a model for the case when amplifiers are operated in automatic power control mode, which is the typical operation mode for optical amplifiers when the network is in steady-state, [2]. We will see that this case leads to a more complex mathematical model; the inherent scaling on the total power translates into coupling between all channels' powers, and OSNR on a particular channel does depend on all other channels' powers.

The following assumption is used for optical amplifiers.

Assumption 6.2 *All spans on each link have equal length and all optical amplifiers are operated in automatic power control mode with the same total power target* $\widehat{P}$ *and have the same gain spectral shape* [2, 85].

This assumption of same gain spectral shape and power target is a reasonable assumption if all spans in a link have the same length, i.e., for uniformly designed optical links, [2]. Then $g_{i,s} = g_i$ for $s = 1, \ldots, N$ (same spectral shape), and by (6.6) the ASE noise generated is the same at each amplifier on the link. In vector form let $\mathbf{g} = [g_1 \ldots g_i \ldots g_m]^T$ or $\mathbf{g} = [g_i]$ denote the vector spectral gain shape of any amplifier on the link. Similarly for each $s = 1, \ \ldots, N$ let $\mathbf{g}^s = [g_1^s \ldots g_i^s \ldots g_m^s]^T$ or $\mathbf{g}^s = [g_i^s]$.

Also by Assumption 6.2 at the output of each span $s = 1, \ldots, N$ one can write

$$\mathbf{1}_m^T \mathbf{p}_s = \widehat{P}, \tag{6.11}$$

and at the output of the link

$$\mathbf{1}_m^T \mathbf{p} = \widehat{P}, \tag{6.12}$$

where $\mathbf{1}_m = [1 \ 1 \ \ldots \ 1]^T$ is the m-dimensional all-ones vector. The above condition is called the *system constraint*. One benefit of keeping a constant total power at the output of the amplifier, or a constant total launching power into next span, is the compensation of variations in fiber-span loss across a link [45]. Moreover, if the total power target is selected to be below the threshold for nonlinear effects, an optimal gain distribution is achieved across the link [85].

Mathematically, in automatic power control mode a scaling of the gain is performed such that the same total power is kept at the output of each amplifier [45]. This can be realized by using a variable optical filter after each amplifier, this filter having the loss coefficient η_s adjusted to achieve the constant total power target $\widehat{P}$ [45]. Thus under Assumption 6.2, the effective gain of an optical amplifier can be written for $s = 1, \dots, N$, $i \in \mathcal{M}$, as

$$g_{i,s} = g_i \eta_s, \tag{6.13}$$

where g_i is the corresponding spectral gain value for channel i, and η_s the filter loss. As we will see next the scaling via η_s introduces coupling between channel powers, and channel OSNR is no longer independent of other channels' powers.

The following result gives the OSNR model for a point-to-point WDM link developed based on (6.2a), (6.2b), (6.11), (6.20), and (6.7a), (6.7b). An earlier similar simple model can be found in [45].

Lemma 6.2 *Let* $\mathbf{u}$, $\mathbf{n}^0$ *be the signal and noise power at the input (Tx) of a point-to-point optical link with N spans and amplifiers in power control mode. Under Assumptions* 6.1 *and* 6.2, *the optical signal power* $\mathbf{p}$ *and the OSNR at Rx* $\mathbf{y}$ (6.3) *are given in vector form as*

$$\mathbf{p} = \text{Diag}(\mathbf{H}_N)\mathbf{u} \tag{6.14}$$

where $\mathbf{H}_N = \frac{\widehat{P}}{[\mathbf{g}^N]^T\mathbf{u}}\mathbf{g}^N$ *and*

$$\mathbf{y} = \mathbf{u}./\big(\mathbf{n}^0 + \Gamma\mathbf{u}\big) \tag{6.15}$$

where $\Gamma = [\Gamma_{i,j}]$ *is the* $(m \times m)$ link system matrix

$$\Gamma = \sum_{r=1}^{N} \big[\text{Diag}(\mathbf{g}^r)\big]^{-1}\frac{\mathbf{ASE}}{\widehat{P}}\big(\mathbf{g}^r\big)^T$$

and Diag($\mathbf{v}$) *denotes the diagonal matrix with the elements of vector* $\mathbf{v}$ *as diagonal entries.*

Component-wise one can write from (6.15)

$$y_i = \frac{u_i}{n_i^0 + \sum_{j\in\mathcal{M}} \Gamma_{i,j}u_j} := \mathbf{f}_i\big(\mathbf{u}, \mathbf{n}^0\big), \tag{6.16}$$

where

$$\Gamma_{i,j} = \sum_{s=1}^{N} \frac{g_j^s}{g_i^s}\frac{ASE_i}{\widehat{P}} \tag{6.17}$$

and $\mathbf{H}_N = [H_{i,N}]$, $i \in \mathcal{M}$

$$H_{i,N} = \frac{g_i^N \widehat{P}}{\sum_{j\in\mathcal{M}} g_j^N u_j}$$

Proof From Lemma 6.1, (6.9a), (6.9b), one can write for $\forall s = 1, \ldots, N$

$$\mathbf{p}_s = \text{Diag}(\mathbf{H}_s)\mathbf{u}, \tag{6.18a}$$

$$\mathbf{n}_s = \text{Diag}(\mathbf{H}_s)\left(\mathbf{n}^0 + \sum_{r=1}^{s}\left[\text{Diag}(\mathbf{H}_r)\right]^{-1}\mathbf{ASE}_r\right), \tag{6.18b}$$

where $\mathbf{ASE}_r = \mathbf{ASE}$, for $r = 1, \ldots, N$ by Assumption 6.2. Using (6.18a) into (6.11) gives

$$\widehat{P} = \mathbf{1}_m^T \,\text{Diag}(\mathbf{H}_s)\mathbf{u} = \mathbf{H}_s^T \mathbf{u} \tag{6.19}$$

Recall that $\mathbf{H}_s = [H_{i,s}]$, $i \in \mathcal{M}$ where $H_{i,s}$ is defined in (6.5) as

$$H_{i,s} = \prod_{q=1}^{s} h_{i,q}$$

From (6.4) and (6.13) $h_{i,q}$ can be written as

$$h_{i,q} = g_i L_q \eta_q = g_i \delta_q, \quad \forall q = 1, \ldots, N, \tag{6.20}$$

where we denoted $\delta_q = L_q \eta_q$ (scalar), or in vector form

$$\mathbf{h}_q = \mathbf{g}\delta_q, \quad \forall q = 1, \ldots, N$$

where $\mathbf{h}_s = [h_{1,s} \ldots h_{i,s} \ldots h_{m,s}]^T = [h_{i,s}]$, $\mathbf{g} = [g_i]$. Substituting (6.20) into the expression above for $H_{i,s}$ yields

$$H_{i,s} = \prod_{q=1}^{s} g_i \delta_q = g_i^s \prod_{q=1}^{s} \delta_q, \quad \forall s = 1, \ldots, N,$$

or, by denoting $\rho_s = \prod_{q=1}^{s} \delta_q$,

$$H_{i,s} = g_i^s \rho_s, \quad \forall s = 1, \ldots, N$$

Then in vector form for $\mathbf{H}_s = [H_{i,s}]$, $i \in \mathcal{M}$ one can write

$$\mathbf{H}_s = \mathbf{g}^s \rho_s, \quad \forall s = 1, \ldots, N \tag{6.21}$$

where the notation $\mathbf{g}^s = [g_i^s]$ was used and ρ_s is scalar.

Now using (6.21) in (6.19) yields

$$\widehat{P} = \rho_s \left[\mathbf{g}^s\right]^T \mathbf{u}, \quad \forall s = 1, \ldots, N$$

so that

$$\rho_s = \frac{\widehat{P}}{[\mathbf{g}^s]^T \mathbf{u}}, \quad \forall s = 1, \ldots, N$$

Hence

$$\mathbf{H}_s = \frac{\widehat{P}}{[\mathbf{g}^s]^T \mathbf{u}} \mathbf{g}^s, \tag{6.22}$$

which when used in (6.18a) for $s = N$ yields (6.14). Note that component-wise one can write

$$H_{i,s} = \frac{g_i^s \widehat{P}}{\sum_{j=1}^m g_j^s u_j}$$

Next, for the second part, observe that using (6.22) one can write for the last term in (6.18b)

$$\left[\mathrm{Diag}(\mathbf{H}_r)\right]^{-1} \mathbf{ASE} = \frac{1}{\widehat{P}} \left[\mathrm{Diag}\left(\mathbf{g}^r\right)\right]^{-1} \mathbf{ASE}\left(\mathbf{g}^r\right)^T \mathbf{u}$$

where the fact that $(\mathbf{g}^r)^T \mathbf{u}$ is a scalar was used. Thus replacing this into (6.18b) yields

$$\mathbf{n}_s = \mathrm{Diag}(\mathbf{H}_s)\left(\mathbf{n}^0 + \Gamma_s \mathbf{u}\right), \tag{6.23}$$

where Γ_s is defined by

$$\Gamma_s = \sum_{r=1}^{s} \frac{1}{\widehat{P}} \left[\mathrm{Diag}\left(\mathbf{g}^r\right)\right]^{-1} \mathbf{ASE}\left(\mathbf{g}^r\right)^T$$

Thus, for OSNR, $\mathbf{y} = \mathbf{p}./\mathbf{n}^{\mathrm{out}}$, (6.3), using $\mathbf{p}$ in (6.14) and $\mathbf{n}^{\mathrm{out}} = \mathbf{n}_N$ in (6.23) for $s = N$, yields

$$\mathbf{y} = \mathbf{p}./\mathbf{n} = \mathbf{u}./\left(\mathbf{n}^0 + \Gamma \mathbf{u}\right)$$

with $\Gamma = \Gamma_N$, or component-wise

$$y_i = OSNR_i = \frac{p_i}{n_i} = \frac{u_i}{n_i^0 + \sum_{j \in \mathcal{M}} \Gamma_{i,j} u_j}$$

and the proof is completed. □

Remark 6.3 This is the OSNR model typically used in optical links. Mathematically the OSNR model in (6.16) is similar to the wireless signal-to-interference ratio (SIR) model [8]. However, it has a richer system structure: Γ with $\Gamma_{i,j}$ defined in (6.17) is a full matrix, with cross-coupling terms, non-zero diagonal elements and all elements dependent on network parameters (e.g., the gain and ASE noise of optical amplifiers). We can rewrite (6.16) as

$$y_i := OSNR_i = \frac{u_i}{X_{-i} + \Gamma_{i,i} u_i}, \tag{6.24}$$

where

$$X_{-i} = n_i^0 + \sum_{j \in \mathcal{M}, j \neq i} \Gamma_{i,j} u_j. \tag{6.25}$$

Hence OSNR is no longer a linear function of channel input power like SIR as in [8]. Furthermore, since optical amplifiers are often cascaded along the link, ASE noise accumulates over many amplifiers and degrades channel OSNR as the number of amplifiers increases, which is reflected by (6.16).

6.4.2 Network Model

Let us look now at the network model. The same assumptions are used and similar notations. Thus consider an optical network with a set $\mathcal{L} = \{1, \ldots, L\}$ of links. A set $\mathcal{M} = \{1, \ldots, m\}$ of channels transmitted over the network. Let $\mathcal{M}_l = \{1, \ldots, m_l\}$ be the set of channels transmitted over link $l \in \mathcal{L}$. Each link l is composed of N_l cascaded optical amplifiers and optical fiber spans and has $\hat{P}_l$ as its constant total power target. We denote by $g_{i,l}$ the gain experienced by channel i due to the spectral gain shape (wavelength dependency) of the amplifier on link l. By $\mathcal{R}_i$ we denote the route of channel i from Tx to Rx. Let the optical fiber from each Tx to the optical switch that is connected be defined as a *virtual optical link* (VOL) (see Fig. 6.7). Thus the set of VOLs, $\mathcal{L}_v$, is equivalent to $\mathcal{M}$. Therefore, the set of all optical links and VOLs is denoted by $\mathcal{L}' = \mathcal{L} \cup \mathcal{L}_v = \{1, \ldots, L'\}$, where $L' = L + m$.

We define two network connection matrices. First, let $A = [A_{i,l}]_{m \times L'}$ denote the *channel transmission matrix*, where

$$A_{i,l} = \begin{cases} 1, & \text{channel } i \text{ uses link or virtual link } l; \\ 0, & \text{otherwise} \end{cases}$$

Similarly, let $B = [B_{k,l}]_{L' \times L}$ denote a *system connection matrix* defined as

$$B_{k,l} = \begin{cases} 1, & \text{link/virtual link } k \text{ is connected to link } l, \\ & \text{i.e., the channel transmission direction between } k \text{ and } l \text{ is } k \to l; \\ 0, & \text{otherwise} \end{cases}$$

Let $A = [A_l]$ with $A_l = [A_{1,l}, \ldots, A_{m,l}]^T$ and $B = [B_l]$ with $B_l = [B_{1,l}, \ldots, B_{L',l}]^T$. The channel transmission matrix and the system connection matrix completely describe channel routing and system interconnection, i.e., the network configuration.

Next we introduce the following notations for channels. Let $\mathbf{u}_l$ denote the m-dimensional input power vector at each link l (all channels), and $\bar{\mathbf{u}}_i$ denotes the L'-dimensional input power vector for each channel (all links), respectively,

$$\mathbf{u}_l = [u_{1,l} \ldots u_{i,l} \ldots u_{m,l}]^T, \quad l = 1, \ldots, L',$$

$$\bar{\mathbf{u}}_i = [u_{i,1} \ldots u_{i,l} \ldots u_{i,L'}]^T, \quad i = 1, \ldots, m$$

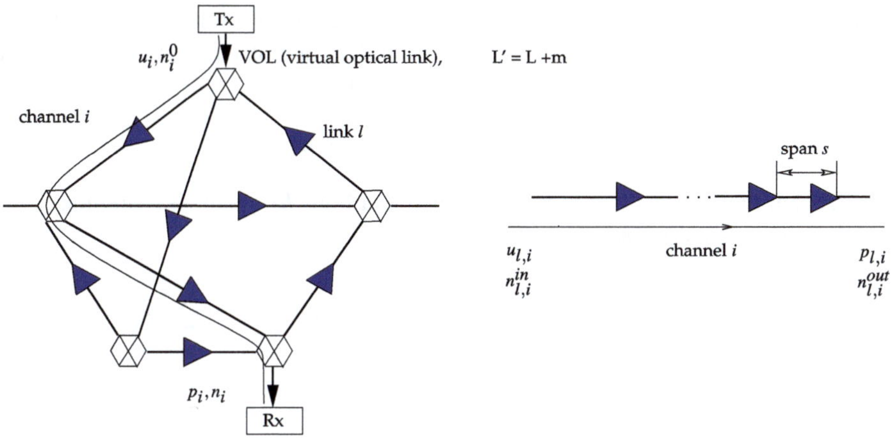

Fig. 6.7 Network configuration: mesh

Similar notations are used for the output powers: $\mathbf{p}_l$ denotes the m-dimensional output power vector at each link l (all channels), and $\overline{\mathbf{p}}_i$ denotes the L'-dimensional output power vector for each channel (all links), respectively.

Based on the network connection matrices, the following result gives interconnection relation for all the channel powers at link l, i.e., how the input channel powers $\mathbf{u}_l$ are obtained from the other links or from the virtual optical links (VOL).

The first case we treat is the end-to-end case where we assume that channel powers are adjustable only at the transmitter sites. Later on in this section we treat the recursive case where channel powers are individually adjustable also at the input of each link by some adjustment factors γ, hence called γ-link. In this first case the channel signal power at the input of link l, denoted by $u_{i,l}$ is identical to the signal power at the output of the preceding link l' of link l on route $\mathcal{R}_i$, i.e.,

$$u_{i,l} = p_{i,l'} \tag{6.26}$$

Lemma 6.4 *The optical signal powers at the input of link l, $\mathbf{u}_l$, $l \in \mathcal{L}$, are given in vector form as*

$$\mathbf{u}_l = V_l^T \overline{\mathbf{p}}, \tag{6.27}$$

where $\overline{\mathbf{p}} = [\overline{\mathbf{p}}_1 \dots \overline{\mathbf{p}}_i \dots \overline{\mathbf{p}}_m]^T$ *and* $\overline{\mathbf{p}}_i = [p_{i,1} \dots p_{i,k} \dots p_{i,L'}]^T$, *with*

$$p_{i,k} = \begin{cases} p_{i,k}, & k \in \mathcal{L} \\ u_i, & k \in \mathcal{L}_v \end{cases}$$

where u_i is the signal power of channel i at Tx, and the matrix V_l is the direct sum of $V_{i,l}$, $V_l = \bigoplus_{i=1}^m V_{i,l}$, where $\forall i = 1, \dots, \mathcal{M}$, $l \in \mathcal{L}$,

$$V_{i,l} = [A_{i,l} B_{1,l} A_{i,1}, \dots, A_{i,l} B_{L',l} A_{i,L'}].$$

Proof For channel i, $i \in \mathcal{M}$ its signal power launched into link l, $l \in \mathcal{L}$ is transmitted either from *one* of previous links of link l or one of virtual links (Txs). Let $k' \in \mathcal{L}'$ be such a link. Then $B_{k',l} = 1$, $A_{i,k'} = 1$ and

$$u_{i,l} = A_{i,l}(B_{k',l} A_{i,k'} p_{i,k'}).$$

Notice that there is *one and only one* k' such that both $B_{k',l}$ and $A_{i,k'}$ are non-zero. So the above equation can be rewritten as

$$u_{i,l} = A_{i,l}\left(B_{k',l} A_{i,k'} p_{i,k'} + \sum_{k \in \mathcal{L}', k \neq k'} B_{k,l} A_{i,k} p_{i,k}\right) = A_{i,l} \gamma_{i,l} \sum_{k \in \mathcal{L}'} B_{k,l} A_{i,k} p_{i,k}$$

where

$$p_{i,k} = \begin{cases} p_{i,k}, & k \in \mathcal{L}, \\ u_i, & k \in \mathcal{L}_v, \end{cases}$$

and u_i is the signal power of channel i at Tx. Since $\mathcal{L}' = \{1, \dots, L'\}$, $u_{i,l}$ can be expressed as

$$u_{i,l} = [A_{i,l} B_{1,l} A_{i,1}, \dots, A_{i,l} B_{L',l} A_{i,L'}] \begin{bmatrix} p_{i,1} \\ \vdots \\ p_{i,L'} \end{bmatrix} = \gamma_{i,l} \cdot V_{i,l} \cdot \bar{\mathbf{p}}_i \tag{6.28}$$

Recalling that $V_l = \bigoplus_{i=1}^m V_{i,l} = \operatorname{Diag}(V_{i,l})$, the vector form of the foregoing yields (6.27). □

Remark 6.5 The matrix $\operatorname{Diag}(V_{i,l}^T)$ is an augmented system configuration matrix, indicating not only the connections between links but also channel routing conditions. From the system control point of view, for the network system with channels transmitted over it, input of this system is the signal power at the input side of each link and the output is the output signal power of each link.

Remark 6.6 Since noise is propagated in the same manner via the optical switch (no O/E conversion), it can be immediately seen that, as in (6.26),

$$n_{i,l}^{\text{in}} = n_{i,l'}^{\text{out}}.$$

Thus the input noise power $n_{i,l}^{\text{in}}$ is given similarly as

$$n_{i,l}^{\text{in}} = A_{i,l} \sum_{k \in \mathcal{L}'} \left(B_{k,l} A_{i,k} n_{i,k}^{\text{out}}\right), \quad \forall i \in \mathcal{M},\ l \in \mathcal{L},$$

where

$$n_{i,k}^{\text{out}} = \begin{cases} n_{i,k}^{\text{out}}, & k \in \mathcal{L}, \\ n_i^0, & k \in \mathcal{L}_v, \end{cases}$$

and in vector form, similar to (6.27), the following holds:

$$\mathbf{n}_l^{\text{in}} = V_l^T \overline{\mathbf{n}}.$$

In the following we present the framework for OSNR modeling in networks. As before, an optical amplified span s on link l is composed of an optical fiber with coefficient, $L_{l,s}$, which is wavelength independent, and an optical amplifier with gain $g_{i,l,s}$. The optical amplifier introduces amplified-spontaneous emission (ASE) noise, denoted by $ASE_{i,l,s}$. Both the gain and ASE noise are wavelength dependent. Let $p_{i,l,s}$ and $n_{i,l,s}$ denote the channel signal and noise power, respectively, at the output of span s on link l (illustrated in Fig. 6.5(c)).

Under similar Assumptions 6.1 and 6.2 and the notations, a constant total power is launched into each span of a link. Under Assumption 6.2, the following relation holds:

$$\sum_{i \in \mathcal{M}_l} p_{i,l} = \widehat{P}_l, \quad \forall l \in \mathcal{L}, \tag{6.29}$$

or in vector notation

$$\mathbf{e}^T \mathbf{p}_l = \widehat{P}_l, \quad \forall l \in \mathcal{L},$$

where $\mathbf{e}$ is an m-dimensional vector such that for each $i \in \mathcal{M}$, $[\mathbf{e}]_i = 1$, if $i \in \mathcal{M}_l$ or $[\mathbf{e}]_i = 0$ else. As before the above condition is called the *system constraint*.

Next we describe a typical optical link extracted from the network and the OSNR at the output of this link.

In Fig. 6.8, $p_{i,l}$ and $n_{i,l}^{\text{out}}$ are the signal and noise power at the output of link l, respectively, and $u_{i,l}$ and $n_{i,l}^{\text{in}}$ are the signal and noise power at the input of link l, respectively, or in vector form $\mathbf{p}_l$ and $\mathbf{u}_l$. The matrix Γ_l is the link system matrix of link l.

At the output of link l with $l \in \mathcal{R}_i$, we denote the OSNR of channel i by $y_{i,l} = \frac{p_{i,l}}{n_{i,l}^{\text{out}}}$, or in vector form

$$\mathbf{y}_l := \mathbf{p}_l ./ \mathbf{n}_l^{\text{out}}$$

Applying directly Lemma 6.2 yields the following result.

Proposition 6.7 *Let* $\mathbf{u}_l$, $\mathbf{n}_l^{\text{in}}$ *be the signal and noise power at the input of a link* l *with* N_l *spans and* m_l *channels. Under Assumptions* 6.1 *and* 6.2, *the optical signal power* $\mathbf{p}_l$ *and the OSNR* $\mathbf{y}_l$ *at the output of the link* l *are given in vector form as*

$$\mathbf{p}_l = \text{Diag}(\mathbf{H}_{N_l})\mathbf{u}_l \tag{6.30}$$

and

$$\mathbf{y}_l = \mathbf{u}_l ./ \left(\mathbf{n}_l^{\text{in}} + \Gamma_l \mathbf{u}_l\right) \tag{6.31}$$

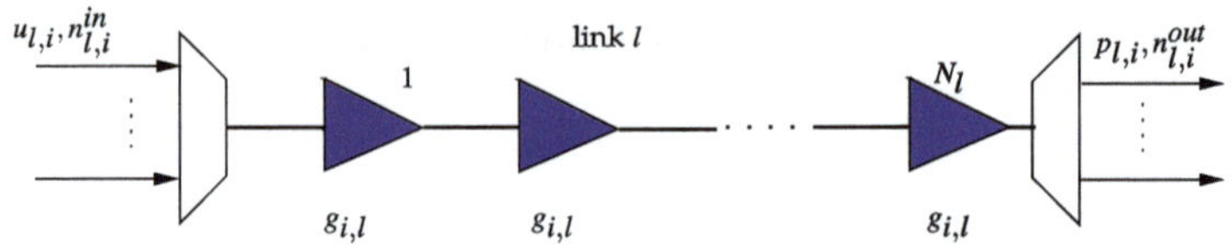

Fig. 6.8 Typical optical link

where $\Gamma_l = [\Gamma_{l_{i,j}}]$ *is the* link l system matrix

$$\Gamma_l = \sum_{r=1}^{N_l} \left[\mathrm{Diag}(\mathbf{g}_l^r)\right]^{-1} \frac{\mathbf{ASE}_l}{\widehat{P}_l} (\mathbf{g}_l^r)^T$$

and $\mathbf{H}_{N_l} = [H_{i,N_l}]$ *with*

$$H_{i,N_l} = \frac{g_{i,l}^{N_l} \widehat{P}_l}{\sum_{j\in\mathcal{M}_l} g_{j,l}^{N_l} u_{j,l}}$$

Component-wise

$$y_{i,l} := \frac{p_{i,l}}{n_{i,l}} = \frac{u_{i,l}}{n_{i,l}^{\mathrm{in}} + \sum_{j\in\mathcal{M}_l} \Gamma_{l_{i,j}} u_{j,l}} \tag{6.32}$$

where

$$\Gamma_{l_{i,j}} = \sum_{s=1}^{N_l} \frac{g_{j,l}^s}{g_{i,l}^s} \frac{ASE_{i,l}}{\widehat{P}_l} \tag{6.33}$$

Based on (6.31) a recursive OSNR model for the network can be obtained.

Lemma 6.8 *Consider a link l and a channel i, $i \in \mathcal{M}_l$, and let l' be the link precedent to link l for channel i along path $\mathcal{R}_i$. Let $y_{i,l}$, and $y_{i,l'}$ denote the channel OSNR at the output of link l and l', respectively. Then the following OSNR recursive relation holds for $\forall i \in \mathcal{M}_l$:*

$$\frac{1}{y_{i,l}} = \frac{1}{y_{i,l'}} + \sum_{j\in\mathcal{M}} A_{j,l} \Gamma_{l_{i,j}} \frac{u_{j,l}}{u_{i,l}} \tag{6.34}$$

Proof Let us write (6.32) with the help of the connection matrix A, such that for $\forall i \in \mathcal{M}_l$

$$y_{i,l} = \frac{u_{i,l}}{n_{i,l}^{\mathrm{in}} + \sum_{j\in\mathcal{M}} A_{j,l} \Gamma_{l_{i,j}} u_{j,l}}$$

Fig. 6.9 End-to-end model

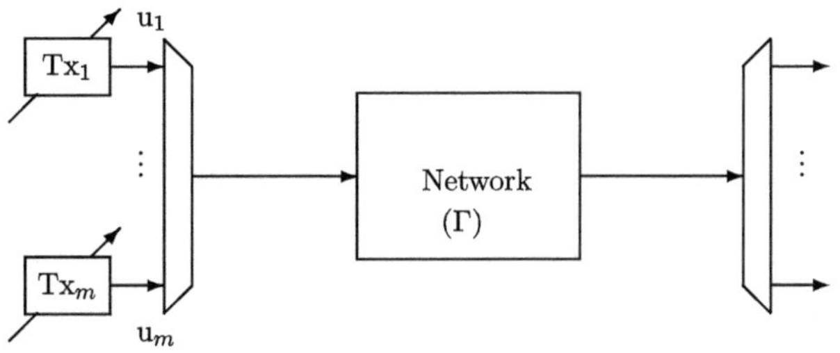

so that

$$\frac{1}{y_{i,l}} = \frac{n_{i,l}^{\text{in}}}{u_{i,l}} + \sum_{j \in \mathcal{M}} A_{j,l} \Gamma_{l_{i,j}} \frac{u_{j,l}}{u_{i,l}}$$

Using (6.26) we have

$$\frac{n_{i,l}^{\text{in}}}{u_{i,l}} = \frac{n_{i,l'}^{\text{out}}}{p_{i,l'}} = \frac{1}{y_{i,l'}} \tag{6.35}$$

the last equality following from the definition of $y_{i,l'}$ (at the output of link l'). Substituting 6.35 into the foregoing completes the proof. □

A similar model as in Lemma 6.2 can be developed for the network end-to-end OSNR albeit with a richer *network system matrix* Γ (Fig. 6.9). The model can be derived in a similar way based on recursive use of propagation relations. The reader is referred to [119] for the full proof of this result as well as for extension to the case when cross-talk at the optical nodes (OXC or OADM) is included (Lemmas 2 and 3 [119]).

Proposition 6.9 (Lemma 2, [119]) *Under Assumptions* 6.1 *and* 6.2, *the OSNR of channel i at Rx, y_i is given as*

$$y_i = \frac{u_i}{n_i^0 + \sum_{j \in \mathcal{M}} \Gamma_{i,j} u_j}, \tag{6.36}$$

where $\Gamma = [\Gamma_{i,j}]$ is the network system matrix with

$$\Gamma_{i,j} = \sum_{l \in \mathcal{R}_i} \frac{\prod_{q \in \mathcal{R}_{j,l}} H_{j,N_q}}{\prod_{q \in \mathcal{R}_{i,l}} H_{i,N_q}} \Gamma_{l_{i,j}},$$

where $\Gamma_{l_{i,j}}$ and H_{i,N_q} are defined in Corollary 6.7 *and $\mathcal{R}_{l,i}$ denotes the set of links on the path $\mathcal{R}_i$ before link l.*

The network OSNR model (6.36) has the same form as the link OSNR model (6.2), but with a more complex system matrix $\Gamma = [\Gamma_{i,j}]$, which we refer to as *network system matrix*. The OSNR model shows that as input power of one channel

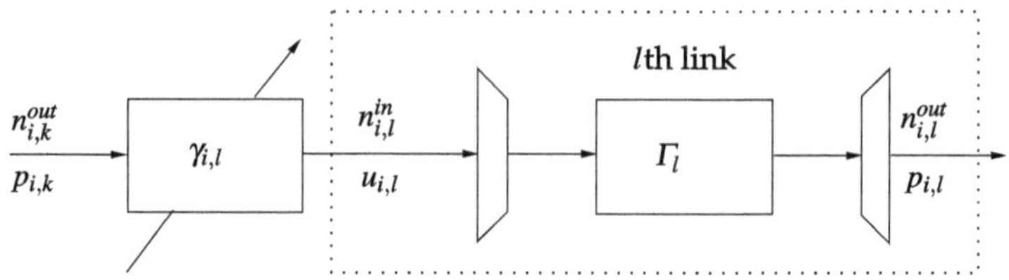

Fig. 6.10 Link l

increases, thereby increasing its OSNR, the noise in the other channels increases, thus decreasing their OSNRs. Based on the OSNR model, we take the channel input power as input to the optical network system and OSNR output and other network measurements as feedback signals to design power control algorithms such that channel OSNRs can be optimized.

Now we consider the second case, whereby in an optical network, channel powers are adjustable not only at Txs but also at optical switches [2]. Thus channel powers can be individually adjusted in the beginning of each optical link via an adjustment parameter for channel i on link l is denoted by $\gamma_{i,l}$, $\gamma_{i,l} \in [\gamma_{\min}, \gamma_{\max}]$ (Fig. 6.10). This is the case called γ-link topology. Then

$$u_{i,l} = \gamma_{i,l} p_{i,l'}, \quad \forall i = 1, \ldots, \mathcal{M}, \ l \in \mathcal{L}$$

where l' is the link precedent to link l for channel i on route $\mathcal{R}_i$.

For $l' = l - 1$ we have the configuration in the following figure (Fig. 6.11), called γ-link topology, whereby each γ-link is composed of cascaded OAs, with adjustable power per channel at each stage.

Similar to the previous results we have the immediate extensions below. First we have the following interconnection result similar to Lemma 6.4.

Lemma 6.10 *The optical signal powers at the input of link l, $\mathbf{u}_l$, $l \in \mathcal{L}$, are given in vector form as*

$$\mathbf{u}_l = \mathrm{Diag}(\boldsymbol{\gamma}_l) V_l^T \overline{\mathbf{p}}, \tag{6.37}$$

where $\boldsymbol{\gamma}_l = [\gamma_{1,l}, \ldots, \gamma_{m,l}]^T$, and V_l, $\overline{\mathbf{p}}$ are defined as in Lemma 6.4.

Proof In this case

$$u_{i,l} = A_{i,l}\gamma_{i,l}(B_{k',l}A_{i,k'}p_{i,k'})$$

or

$$u_{i,l} = A_{i,l}\gamma_{i,l}\left(B_{k',l}A_{i,k'}p_{i,k'} + \sum_{k\in\mathcal{L}',k\neq k'} B_{k,l}A_{i,k}p_{i,k}\right) = A_{i,l}\gamma_{i,l}\sum_{k\in\mathcal{L}'} B_{k,l}A_{i,k}p_{i,k}$$

and the proof follows immediately as in Lemma 6.4. □

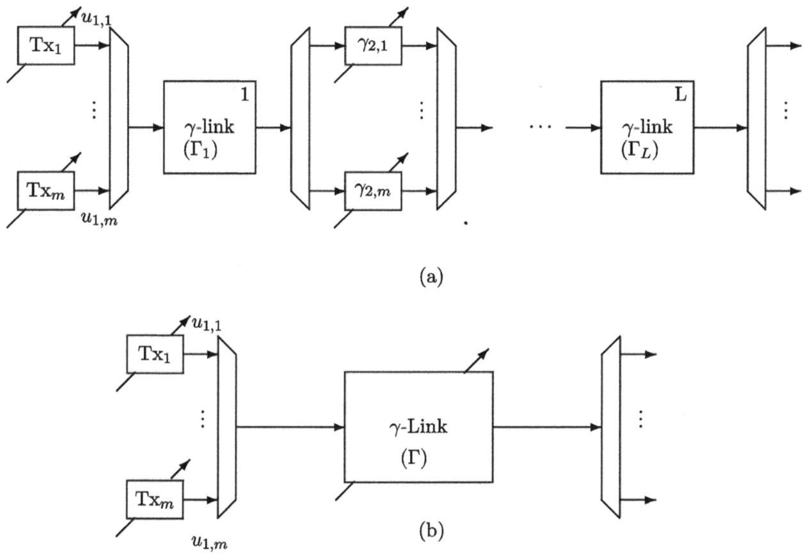

Fig. 6.11 A γ-link topology

Now, following Remark 6.6, since the adjustment factors affect both input signal and noise power simultaneously, one can write

$$n_{i,l}^{\text{in}} = \gamma_{i,l} n_{i,l'}^{\text{out}}, \quad \forall i = 1, \dots, \mathcal{M}, \; l \in \mathcal{L}$$

That is, for noise power a relation similar to (6.37) holds, i.e.,

$$\mathbf{n}_l^{\text{in}} = \text{Diag}(\boldsymbol{\gamma}_l) V_l^T \overline{\mathbf{n}}$$

Moreover, let us note that

$$\frac{n_{i,l}^{\text{in}}}{u_{i,l}} = \frac{\gamma_{i,l} n_{i,l'}^{\text{out}}}{\gamma_{i,l} p_{i,l'}} = \frac{1}{y_{i,l'}}$$

Thus one can see that in this second case with $\gamma_{i,l}$ adjustment factors (γ-link), the same recursive OSNR model as in (6.35) and Lemma 6.8 is valid on a per-link basis.

6.4.3 Link Capacity Constraint

Recall that in Sect. 6.2 we mentioned that nonlinear effects on optical fiber along links limit the total input power. After each span of optical fiber, the total power is influenced by the gain distribution of optical amplifiers [33]. Under Assumption 6.2, the same total power is launched into each span of a link. This leads to a uniform total power distribution along the link, which minimizes the effects of optical non-linearity [85].

However, this optimal gain distribution of optical amplifiers is not applicable to limit nonlinear effects on the optical fiber at the beginning of each link, i.e., the fiber segment between Txs (or optical switches) and the first optical amplifier along the link. Since the effects of the nonlinearity can be reduced by lowering the total launched power, the following condition is imposed as the *coupled power constraint*, also called the *link capacity constraint* in optical networks:

$$\sum_{i \in \mathcal{M}_l} u_{i,l} \leq \hat{P}_l, \quad \forall l \in \mathcal{L} \tag{6.38}$$

where $u_{i,l}$ is the signal power of channel i at the input of link l. When channel i is transmitted from its associated Tx directly to link l, $u_{i,l} = u_i$, where u_i is the signal power of channel i at Tx. Otherwise, due to the power propagation along links and among networks, $u_{i,l}$ is a function of u_i, or u where u is defined as the signal power vector of all channels at Tx. This is similar to capacity constraints in flow control [13]. However, capacity constraints in flow control are unchanged along links, while link coupled power constraints in optical networks are propagated along links. The foregoing (6.38) can be rewritten as

$$g_l(u) = \sum_{i \in \mathcal{M}_l} u_{i,l} - \hat{P}_l \leq 0, \quad \forall l \in \mathcal{L} \tag{6.39}$$

We are interested in the convexity of the constraint (6.39).

Example 6.11 In a single point-to-point WDM fiber link (Fig. 6.6), channel i is transmitted from its associated Tx directly to the link. Then one can drop the link index and (6.39) is reduced to

$$g(u) = \sum_{i \in \mathcal{M}} u_i - \widehat{P} \leq 0. \tag{6.40}$$

The function $g(u)$ in the single link case is always convex.

In optical networks, the coupled power constraints are propagated along links, and due to this, convexity of (6.39) is not automatically ensured, as shown in the next example.

Example 6.12 Consider the network shown in Fig. 6.12 with three links ($L = 3$) and three channels ($m = 3$) transmitted over links. Each link l has N_l spans and a total power target $\widehat{P}_l$. Let us study the convexity of the constraint on the third link, $g_3(u)$:

$$g_3(u) = u_{3,1} + u_{3,2} - \widehat{P}_3$$

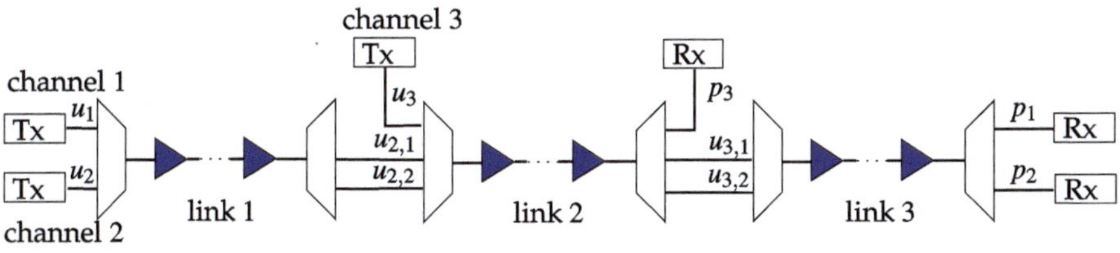

Fig. 6.12 Three-link topology with three channels

By using (6.26), i.e., $u_{i,l}$ is identical to $p_{i,l'}$ with l' being the preceding link of l on route $\mathcal{R}_i$, we have

$$g_3(u) = p_{2,1} + p_{2,2} - \widehat{P}_3$$

Then by using the system constraint (6.29) on link 2, we get

$$g_3(u) = p_{2,1} + p_{2,2} - \widehat{P}_3 = -p_{2,3} + (\widehat{P}_2 - \widehat{P}_3)$$

By using (6.14) in Lemma 6.2 and (6.26), we have

$$p_{2,3} = \frac{\widehat{P}_2 (G_{2,3})^{N_2} u_3}{(G_{2,1})^{N_2} u_{2,1} + (G_{2,2})^{N_2} u_{2,2} + (G_{2,3})^{N_2} u_3},$$

$$u_{2,i} = p_{i,1} = \frac{\widehat{P}_1 (G_{i,1})^{N_1} u_i}{(G_{1,1})^{N_1} u_1 + (G_{1,2})^{N_1} u_2}, \quad i = 1, 2$$

As can be easily seen $g_3(u)$ is a more intricate function of u. Moreover, when computing the Hessian matrix of $g_3(u)$, the sign of $\frac{\partial^2 g_3(u)}{\partial u_1^2}$ and $\frac{\partial^2 g_3(u)}{\partial u_2^2}$ may be different. Thus the Hessian matrix $\nabla^2 g_3(u)$ may not be positive semidefinite, that is, $g_3(u)$ may not remain convex independently of link parameters.

6.5 Notes

This chapter provided an overview of basic background on transmission in optical networks and on general topologies to be studied, as well as modeling of OSNR. The reader is referred to [2] and [126] for more in-depth material. The concepts of OSNR and link power capacity constraint in optical networks are introduced as a preface to the remaining chapters.

Chapter 7
Games in Point-to-Point Topologies

Abstract This chapter provides the basic formulation of a game framework towards solving the OSNR optimization problem in optical networks. We restrict the analysis to single point-to-point optical links, as the simplest network topology. A Nash game played among channels is employed towards maximizing OSNR firstly without coupled link capacity constraint. Then for incorporating the coupled power constraint, two approaches are considered—an indirect and a direct one, based on Lagrangian pricing and duality extension. Sufficient conditions are derived for the existence and uniqueness of an NE solution for both approaches. Two convergent iterative algorithms are developed towards finding the NE solution.

7.1 Game Formulation

Noncooperative game theory represents a powerful tool that can capture strategy interactions among self-interested players. Game-theoretic models have started to be used recently in networks [163]–[8] in the context of dynamic resource allocation, as an alternative to traditional system-wide optimization [164]–[68]. In large-scale networks decisions are often made by users independently [78], each according to its own performance objective. This is also appropriate for large-scale optical networks, where it is difficult to maintain a centralized system for transmitting real-time information between all channels, and cooperation among channels is impractical. This makes noncooperative game theory a suitable framework [20, 96]. A game-theoretic approach can be employed to optimize channel OSNR. The interest in such an approach is motivated by the departure from the assumption of cooperation among channels, particularly in large-scale optical networks. Here too decisions are often made independently by channels according to the users' own performance objectives. Moreover, a beneficial feature of a game formulation is that it leads itself to iterative distributed algorithms for the computation of an NE solution.

This problem belongs to a subclass of resource allocation in general communication networks [4, 134]. In optical network systems, a signal over the same fiber link can be regarded as an interfering noise for others, which leads to OSNR degradation. A satisfactory OSNR at Rx for each channel may be achieved by regulating the input power per channel at Tx. As a first step towards solving the OSNR optimization problem in optical networks, in this chapter we study such a problem in

L. Pavel, *Game Theory for Control of Optical Networks*,
Static & Dynamic Game Theory: Foundations & Applications,
DOI 10.1007/978-0-8176-8322-1_7,

a *single point-to-point optical link*. A Nash game played among channels is employed towards maximizing OSNR firstly without coupled link capacity constraint (Sect. 7.2). Channel optical signal-to-noise ratio (OSNR) optimization is formulated as an m-player noncooperative game. Conditions for the existence and uniqueness of the game equilibrium solution are given. We discuss an iterative algorithm for power control, which uses only channel feedback measurements and which is shown to converge to the equilibrium solution [115, 118]. Then for incorporating this link capacity coupled constraint, two approaches are considered: an indirect penalty-based approach (Sect. 7.3) and a direct Lagrangian pricing approach (Sect. 7.4). Sufficient conditions are derived for the existence and uniqueness of an NE solution for both approaches. Two convergent iterative algorithms are developed towards finding the NE solution.

7.2 Games Without Coupled Constraints

7.2.1 Utility and Nash Equilibrium Solution

The OSNR game problem is similar to power control via noncooperative game approaches in wireless networks, a topic which has been explored extensively in works such as [8, 133]. The problem needs to be formulated and addressed separately for optical networks, as there are several differences when compared to wireless networks [8]. In wireless networks, channels are characterized by uncoupled loss/gain with single Tx to Rx links. Optical networks have cascaded amplified spans, accumulation and self-generation of optical noise, cross-talk and possible coupling, and saturation [2].

Considering these differences and physical features of optical networks, relevant questions that we address in this section are: what is a natural way of formulating a tractable OSNR optimization game in optical links and then networks of general topology? How is this game compared to the SIR optimization game in wireless networks? Is there a natural decomposition that leads to iterative algorithms with provable convergence, decentralized with respect to channels?

We concentrate on a basic game formulation for an end-to-end optical link, i.e., a point-to-point link topology. Conditions for uniqueness of the Nash equilibrium as well as proof techniques and conditions for selecting pricing parameters are different than in the wireless case. This chapter focuses on a point-to-point link topology and it provides a starting point for the next chapter where optical network topologies are considered.

We study a point-to-point WDM fiber link shown in Fig. 6.6 in Chap. 6. We assume that reconfiguration is finished, i.e., channels will not be added or dropped while performing adjustments. A set $\mathcal{M} = \{1, \dots, m\}$ of channels are transmitted over the link. The link consists of N cascaded spans of optical fiber each followed by an OA (Fig. 7.1). All OAs have the same gain spectral shape and the gain value for channel i is g_i. We denote u_i and n_i^0 the signal power and noise power of channel

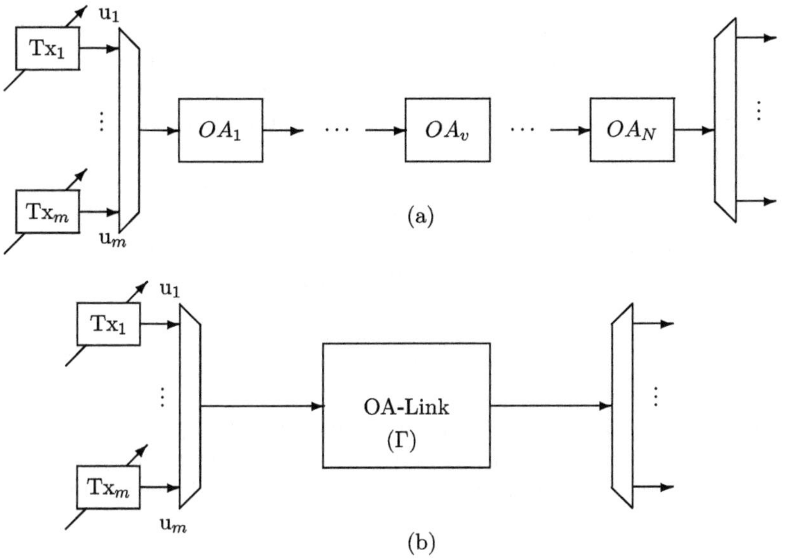

Fig. 7.1 A point-to-point link

$i \in \mathcal{M}$ at Tx, respectively. Similarly, we denote p_i and n_i the signal power and noise power of channel $i \in \mathcal{M}$ at Rx, respectively. Let $\mathbf{u} = [u_1, \dots, u_m]^T = [u_i]$ denote the vector form of the signal power at Tx. Equivalently, we write $\mathbf{u} = (\mathbf{u}_{-i}, u_i)$ with $\mathbf{u}_{-i} = [u_1, \dots, u_{i-1}, u_{i+1}, \dots, u_m]^T$ in some context to represent the same vector $\mathbf{u}$. The signal power at Tx is typically limited for every channel. That is, u_i is in a bounded set $\Omega_i = [0, u_{\max}]$ where $u_{\max}$ is a positive constant. We use Ω to represent the Cartesian product of Ω_i, $i \in \mathcal{M}$, $\Omega = \Omega_1 \times \cdots \times \Omega_m$. We also let $\Omega_{-i} = \Omega_1 \times \cdots \times \Omega_{i-1} \times \Omega_{i+1} \times \cdots \times \Omega_m$. Thus $u_i \in \Omega_i$ and $\mathbf{u} \in \Omega$.

The OSNR of channel i at Rx denoted by y_i, $y_i = \frac{p_i}{n_i}$, is given as in (6.16), i.e.,

$$y_i = \frac{u_i}{n_i^0 + \sum_{j \in \mathcal{M}} \Gamma_{i,j} u_j}, \tag{7.1}$$

with Γ as in (6.17). Equivalently, (7.1) can be rewritten as

$$y_i = \frac{u_i}{X_{-i} + \Gamma_{i,i} u_i} \quad \text{with } X_{-i} = \sum_{j \in \mathcal{M}, j \neq i} \Gamma_{i,j} u_j + n_{0,i} \tag{7.2}$$

where X_{-i} denotes the total interference on channel i due to other channels' power.

The OSNR model reflects that the signal power of one channel can be regarded as an interfering noise for others, which leads to OSNR degradation. Regulating the optical powers at Tx, i.e., allocating optical power as a resource among channels, aims to achieve a satisfactory OSNR for each channel at Rx.

Let us now formulate a game-theoretic approach to solve an OSNR optimization problem from the user optimization point of view, based on this OSNR model.

Specifically we formulate a Nash game where a player is identified with a channel. The objective of each player is to maximize its utility function related to individual channel OSNR. Each channel adjusts its power towards this goal in the presence of all other channels. The game settles at an equilibrium when no channel can improve its utility unilaterally.

Each channel i, $i \in \mathcal{M}$ is a player that minimizes its own cost function J_i (or maximizes its utility function), by adjusting its transmission power, in response to the other channels' (players') actions. Such a game is denoted by $\mathcal{G}(\mathcal{M}, \Omega_i, J_i)$. This means that the relevant concept is the noncooperative Nash equilibrium (NE) (see Chap. 4), i.e., a vector $\mathbf{u}^*$ such that

$$J_i\left(\mathbf{u}_{-i}^*, \mathbf{u}_i^*, \right) \leq J_i\left(\mathbf{u}_{-i}^*, u_i\right), \quad \forall i \in \mathcal{M},$$

for any given $\mathbf{u}_{-i}^*$. Recall that NE optimality means that no player has an incentive to change its action, since no further individual improvement in its cost is possible.

Let us consider that each individual cost function J_i is defined as the difference between a *pricing function* $\mathcal{P}_i$ and a *utility function* $\mathcal{U}_i$

$$J_i(\mathbf{u}) = \mathcal{P}_i(\mathbf{u}) - \mathcal{U}_i(\mathbf{u}) \tag{7.3}$$

The utility $\mathcal{U}_i$ is related to channel's performance criteria, while the pricing term $\mathcal{P}_i$ is used to penalize a channel for using too large an action (power). Minimizing the cost function is equivalent to maximizing the net utility, i.e., the difference between utility and pricing.

In general a pricing mechanism is known to improve the NE efficiency, and a linear pricing is the simplest one [133]. For such a simple linear pricing term each channel minimizes the cost function J_i:

$$J_i(\mathbf{u}_{-i}, u_i) = \alpha_i u_i - \mathcal{U}_i(\mathbf{u}_{-i}, u_i), \quad u_i \in [0, u_{\max}] \tag{7.4}$$

where channel power is limited to be below some $u_{\max}$ threshold on transmitter power. This linear pricing term in the individual channel input signal power can be interpreted as the "price" a channel pays for using the system resources [133], reflecting the fact that increasing one channel's power degrades the OSNR of all other channels. This price can be set by the network/link and dictated to each channel. From a system performance point of view, this linear pricing term limits the interference of this channel on the other ones [134]. Intuitively speaking, a large α_i means that a large penalty is put by the network on the use of power.

With OSNR as a performance criterion, the utility $\mathcal{U}_i$ should reflect a channel's preference for high OSNR or for low OSNR degradation. We define an OSNR-like channel utility function $\mathcal{U}_i(\mathbf{u})$ that reflects the channel's preference for maximizing OSNR_i, y_i. We consider a utility function appropriately defined for optical networks, a generalization of the one used in the wireless case. The following assumptions are used.

(A.ii.1) The utility function $\mathcal{U}_i(\mathbf{u})$ is a continuously differentiable function in u_i, monotone increasing and strictly concave in u_i.

Fig. 7.2 $\mathcal{U}_i$ vs. y_i

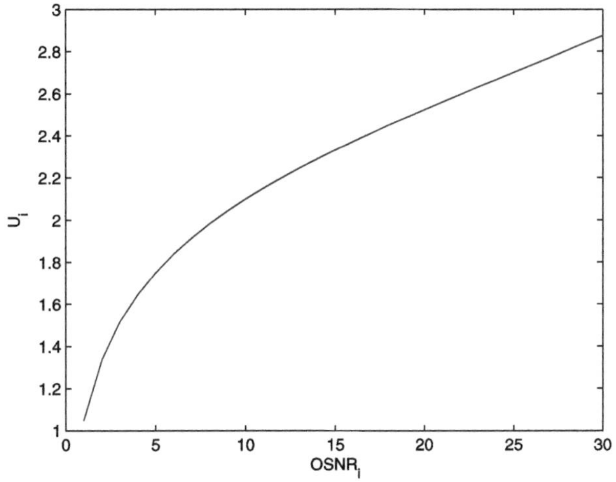

(A.ii.2) $u_i = 0$, $u_i = u_{\max}$ are not solutions to the minimization of the cost function J_i.

We construct a utility function that satisfies (A.ii.1), (A.ii.2). Note that OSNR_i, i.e., y_i in (7.2), is a strictly increasing function with respect to u_i, and tends to $1/\Gamma_{i,i}$, for infinite channel power. It is apparent that relation (7.2) bears a striking similarity with the wireless SIR model, [8], even though a different physical mechanism is present (ASE noise accumulation). The system matrix in the SIR model [8] has a special structure with equal rows, which is instrumental in obtaining uniqueness results. In contrast, for optical networks, Γ is a full general structure matrix, with coupling due to all channels and all spans. Moreover, OSNR_i is no longer a linear function of u_i and a direct logarithmic utility function of associated SIR as in the wireless case cannot be applied. For the general full matrix Γ, one can define a more general utility function $\mathcal{U}_i(\mathbf{u})$ here chosen to be a logarithmic function of the associated channel's OSNR, $y_i(\mathbf{u})$,

$$\mathcal{U}_i(\mathbf{u}) = \beta_i \ln\left(1 + a_i \frac{y_i(\mathbf{u})}{1 - \Gamma_{i,i} y_i(\mathbf{u})}\right), \quad \forall i \in \mathcal{M} \tag{7.5}$$

where $a_i, \beta_i > 0$ are channel-dependent parameters, a_i a scaling parameter for flexibility and β_i quantifying the desire to maximize the OSNR, respectively.

The utility function $\mathcal{U}_i$ defined in (7.5) is monotonically increasing in y_i, so that maximizing utility is related to maximizing channel OSNR, y_i. This relationship is illustrated in Fig. 7.2. Equivalently, using (7.2)

$$\mathcal{U}_i(\mathbf{u}_{-i}, u_i) = \beta_i \ln\left(1 + a_i \frac{u_i}{X_{-i}}\right) \tag{7.6}$$

where X_{-i} is given as in (7.2), as a function of the full system-matrix Γ. This dependence will be instrumental in the following results. Therefore, the cost function to be minimized, (7.4), is given as

$$J_i(\mathbf{u}_{-i}, u_i) = \alpha_i u_i - \beta_i \ln\left(1 + a_i \frac{u_i}{X_{-i}}\right) \tag{7.7}$$

Herein α_i is the price set by the network/link and β_i is set by each channel, respectively. These act as weighting factors, quantifying the trade-off between pricing (penalizing a channel for using large power) and the desire to maximize its utility.

From (7.6) it follows immediately that $\mathcal{U}_i$ satisfies (A.ii.1), i.e., the utility function $\mathcal{U}_i(\mathbf{u}_{-i}, u_i)$ is twice continuously differentiable in its arguments, monotonically increasing and strictly concave in u_i. Using (7.7), it can be shown that there exists a non-empty interval from which to select β_i/α_i, such that (A.ii.2) holds. Thus α_i, β_i are selected such that an NE solution is an interior point of the action set, and hence is an *inner NE* (see remark after Definition 4.1 in Chap. 4). This is what we assume in the following.

One can note that this is only a particular utility function that has certain nice properties being logarithmic, in that it allows for closed-form expression of the Nash equilibrium solution. Other examples can be given such as a linear function in OSNR. A logarithmic function is analytically useful and moreover is widely used as a utility function in flow control [7, 67, 139] and power control [3, 6, 8] for general communication networks. In some cases, the logarithmic utility function is intimately associated with the concept of proportional fairness [67].

Let us recall from Chap. 4 that a point $\mathbf{u}^*$ is an NE when $\mathbf{u}_i^*$ solves individual optimization problem J_i, given all channels on its path have equilibrium power levels, $\mathbf{u}_{-i}^*$. Existence of an NE solution depends on existence of a common intersection point for all players' reaction curves, while uniqueness depends on the particular problem. The following result characterizes the Nash equilibrium (NE) solution.

Theorem 7.1 *Consider $\mathcal{G}(\mathcal{M}, \Omega_i, J_i)$ with individual cost functions J_i, (7.7). This game admits a unique NE solution $\mathbf{u}^*$ if a_i are selected such that*

$$\sum_{j \neq i} \Gamma_{i,j} < a_i, \quad \forall i \in \mathcal{M} \tag{7.8}$$

The optimal NE solution $\mathbf{u}^*$ is inner and

$$\mathbf{u}^* = \widetilde{\Gamma}^{-1} \tilde{\mathbf{b}} \tag{7.9}$$

where $\widetilde{\Gamma} = [\widetilde{\Gamma}_{i,j}]$ and $\tilde{\mathbf{b}} = [\tilde{b}_i]$ are defined as

$$\widetilde{\Gamma}_{i,j} = \begin{cases} a_i, & j = i \\ \Gamma_{i,j}, & j \neq i \end{cases} \qquad \tilde{b}_i = \frac{a_i \beta_i}{\alpha_i} - n_{0i}$$

and $\Gamma_{i,j}$ being the link system matrix.

This result shows that while similar in some respects to wireless case, [8], for optical networks a more general uniqueness condition is needed for uniqueness of the NE solution.

Proof From (7.4, 7.7) and (A.ii.1) it follows directly that $\frac{\partial^2 J_i}{\partial {u_i}^2} > 0$. Since the cost function J_i is strictly convex in u_i, there exists a minimizing u_i^*, for any given $\mathbf{u}_{-i}$, such that

$$J(\mathbf{u}_{-i}, u_i^*) < J(\mathbf{u}_{-i}, u_i), \quad \forall u_i \neq u_i^*$$

on the closed and bounded (compact) set $[0, u_{\max}]$. Furthermore by (A.ii.2) u_i^* is inner.

To find u_i^* we solve the necessary conditions $\frac{\partial J_i}{\partial u_i} = 0$. From (7.7) one obtains

$$a_i \mathbf{u}_i^* + X_{-i}^* = \frac{a_i \beta_i}{\alpha_i}, \quad \forall i \tag{7.10}$$

which defines the ith player's reaction curve or reaction function r_i.

Note now that a linear reaction function greatly simplifies the closed-form expression of the NE. From Chap. 4 and Theorem 4.3 in [20], the vector solution of (7.10) is an NE solution to the m-player game. Use the definition of X_{-i}^*, (7.2), to rewrite (7.10) as

$$a_i \mathbf{u}_i^* + \sum_{j \neq i} \Gamma_{i,j} \mathbf{u}_j^* = \frac{a_i \beta_i}{\alpha_i} - n_{0,i}, \quad \forall i$$

Equivalently, in matrix form, this is written as

$$\tilde{\Gamma} \mathbf{u}^* = \tilde{\mathbf{b}} \tag{7.11}$$

where matrix $\tilde{\Gamma}$ and vector $\tilde{\mathbf{b}}$ are defined as $\tilde{\Gamma} = [\tilde{\Gamma}_{i,j}]$ and $\tilde{\mathbf{b}} = [\tilde{b}_i]$ with

$$\tilde{\Gamma}_{i,j} = \begin{cases} a_i, & j = i \\ \Gamma_{i,j}, & j \neq i \end{cases} \qquad \tilde{b}_i = \frac{a_i \beta_i}{\alpha_i} - n_{0,i}$$

Therefore a unique NE solution $\mathbf{u}^*$ exists if the matrix $\tilde{\Gamma}$ is invertible. Recall that Γ is a positive-entry matrix. If (7.8) holds, then $\tilde{\Gamma}$, (7.11), is strictly diagonally dominant; from Gershgorin's Theorem [59], it follows that $\tilde{\Gamma}$ is invertible, and the unique NE solution is $\mathbf{u}^* = \tilde{\Gamma}^{-1} \tilde{\mathbf{b}}$. □

Remark 7.2 For a given Γ, the a_i factors can be selected such that the diagonal dominance condition (7.8) holds on $\tilde{\Gamma}$. If Γ is itself diagonal dominant, then a possible choice is $a_i = \Gamma_{i,i}$. Once Γ changes (e.g. upon reconfiguration) these factors can be adjusted to satisfy (7.8).

7.2.2 Iterative Algorithm

In this section we discuss a decentralized iterative algorithm developed by assuming that all players follow best response strategy, i.e., based on best response dynamics. It can be shown that this algorithm converges to the NE solution if the diagonal dominance condition in Theorem 7.1 holds. Other control update laws can be considered such as gradient dynamics, etc. (see Chap. 4) with similar results. Consider

$$u_i(n+1) = \frac{\beta_i}{\alpha_i} - \frac{X_{-i}(n)}{a_i}, \quad \forall i \tag{7.12}$$

as recursive relation for updating transmitter power level, based on (7.10). Thus (7.12) corresponds to a parallel adjustment scheme (PUA) based on best response (BR) (see Chap. 4 and [20]), whereby each player responds optimally to the previously selected action of the other players. Relation (7.12) requires the total interference factor X_{-i}, which from (7.2) depends on all channel powers, u_j, and all channel gains, i.e., centralized information. However, using (7.2) one can express (7.12) in terms of y_i, i.e,

$$u_i(n+1) = \frac{\beta_i}{\alpha_i} - \frac{1}{a_i}\left(\frac{1}{y_i(n)} - \Gamma_{i,i}\right)u_i(n) \tag{7.13}$$

where n denotes the iteration number. This corresponds to a decentralized algorithm as in [46, 72], since the only information feedback is the individual channel y_i, which can be measured in real-time, and the channel "gain", $\Gamma_{i,i}$.

The following result gives convergence conditions for this algorithm.

Lemma 7.3 *If* (7.8) *holds, then algorithm* (7.13) *converges to the NE solution.*

Proof Let $e_i(n) = u_i(n) - \mathbf{u}_i^*$ where $\mathbf{u}^* = [u_i^*]$ is the NE solution. Using (7.10, 7.12, 7.13) yields

$$e_i(n+1) = -\frac{1}{a_i}\sum_{j \neq i} \Gamma_{i,j} e_j(n) \tag{7.14}$$

so that

$$\left\|\mathbf{e}(n+1)\right\|_\infty = \max_i \left|e_i(n+1)\right| \leq \max_i \left(\frac{1}{a_i}\sum_{j \neq i} \Gamma_{i,j}\left|e_j(n)\right|\right)$$

where the fact that Γ is component-wise positive was used.

Using $|e_j(n)| \leq \|\mathbf{e}(n)\|_\infty$, $\forall j$, yields

$$\left\|\mathbf{e}(n+1)\right\|_\infty \leq \max_i \left(\frac{1}{a_i}\sum_{j \neq i} \Gamma_{i,j}\right)\left\|\mathbf{e}(n)\right\|_\infty$$

Using (7.8) on the right-hand side of the foregoing yields

$$\left\|\mathbf{e}(n+1)\right\|_\infty < \left\|\mathbf{e}(n)\right\|_\infty, \quad \forall n = 0, 1, 2, \ldots$$

By a contraction mapping argument this shows that $\mathbf{e}(n) \to 0$ and $\mathbf{u}(n) \to \mathbf{u}^*$. □

Next we present some discussion on parameter selection via pricing strategies.

(a) Proportional Pricing This is the case when pricing parameters α_i are selected to be proportional to the system matrix entry for channel i, $\Gamma_{i,i}$, i.e., $\alpha_i = \Gamma_{i,i} k_i$, and all β_i are equal to 1. It can be shown that the scaling factors k_i can be selected such that all channels achieve some desired OSNR level, γ_i^*. Using (7.2) and (7.10) for the NE solution, one can write

$$\frac{1}{\gamma_i^*}\mathbf{u}_i^* = (\Gamma_{i,i} - a_i)\mathbf{u}_i^* + \frac{a_i}{\Gamma_{i,i}}\frac{1}{k_i}, \quad \forall i$$

In matrix vector form this yields

$$\Sigma\mathbf{u}^* = \mathbf{v} \quad \text{with } \Sigma = \operatorname{diag}(\varepsilon_1, \ldots, \varepsilon_i, \ldots \varepsilon_m), \ \mathbf{v} = [v_i] \tag{7.15}$$

where $\varepsilon_i = \frac{1}{\gamma_i^*} + a_i - \Gamma_{i,i}$, and $v_i = \frac{a_i}{\Gamma_{i,i}}\frac{1}{k_i}$. Using the NE solution, (7.11), $\mathbf{u}^* = \tilde{\Gamma}^{-1}\tilde{\mathbf{b}}$, and $\tilde{\mathbf{b}} = \mathbf{v} - \mathbf{n}_0$, yields after some manipulation

$$\left(I - \tilde{\Gamma}\Sigma^{-1}\right)\mathbf{v} = \mathbf{n}_0 \tag{7.16}$$

assuming that $\varepsilon_i \neq 0$. Then if $\rho(\tilde{\Gamma}\Sigma^{-1}) < 1$, where ρ is the matrix spectral radius, equation (7.16) has a unique solution $\mathbf{v}$. Moreover, for $\varepsilon_i = \varepsilon \neq 0$, using (7.15), one can rewrite this condition as an upper bound condition on the desired OSNR level γ_i^*, i.e.,

$$\gamma_i^* < \frac{1}{\rho(\tilde{\Gamma}) + (\Gamma_{i,i} - a_i)}, \quad \forall i \tag{7.17}$$

Therefore, if (7.17) holds $\mathbf{v}$ can be uniquely found as in (7.16), and hence the k_i factors can be found such that the desired OSNR level γ_i^* is achieved. This corresponds to a centralized pricing strategy, and as in [8], shows the trade-off between "gain" of the system matrix, in terms of the spectral radius, and the level of OSNR achieved. Unlike [8], for the general model Γ these parameters and the upper bound for the desired OSNR level are different for each channel.

(b) Decentralized Pricing For the case when β_i can be adjusted individually, one can show that if β_i satisfies the lower and upper bounds

$$\beta_i > \frac{1 + (a_i - \Gamma_{i,i})\gamma_i^*}{1 - \Gamma_{i,i}\gamma_i^*}\frac{\alpha_i}{a_i}X_{-i}, \qquad \beta_i < \frac{\alpha_i}{a_i}\left[u_{\max}\left(a_i + \sum_{j\neq i}\Gamma_{i,j}\right) + n_0\right], \quad \forall i$$

then each channel will achieve at least γ_i^*, i.e., $y_i > \gamma_i^*$, with $u_i \le u_{\max}$ at each iteration.

(c) Minimum OSNR Level Finally, consider the case when all $u_i = u_{\max}$ and $y_i \geq \gamma_{\min}$. Using the OSNR model in Chap. 6, it can be shown that a sufficient condition for this is

$$\gamma_{\min} \leq \frac{1}{\max_{i \in \mathcal{M}}(\sum_{j \in \mathcal{M}} \Gamma_{i,j})}$$

which shows the trade-off between minimum achievable OSNR, $\gamma_{\min}$, and the norm (gain) of system matrix Γ. Unlike the wireless case, [8], the system matrix Γ plays an essential role, i.e., a condition independent of the system matrix cannot be obtained.

As long as coupling constraints are not considered these point-to-point link results are directly extendable to end-to-end games in optical network topologies (see Chap. 8). In the next section we show how distributed links can be treated.

7.3 Games with Coupled Constraints: Indirect Penalty Approach

As we saw in Chap. 6 in optical networks cascaded optical amplified spans are present, as well as accumulation and self-generation of optical ASE noise. These realistic physical features have been considered in OSNR modeling (see Chap. 6). In order for game methods to be practical, they must incorporate realistic constraints arising from the underlying network systems. One such important constraint is the *link capacity constraint* (see Sect. 6.4.3 in Chap. 6), which is a coupled power constraint on channels that share an optical fiber. As we recall, the total power launched into the fiber needs to be restricted below the nonlinearity threshold [2, 85]. This constraint is satisfied at intermediary amplifier sites, which are operated in the automatic power control (APC) mode, but not at Txs. Thus unlike the previous section, herein we consider a Nash game with coupled constraint.

Thus, incorporating the link capacity constraint (6.38), the action space $\overline{\Omega}$ is

$$\overline{\Omega} = \left\{ \mathbf{u} \in \Omega \,\middle|\, \sum_{j \in \mathcal{M}} u_j - \widehat{P} \leq 0 \right\} \tag{7.18}$$

This $\overline{\Omega}$ is coupled in the sense that one player's action affects the feasible action sets of the other players. The feasible action set for each channel i is the projection set

$$\widehat{\Omega}_i(\mathbf{u}_{-i}) = \left\{ \xi \in \Omega_i \,\middle|\, \sum_{j \in \mathcal{M}, j \neq i} u_j + \xi - \widehat{P} \leq 0 \right\} \tag{7.19}$$

The first approach we consider in order to treat such a game is based on incorporating the coupled constraint into each cost function, thus in effect an indirect approach. Part of the results in this section are based mainly on [105, 108, 110].

Let us consider then that for each channel $i \in \mathcal{M}$ an individual cost function $\widehat{J}_i : \overline{\Omega} \to \mathbb{R}$ is assigned which is defined as the difference between a *pricing function* $\widehat{\mathcal{P}}_i : \overline{\Omega} \to \mathbb{R}$ and the same *utility function* as before , $\mathcal{U}_i : \overline{\Omega} \to \mathbb{R}$:

$$\widehat{J}_i = \widehat{\mathcal{P}}_i - \mathcal{U}_i \tag{7.20}$$

Specifically, the utility function $\mathcal{U}_i$ is the same as in (7.5), while the new pricing function consists of two terms: a linear pricing term (as in Sect. 7.2.1) and a new regulation (penalty) term, i.e.,

$$\widehat{\mathcal{P}}_i(\mathbf{u}_{-i}, u_i) = \alpha_i u_i + \frac{1}{\widehat{P} - \sum_{j \in \mathcal{M}} u_j}, \quad \forall i \in \mathcal{M}, \tag{7.21}$$

where $\alpha_i > 0$ is a pricing parameter determined by the system.

The regulation term is constructed by considering the link capacity constraint. It penalizes any violation of the constraint as following: the regulation term tends to infinity when the total power approaches the total power target $\widehat{P}$, so the pricing function $\widehat{\mathcal{P}}_i(\mathbf{u}_{-i}, u_i)$ increases without bound. Hence the system resource is preserved by forcing all channels to decrease their input powers and indirectly satisfies the link capacity constraint.

Thus the m-player Nash game is defined in terms of the new cost functions $\widehat{J}_i(\mathbf{u}_{-i}, u_i)$, $i \in \mathcal{M}$, played within the action space $\overline{\Omega}$. This game is denoted by $\mathcal{G}(\mathcal{M}, \widehat{\Omega}_i, \widehat{J}_i)$.

7.3.1 Nash Equilibrium Solution

A solution $\mathbf{u}^*$ of $\mathcal{G}(\mathcal{M}, \widehat{\Omega}_i, \widehat{J}_i)$ is called an NE solution in the sense of Definition 5.2 for coupled games. If in addition, such a solution is not on the boundary of the action space $\overline{\Omega}$, then it is an *inner NE solution.*

Notice that

$$\frac{1}{\widehat{P} - \sum_{j \in \mathcal{M}} u_j} \to \infty \quad \text{as } u_i \to \widehat{P} - \sum_{j \in \mathcal{M}, j \neq i} u_j$$

Thus the points on the hyperplane $\{\mathbf{u} \in \mathbb{R}^m \mid \sum_{j \in \mathcal{M}} u_j = \widehat{P}\}$ are not NE solutions of $\mathcal{G}(\mathcal{M}, \widehat{\Omega}_i, \widehat{J}_i)$. In addition, since $u_i = 0$ means channel i is inactive in the link, an NE solution $\mathbf{u}^*$ with a zero component, say, $u_1^* = 0$, of $\mathcal{G}(\mathcal{M}, \widehat{\Omega}_i, \widehat{J}_i)$ implies that channel 1 does not have any effect on the game. So in this case, the game is equivalent to the one in which $(m-1)$ channels play and the NE solution to the $(m-1)$-player Nash game does not have a zero component. In this sense, we assume that any NE solution $\mathbf{u}^*$ to the m-player OSNR Nash game does not have zero components, thus is always inner.

The following result provides sufficient conditions for existence and uniqueness of an inner NE solution.

Theorem 7.4 (See [105]) *$\mathcal{G}(\mathcal{M}, \widehat{\Omega}_i, \widehat{J}_i)$ admits a unique inner NE solution* $\mathbf{u}^*$ *if* $\forall i \in \mathcal{M}$,

$$a_i > \sum_{j \neq i, j \in \mathcal{M}} \Gamma_{i,j}, \tag{7.22}$$

$$\beta_i < \frac{\beta_{\min}}{\sum_{j \neq i, j \in \mathcal{M}} \frac{\Gamma_{j,i}}{a_j}}, \tag{7.23}$$

$$\alpha_i > \alpha_{\max} \sqrt{\sum_{j \neq i, j \in \mathcal{M}} \frac{\beta_i}{\beta_j} \frac{\Gamma_{j,i}}{a_j}}, \tag{7.24}$$

where $\beta_{\min} = \min_{j \in \mathcal{M}} \beta_j$ *and* $\alpha_{\max} = \max_{j \in \mathcal{M}} \alpha_j$.

Proof (*Existence*) The action space $\overline{\Omega}$ is a compact and convex set with a non-empty interior. Each cost function $\widehat{J}_i(\mathbf{u}_{-i}, u_i)$ is continuous and bounded and the first and second partial derivatives of $\widehat{J}_i(\mathbf{u}_{-i}, u_i)$ with respect to u_i are well defined on $\overline{\Omega}$ except the hyperplane $\{\mathbf{u} \in \mathbb{R}^m \mid \sum_{j \in \mathcal{M}} u_j = \widehat{P}\}$, given as

$$\frac{\partial \widehat{J}_i(\mathbf{u})}{\partial u_i} = \alpha_i + \frac{1}{(\widehat{P} - \sum_{j \in \mathcal{M}} u_j)^2} - \frac{\beta_i a_i}{X_{-i} + a_i u_i}, \quad \forall i \in \mathcal{M} \tag{7.25}$$

$$\frac{\partial^2 \widehat{J}_i(\mathbf{u})}{\partial u_i^2} = \frac{2}{(\widehat{P} - \sum_{j \in \mathcal{M}} u_j)^3} + \frac{\beta_i a_i^2}{(X_{-i} + a_i u_i)^2}, \quad \forall i \in \mathcal{M} \tag{7.26}$$

It follows that $\frac{\partial^2 \widehat{J}_i(\mathbf{u})}{\partial u_i^2}$ is positive and therefore $\widehat{J}_i(\mathbf{u}_{-i}, u_i)$ is strictly convex in u_i. Since each point on the hyperplane $\{\mathbf{u} \in \mathbb{R}^m \mid \sum_{j \in \mathcal{M}} u_j = \widehat{P}\}$ is not an NE solution, by Proposition 5.3, the game with the action space $\overline{\Omega}$ admits an NE solution, which is the inner NE solution of $\mathcal{G}(\mathcal{M}, \widehat{\Omega}_i, \widehat{J}_i)$.

Since an NE solution is inner, it follows from Proposition 4.6 that (4.6), $\frac{\partial \widehat{J}_i(\mathbf{u})}{\partial u_i} = 0$, $\forall i \in \mathcal{M}$, holds. The vector form is $\overline{\nabla} \widehat{J}(\mathbf{u}) = 0$, where $\overline{\nabla} \widehat{J}(\mathbf{u})$ is defined as in (A.3).

(*Uniqueness*) The uniqueness part is proved by contradiction and involves rather long manipulations. The reader is referred to [105, 110] for the detailed steps. □

Based on the results in Theorem 7.4, let us briefly discuss parameter selection strategies. The link sets fixed channel prices and each channel decides its willingness β_i to obtain a satisfactory OSNR, y_i, denoted by $\beta_i(y_i)$. From the necessary condition, $\frac{\partial \widehat{J}_i(\mathbf{u})}{\partial u_i} = 0$, using (7.25) one obtains β_i as a function of y_i to be

$$\beta_i(y_i) = \frac{\alpha_i}{a_i}\big(X_{-i} + a_i u_i(y_i)\big) + \frac{X_{-i} + a_i u_i(y_i)}{a_i(\widehat{P} - \sum_{j \in \mathcal{M}, j \neq i} u_j - u_i(y_i))^2}, \tag{7.27}$$

where

$$u_i(y_i) = \frac{X_{-i}}{1/y_i - \Gamma_{i,i}}$$

For a given lower OSNR bound $\widehat{\gamma}_i$, one can show that if β_i is adjusted to satisfy the lower bound

$$\beta_i > \frac{\alpha_i}{a_i} \frac{1 + (a_i - \Gamma_{i,i})\widehat{\gamma}_i}{1 - \Gamma_{i,i}\widehat{\gamma}_i} X_{-i} + \frac{(1 - \Gamma_{i,i}\widehat{\gamma}_i)(1 + (a_i - \Gamma_{i,i})\widehat{\gamma}_i)}{a_i(\widehat{P} - \sum_{j \in \mathcal{M}, j \neq i} u_j - (\widehat{P} - \sum_{j \in \mathcal{M}, j \neq i} u_j \Gamma_{i,i} + X_{-i})\widehat{\gamma}_i)^2} X_{-i}, \tag{7.28}$$

so that each channel has $y_i > \widehat{\gamma}_i$.

Theorem 7.4 shows the existence of a unique inner NE solution under certain conditions, but without giving an explicit expression of this solution. Some properties of the NE solution, which directly lead to the development of two iterative algorithms towards finding the NE solution numerically, are characterized in the following.

Consider the necessary conditions $\frac{\partial \widehat{J}_i(\mathbf{u})}{\partial u_i} = 0$, i.e.,

$$\alpha_i + \frac{1}{(\widehat{P} - \sum_{j \in \mathcal{M}} u_j)^2} = \frac{\beta_i a_i}{X_{-i} + a_i u_i}, \quad \forall i \in \mathcal{M} \tag{7.29}$$

The unique inner NE solution of $\mathcal{G}(\mathcal{M}, \widehat{\Omega}_i, \widehat{J}_i)$ satisfies (7.29).

Alternatively, the necessary condition can be written in terms of reaction functions. Recall that the action space $\overline{\Omega}$ is compact and convex and the cost function $\widehat{J}_i(\mathbf{u}_{-i}, u_i)$ is twice continuously differentiable over $\overline{\Omega}$ in its arguments and strictly convex in u_i. A reaction function r_i for channel i is defined as

$$r_i(\mathbf{u}_{-i}) := \arg\min_{u_i \in \widehat{\Omega}_i(\mathbf{u}_{-i})} \widehat{J}_i(\mathbf{u}_{-i}, u_i), \quad \forall i \in \mathcal{M} \tag{7.30}$$

Then it follows that for every given $\mathbf{u}_{-i}$, $r_i(\mathbf{u}_{-i})$ is single-valued. Furthermore, by Berge's Maximum Theorem (Theorem A.17), $r_i(\mathbf{u}_{-i})$ is continuous. Thus, there is a unique point, $u_i = r_i(\mathbf{u}_{-i})$ for every $i \in \mathcal{M}$. In vector form,

$$\mathbf{u} = r(\mathbf{u}), \tag{7.31}$$

where $r := [r_1(u_{-1}), \dots, r_m(u_{-m})]^T$.

As discussed in Chap. 4, an NE solution $\mathbf{u}^* \in \overline{\Omega}$ satisfies component-wise

$$u_i^* = r_i(\mathbf{u}_{-i}^*), \quad \forall i \in \mathcal{M}, \tag{7.32}$$

or $\mathbf{u}^* = r(\mathbf{u}^*)$. Unlike the case with no coupling constraints in Sect. 7.2.1 where a closed-form expression was found, here the reaction function r_i is highly nonlinear and the NE solution is analytically intractable.

The reaction function is in fact the implicit solution of (7.29). Notice that over $\overline{\Omega}$, the left-hand side of (7.29) is a monotonically increasing function with respect to u_i, while the right-hand side of (7.29) is monotonically decreasing. Therefore, for every $i \in \mathcal{M}$, there exists a unique intersection between the left-hand side and the right-hand side of (7.29), which is the NE solution $\mathbf{u}^*$. One can get insights into the NE solution's characteristics by studying the reaction function.

Lemma 7.5 (See [110]) *The reaction function* $r(\mathbf{u})$ *has the following properties*:

- $r(\mathbf{u}) \geq 0$ (non-negativity);
- *If* $\mathbf{u} > \mathbf{u}'$, *then* $r(\mathbf{u}) < r(\mathbf{u}')$ (monotonicity).

It follows from Lemma 7.5 that the reaction function $r(\mathbf{u})$ is not *standard*. As defined in [163], a function $I(p)$ is standard if for all $p \geq 0$, the following properties are satisfied: $I(p) > 0$; if $p \geq p'$, then $I(p) \geq I(p')$; for all $\alpha > 1$, $\alpha I(p) > I(\alpha p)$. Such a property typically helps in developing iterative algorithms.

7.3.2 Iterative Algorithm and Convergence Analysis

In this section we present two iterative algorithms towards finding the unique inner NE solution: a parallel update algorithm (PUA) and a gradient algorithm (GA). PUA is developed using the reaction functions and GA is developed based on the gradient descent method.

In the parallel update algorithm each channel i updates its input power u_i based on its associated reaction function $r_i(\mathbf{u}_{-i})$. Let

$$\{\mathbf{u}(n)\} := \{\mathbf{u}(0), \mathbf{u}(1), \ldots, \mathbf{u}(n), \ldots\}$$

denote the sequence of channel input power vectors, where $\mathbf{u}(0)$ is the initial channel input power vector. Then at each iteration time $(n+1)$, channel input power $u_i(n+1)$ is updated by

$$\text{(PUA)} \quad u_i(n+1) = r_i\big(\mathbf{u}_{-i}(n)\big), \quad \forall i \in \mathcal{M}, \tag{7.33}$$

or equivalently, in a vector form,

$$\mathbf{u}(n+1) = r\big(\mathbf{u}(n)\big) \tag{7.34}$$

If the reaction function $r(\mathbf{u})$ were standard, PUA would be the *standard power control algorithm* [163] and the convergence results in the synchronous case could be directly applied to PUA. However, this is not the case here and convergence is really difficult to prove. Results have been obtained only for the case of 2-players ($m = 2$) and extension to m-player case, $m > 2$ is an open problem [110].

Remark 7.6 PUA may not converge when more than two channels ($m > 2$) exist. The intuitive reason is that each channel updates its optical power only based on instant costs and parameters, ignoring future implications of its action. Therefore, at some iterations the total optical power of all other channels will exceed the target power $\widehat{P}$ when $m > 2$. To overcome these power fluctuations, one can use a relaxed PUA, in which a relaxation parameter is used to determine the step size that each channel takes towards finding the NE solution at each iteration step. For example, (7.33) can be modified as

$$u_i(n+1) = (1-\mu_i)u_i(n) + \mu_i r_i\big(\mathbf{u}_{-i}(n)\big),$$

where the coefficient $0 < \mu_i < 1$ and it can be shown that the relaxed PUA converges if μ_i is restricted.

PUA is developed by using the reaction function $r(\mathbf{u})$ but convergence can be shown only for $m = 2$. As alternative a gradient algorithm (GA) can be designed based on the gradient descent method [24]. Convergence of this algorithm can be proved even when $m > 2$. For convenience, particularly for the simplification of convergence proof, we follow the approach in [107, 109] and develop the algorithm in the continuous-time domain.

We consider a model where each channel uses a gradient algorithm (GA) to update its power, given as

$$\dot{u}_i(t) = \frac{du_i}{dt} = -\mu\left(\frac{\partial \widehat{J}_i(\mathbf{u}_{-i}, u_i)}{\partial u_i(t)}\right), \quad \forall i \in \mathcal{M}, \tag{7.35}$$

where t is the continuous-time variable and the coefficient $\mu > 0$.

Using (7.3) and (7.25), one can rewrite (7.35) as

$$\text{(GA)} \quad \dot{u}_i(t) = -\mu\left(\alpha_i + \frac{1}{(\widehat{P} - \sum_{j\in\mathcal{M}} u_j(t))^2} - \frac{\beta_i a_i}{\sum_{k=1}^m \tilde{\Gamma}_{i,k} u_k(t)}\right) \tag{7.36}$$

where

$$\tilde{\Gamma}_{i,k} = \begin{cases} \Gamma_{i,k}, & k \neq i \\ a_i, & k = i \end{cases}$$

Recalling the definition of OSNR, one can rewrite (7.36) as

$$\dot{u}_i(t) = -\mu\left(\alpha_i + \frac{1}{(\widehat{P} - \sum_{j\in\mathcal{M}} u_j(t))^2} - \frac{\beta_i a_i}{(\frac{1}{y_i(t)} + a_i - \Gamma_{i,i})u_i(t)}\right) \tag{7.37}$$

Hence, at each iteration, individual channels need only the sum of power of all channels and local measurements, namely, its own power and current OSNR level. Thus the iterative algorithm (7.37) is distributed.

Next let us define a set $\overline{\Omega}_\delta$, which is a subset of the action space $\overline{\Omega}$. We slightly modify the bounded set $\Omega_i = [0, u_{\max}]$ to be

$$\widetilde{\Omega}_i = [u_{\min}, u_{\max}], \quad \forall i \in \mathcal{M}$$

where $0 < u_{\min} < u_{\max}$. The Cartesian product of $\widetilde{\Omega}_i$ is given by Ω_δ. The set $\overline{\Omega}_\delta$ is defined as

$$\overline{\Omega}_\delta = \left\{ \mathbf{u} \in \Omega_\delta \;\middle|\; \sum_{j \in \mathcal{M}} u_j - \widehat{P} \le 0 \right\}$$

We set $u_{\min}$ to be sufficiently small such that Theorem 7.4 holds, i.e., $u_i^* > u_{\min}$ for all $i \in \mathcal{M}$ and

$$\frac{\partial \widehat{J}_i}{\partial u_i}(\mathbf{u}_{-i}, u_i)\bigg|_{u_i = u_{\min}} < 0, \quad \forall \mathbf{u}_{-i},\ \forall i \in \mathcal{M} \tag{7.38}$$

Notice that because of the capacity constraint, the following statement is always true:

$$\frac{\partial \widehat{J}_i}{\partial u_i}(\mathbf{u}_{-i}, u_i)\bigg|_{u_i = \widehat{P} - \sum_{j \in \mathcal{M}, j \ne i} u_j} > 0, \quad \forall \mathbf{u}_{-i},\ \forall i \in \mathcal{M} \tag{7.39}$$

Thus the set $\overline{\Omega}_\delta$ is invariant under the algorithm (7.36). That is, the trajectory remains inside $\overline{\Omega}_\delta$. It follows that the trajectory lies in $\overline{\Omega}$ if the initial state is in $\overline{\Omega}_\delta$. Moreover, the equilibrium of (7.36) in $\overline{\Omega}_\delta$ is the unique NE solution $\mathbf{u}^*$ of $\mathcal{G}(\mathcal{M}, \widehat{\Omega}_i, \widehat{J}_i)$.

The following result proves the convergence of algorithm (7.36).

Theorem 7.7 *For $\mathcal{G}(\mathcal{M}, \widehat{\Omega}_i, \widehat{J}_i)$, let the initial condition $u(0) \in \overline{\Omega}_\delta$. Then the update scheme* (7.35) *converges to the NE solution $\mathbf{u}^*$ if*

$$a_{\min} > \left(\frac{2\widehat{P}}{u_{\min}}\right)^2 \max_{i \in \mathcal{M}} \sum_{j \in \mathcal{M}, \ne i} \Gamma_{j,i}, \tag{7.40}$$

$$\beta_i > S \cdot \beta_{\max}, \tag{7.41}$$

where $a_{\min} = \min_{j \in \mathcal{M}} a_j$, $\beta_{\max} = \max_{j \in \mathcal{M}} \beta_j$ and

$$S = \left(\frac{2\widehat{P}}{u_{\min}}\right)^2 \frac{\max_{i \in \mathcal{M}} \sum_{j \in \mathcal{M}, j \ne i} \Gamma_{j,i}}{a_{\min}} \tag{7.42}$$

Proof The proof is based on a Lyapunov approach. Let $\phi_i(\mathbf{u}) := \dot{u}_i(t)$, where $\dot{u}_i(t)$ is defined in (7.35) or (7.36), and define a candidate Lyapunov function,

$$V(\mathbf{u}) := \frac{1}{2} \sum_{i \in \mathcal{M}} \phi_i^2(\mathbf{u}) \tag{7.43}$$

Note that $V(\mathbf{u})$ is restricted to the set, $\overline{\Omega}_\delta$. Because of the uniqueness of the NE solution $\mathbf{u}^*$, $\phi_i(\mathbf{u}) = 0, \forall i \in \mathcal{M}$, if and only if $u = \mathbf{u}^*$. Therefore, $V(\mathbf{u})$ is strictly positive for all $u \ne \mathbf{u}^*$.

From (7.36), taking the second derivative of u_i with respect to time t, yields

$$\ddot{u}_i(t) = -\mu \sum_j \left(\frac{2}{(\widehat{P} - \sum_{j\in\mathcal{M}} u_j(t))^3} + \frac{\beta_i a_i \tilde{\Gamma}_{i,j}}{(\sum_{k\in\mathcal{M}} \tilde{\Gamma}_{i,k} u_k(t))^2} \right) \phi_j(\mathbf{u}) \quad (7.44)$$

It is obvious that $\dot{\phi}_i(\mathbf{u}) = \ddot{u}_i(t)$.

Taking the derivative of $V(\mathbf{u})$, (7.43), with respect to time t, yields

$$\begin{aligned} \dot{V}(\mathbf{u}) &= \sum_{i\in\mathcal{M}} \phi_i(\mathbf{u}) \cdot \dot{\phi}_i(\mathbf{u}) \\ &= -\mu \sum_{i\in\mathcal{M}} \phi_i(\mathbf{u}) \sum_{j\in\mathcal{M}} \left(\frac{2}{(\widehat{P} - \sum_{j\in\mathcal{M}} u_j(t))^3} + \frac{\beta_i a_i \tilde{\Gamma}_{i,j}}{(\sum_{k\in\mathcal{M}} \tilde{\Gamma}_{i,k} u_k(t))^2} \right) \phi_j(\mathbf{u}), \end{aligned} \quad (7.45)$$

where (7.44) is used. For simplicity, let

$$a(t) = \frac{2}{(\widehat{P} - \sum_{j\in\mathcal{M}} u_j(t))^3}, \qquad b_i(t) = \frac{\beta_i a_i}{(\sum_{k\in\mathcal{M}} \tilde{\Gamma}_{i,k} u_k(t))^2},$$

$$\theta_{ij}(t) = a(t) + b_i(t)\tilde{\Gamma}_{i,j}$$

Therefore, (7.45) is rewritten as

$$\begin{aligned} \dot{V}(\mathbf{u}) &= -\mu \sum_{i\in\mathcal{M}} \sum_{j\in\mathcal{M}} \big(a(t) + b_i(t)\tilde{\Gamma}_{i,j}\big) \phi_j(\mathbf{u}) \phi_i(\mathbf{u}) \\ &= -\mu \sum_{i\in\mathcal{M}} \phi_i(\mathbf{u}) \sum_{j\in\mathcal{M}} \theta_{ij}(t) \phi_j(\mathbf{u}) \\ &= -\mu \phi^T(\mathbf{u}) \Theta(t) \phi(\mathbf{u}), \end{aligned} \quad (7.46)$$

where vector $\phi = [\phi_i]$ and matrix $\Theta(t) = [\theta_{ij}(t)]_{m\times m}$ are appropriately defined. If $\Theta(t)$ is uniformly positive definite then $\dot{V}(\mathbf{u})$ in (7.46) is negative definite, and the system is asymptotically stable by Lyapunov's stability theorem [69]. This is the main idea of the proof and the reader is referred to [107] for the detailed steps. Then $\phi_i(u(t)) = \dot{u}_i(t) \to 0, \forall i \in \mathcal{M}$ which implies that $u_i(t)$ converges to the unique NE solution point $\mathbf{u}^*$. □

A gradient algorithm has been studied for stability in a congestion control game [4, 5] where a special structure matrix is automatically uniformly positive definite in the proof. In contrast, herein the corresponding matrix $\Theta(t)$ is more general. System dependent conditions are required for Θ to be positive definite.

Compared the conditions (7.40) and (7.41) for the stability of GA with the sufficient conditions (7.22)–(7.24) for the existence of a unique NE solution, it readily follows that β_i is not only upper-bounded by (7.23), but also lower-bounded by (7.41).

As we have seen in this section a relatively tedious process has to be followed to treat coupled constraints indirectly, with respect to both NE existence and uniqueness conditions, as well as with respect to NE computation (algorithm). In the next section we will use a direct approach to treat these constraints, which, as we will see, offers reduced complexity.

7.4 Games with Coupled Constraints: Lagrangian Pricing Approach

7.4.1 Lagrangian Pricing and Duality Extension

In this section we consider Nash games with coupled constraints and their solvability by a direct approach based on Lagrangian extension. Specifically we apply directly the procedure developed in Chap. 5. We note that this direct approach has advantages compared to the indirect one in the previous section. As we saw the indirect, penalty-based approach does not allow for an analytically tractable NE solution and developing decentralized algorithms is not immediate.

As in Sect. 7.2, each channel $i \in \mathcal{M}$ minimizes the original individual cost function $J_i : \overline{\Omega} \to \mathbb{R}$, with J_i defined in (7.4), not the penalized cost $\widehat{J}_i$ as used the previous section. The cost J_i is composed of a linear pricing and a channel utility $\mathcal{U}_i$ related to OSNR maximization defined as before. Unlike Sect. 7.2, where coupled constraints were not considered, here the action set is coupled as in (7.18), i.e.,

$$\overline{\Omega} = \left\{ \mathbf{u} \in \Omega \mid \mathbf{1}^T \mathbf{u} \le \widehat{P} \right\} \tag{7.47}$$

and the action set of channel $i \in \mathcal{M}$ is the projection of $\overline{\Omega}$ on channel i's direction, namely as in (7.19):

$$\widehat{\Omega}_i(\mathbf{u}_{-i}) = \left\{ x_i \in \Omega_i \;\middle|\; \sum_{j \neq i}^{m} u_j + x_i \le \widehat{P} \right\} \tag{7.48}$$

We denote the game by $\mathcal{G}(\mathcal{M}, \widehat{\Omega}_i, J_i)$, which is in the class of m-player Nash games with coupled utilities and coupled constraints. It follows from Sect. 7.2 that J_i is continuously differentiable in its arguments and convex in u_i. Note that the overall coupled action space $\overline{\Omega}$ is compact and convex as well. Then from Proposition 5.3, $\mathcal{G}(\mathcal{M}, \widehat{\Omega}_i, J_i)$ admits an NE solution. Moreover, due to the coupled constraint (7.47), solving directly for an NE solution of this game requires coordination among possibly all channels and is impractical as seen in the previous section.

In the following, we use Lagrangian extension and decomposition results in Chap. 5 as a natural way to obtain a hierarchical decomposition and compute an NE solution of $\mathcal{G}(\mathcal{M}, \widehat{\Omega}_i, J_i)$. For this game, using the notations as defined in Chaps. 4 and 5, consider the separable NG cost function $\widetilde{J}(\mathbf{u}; \mathbf{x})$, (4.7),

$$\widetilde{J}(\mathbf{u}; \mathbf{x}) = \sum_{i=1}^{m} J_i(\mathbf{u}_{-i}, x_i),$$

the separable augmented constraint $\widetilde{g}(\mathbf{u}; \mathbf{x})$, (5.8),

$$\widetilde{g}(\mathbf{u}; \mathbf{x}) = \sum_{i=1}^{m} g(\mathbf{u}_{-i}, x_i),$$

together with the augmented Lagrangian function $\widetilde{L}(\mathbf{u}; \mathbf{x}; \boldsymbol{\mu})$, (5.11),

$$\widetilde{L}(\mathbf{u}; \mathbf{x}; \boldsymbol{\mu}) = \widetilde{J}(\mathbf{u}; \mathbf{x}) + \boldsymbol{\mu}^T \widetilde{g}(\mathbf{u}; \mathbf{x})$$

and the dual cost function $D(\boldsymbol{\mu})$, (5.17),

$$D(\boldsymbol{\mu}) = \widetilde{L}\big(\mathbf{u}^*; \mathbf{u}^*; \boldsymbol{\mu}\big), \tag{7.49}$$

where $\mathbf{u}^*$ is such that $\mathbf{u} = \mathbf{u}^*$ satisfies $\mathbf{u} = \arg\min_{\mathbf{x}\in\Omega} \widetilde{L}(\mathbf{u}; \mathbf{x}; \boldsymbol{\mu})$. In a fixed-point notation,

$$D(\boldsymbol{\mu}) = \Big[\min_{\mathbf{x}\in\Omega} \widetilde{L}(\mathbf{u}; \mathbf{x}; \boldsymbol{\mu})\Big]\Big|_{\arg\min_{\mathbf{x}\in\Omega} \widetilde{L}=\mathbf{u}},$$

or

$$D(\mu) := \Big[\min_{\mathbf{x}\in\Omega} \widetilde{L}(\mathbf{u}; \mathbf{x}; \boldsymbol{\mu})\Big]\Big|_{\mathbf{x}=\mathbf{u}}$$

Then $\mathcal{G}(\mathcal{M}, \widehat{\Omega}_i, J_i)$ is related to a constrained minimization of $\widetilde{J}$, (4.7), (5.8), with respect to the second argument x, which admits a fixed-point solution. In particular individual components of a solution $\mathbf{u}^*$ to this constrained minimization constitute an NE solution to $\mathcal{G}(\mathcal{M}, \widehat{\Omega}_i, J_i)$ in the sense of Definition 5.2. From Remark 5.6, we know that $\mathbf{u}^*$ is obtained by first minimizing $\widetilde{L}(\mathbf{u}; \mathbf{x}; \boldsymbol{\mu})$ with respect to x, $x(\mathbf{u})$. The next step involves finding a fixed-point solution $x(\mathbf{u}^*) = \mathbf{u}^*$ which depends on $\boldsymbol{\mu}$, $\mathbf{u}^*(\boldsymbol{\mu})$. An NE solution–Lagrange multiplier pair $(\mathbf{u}^*(\boldsymbol{\mu}^*), \boldsymbol{\mu}^*)$ is obtained if ${\boldsymbol{\mu}^*}^T g(\mathbf{u}^*(\boldsymbol{\mu}^*)) = 0$ and $\boldsymbol{\mu}^* \geq 0$.

Thus $D(\boldsymbol{\mu})$ is related to the minimization of the associated Lagrangian function $\widetilde{L}(\mathbf{u}; \mathbf{x}; \boldsymbol{\mu})$. For $\mathcal{G}(\mathcal{M}, \widehat{\Omega}_i, J_i)$ with convex coupled constraints, Proposition 5.9 yields a decomposition into a lower-level Nash game with no coupled constraints and a higher-level problem for pricing. Thus based on Proposition 5.9 applied to the OSNR game $\mathcal{G}(\mathcal{M}, \widehat{\Omega}_i, J_i)$, with separable NG cost function and linear constraints, one has the following decomposition result.

Proposition 7.8 *Consider the coupled OSNR Nash game $\mathcal{G}(\mathcal{M}, \widehat{\Omega}_i, J_i)$ with cost functions $J_i(\mathbf{u}_{-i}, u_i)$, (7.4), subject to the linear constraint (7.47), i.e., over $\widehat{\Omega}_i$.*

Then the dual cost function $D(\mu)$, (7.49), *can be decomposed as*

$$D(\mu) = \sum_{i=1}^{m} \overline{L}_i\big(\mathbf{u}^*_{-i}, u_i^*(\mu), \boldsymbol{\mu}\big) + \sum_{i=1}^{m} \mu\big(\mathbf{e}^T \mathbf{u}^*_{-i} - \widehat{P}\big) \tag{7.50}$$

where $\mathbf{u}^*(\mu) = [u_i^*(\mu)]$ *is an NE solution to the Nash game* $\mathcal{G}(\mathcal{M}, \Omega_i, \overline{L}_i)$ *with cost functions* $\overline{L}_i$ (7.51) *and no coupled constraints.*

$$\overline{L}_i(\mathbf{u}_{-i}, x_i, \boldsymbol{\mu}) = J_i(\mathbf{u}_{-i}, x_i) + \mu x_i, \quad i \in \mathcal{M} \tag{7.51}$$

Proof Each cost function J_i, (7.4), is continuously differentiable and convex in u_i and the constraints are linear so that one can apply Proposition 5.9. The linear constraints (7.47) are rewritten in a two-argument form

$$g(\mathbf{u}_{-i}, x_i) = \mathbf{e}^T \mathbf{u}_{-i} + x_i - \widehat{P} \tag{7.52}$$

where $\mathbf{e} = [1, \dots, 1]^T$ is the $(m-1) \times 1$ all ones vector. Then by using Proposition 5.9, $D(\boldsymbol{\mu})$ can be decomposed as

$$D(\boldsymbol{\mu}) = \sum_{i=1}^{m} L_i\big(\mathbf{u}^*_{-i}(\boldsymbol{\mu}), \mathbf{u}_i^*(\boldsymbol{\mu}), \boldsymbol{\mu}\big)$$

where using (7.52) L_i is given here as

$$L_i(\mathbf{u}_{-i}, x_i, \boldsymbol{\mu}) = J_i(\mathbf{u}_{-i}, x_i) + \mu x_i + \mu\big(\mathbf{e}^T \mathbf{u}_{-i} - \widehat{P}\big) \tag{7.53}$$

Recall that in $D(\mu)$ in Proposition 5.9 one has to minimize first with respect to x_i on the right-hand side, and then solve for a fixed-point solution. From (7.53) it can be seen that only the first two terms depend on x_i. Hence, substituting for $L_i(\mathbf{u}_{-i}, x_i, \boldsymbol{\mu})$, (7.53), on the right-hand side of $D(\mu)$ and isolating the terms that are independent of x_i, yields

$$D(\mu) = \sum_{i=1}^{m} \min_{x_i \in \Omega_i} \overline{L}_i(\mathbf{u}_{-i}, x_i, \mu)|_{x_i = u_i} + \sum_{i=1}^{m} \mu\big(\mathbf{e}^T \mathbf{u}_{-i} - \widehat{P}\big)$$

where $\overline{L}_i$ is defined as in corollary statement (7.51). A fixed-point solution $\mathbf{u}^* = [u_i^*]$ to the set of m optimizations on the right-hand side of the foregoing is an NE solution to the Nash game with cost functions, $\overline{L}_i$, (7.51), and the last part of the claim follows. □

Proposition 7.8 leads to a hierarchical decomposition of $\mathcal{G}(\mathcal{M}, \widehat{\Omega}_i, J_i)$ into a lower-level modified Nash game $\mathcal{G}(\mathcal{M}, \Omega_i, \overline{L}_i)$ with cost functions $\overline{L}_i$ (7.51) and no coupled constraints, and a higher-level optimization problem used for coordination. This decomposition is computationally simpler as shown below. For a given pricing μ, the lower-level game admits a closed-form explicit solution. Specifically using (7.51, 7.4, 7.7) we see that $\overline{L}_i$ satisfies

$$\overline{L}_i(\mathbf{u}_{-i}, x_i, \mu) = (\alpha_i + \mu)u_i - \beta_i \ln\left(1 + a_i \frac{u_i}{X_{-i}}\right)$$

i.e., $\overline{L}_i$ is the same as J_i, (7.7) for α_i replaced by $\alpha_i + \mu$, $\forall i$. Therefore, for each given μ, the NE solution $\mathbf{u}^*(\mu)$ to the lower-level game $\mathcal{G}(\mathcal{M}, \Omega_i, \overline{L}_i)$ with cost $\overline{L}_i$ is unique and can be obtained as in Theorem 7.1 as

$$\mathbf{u}^*(\mu) = \widetilde{\Gamma}^{-1}\big(\mathrm{Diag}\big[1./(\alpha + \mu)\big]\mathbf{b}_0 - \mathbf{n}_0\big) \tag{7.54}$$

where $\mathbf{b}_0 = [a_i\beta_i]$, $\mathbf{n}_0 = [n_{0i}]$ and $\mathrm{Diag}[1./(\alpha + \mu)] = \mathrm{Diag}([1/(\alpha_i + \mu)])$. This is formalized in the following result.

Proposition 7.9 *For each given $\mu \geq 0$, $\mathcal{G}(\mathcal{M}, \Omega_i, \overline{L}_i)$ admits an NE solution if a_i in $\overline{L}_i$ (7.51) satisfies*

$$\sum_{j\neq i} \Gamma_{i,j} < a_i, \quad \forall i \in \mathcal{M} \tag{7.55}$$

Then the inner NE solution $\mathbf{u}^*(\boldsymbol{\mu})$ is unique and given as

$$\mathbf{u}^*(\boldsymbol{\mu}) = \widetilde{\Gamma}^{-1}\widetilde{b}(\boldsymbol{\mu}),$$

where $\widetilde{\Gamma} = [\widetilde{\Gamma}_{i,j}]$ and $\widetilde{b}(\boldsymbol{\mu}) = [\widetilde{b}_i(\boldsymbol{\mu})]$ are defined as

$$\widetilde{\Gamma}_{i,j} = \begin{cases} a_i & j = i, \\ \Gamma_{i,j} & j \neq i, \end{cases} \qquad \widetilde{b}_i(\boldsymbol{\mu}) = \frac{a_i\beta_i}{\alpha_i + \boldsymbol{\mu}} - n_i^0, \tag{7.56}$$

with $\Gamma_{i,j}$ being the system matrix.

Proof The result can be proved by applying Proposition 4.5. We rewrite $\overline{L}_i$ (7.51) as

$$\overline{L}_i(\mathbf{u}_{-i}, u_i, \boldsymbol{\mu}) = (\alpha_i + \boldsymbol{\mu})u_i - \beta_i \ln\left(1 + \frac{a_i u_i}{n_i^0 + \sum_{j\neq i} \Gamma_{i,j} u_j}\right), \quad \forall i \in \mathcal{M} \tag{7.57}$$

It can be seen that for any given $\boldsymbol{\mu} \geq 0$, $\overline{L}_i$ is jointly continuous in all its arguments and

$$\frac{\partial^2 \overline{L}_i}{\partial u_i^2} > 0, \quad \forall u_i \in \Omega_i$$

It follows that $\overline{L}_i$ is strictly convex in u_i. Recall that for each $i \in \mathcal{M}$, Ω_i is a closed, bounded, and convex subset of $\mathbb{R}$. Then by Proposition 4.5, $\mathcal{G}(\mathcal{M}, \Omega_i, \overline{L}_i)$ admits an NE solution.

An inner NE solution can be found by solving the necessary conditions, $\frac{\partial \overline{L}_i}{\partial u_i} = 0$. From (7.57), one can obtain

$$a_i u_i^*(\boldsymbol{\mu}) + \sum_{j \neq i} \Gamma_{i,j} u_j^*(\boldsymbol{\mu}) = \frac{a_i \beta_i}{\alpha_i + \boldsymbol{\mu}} - n_i^0, \quad \forall i \in \mathcal{M}$$

Equivalently, in a matrix form, this is written as

$$\widetilde{\Gamma} \cdot \mathbf{u}^*(\boldsymbol{\mu}) = \widetilde{b}(\boldsymbol{\mu}), \tag{7.58}$$

where matrix $\widetilde{\Gamma}$ and vector $\widetilde{b}(\boldsymbol{\mu})$ are defined in (7.56). Therefore a unique solution of (7.58) exists if the matrix $\widetilde{\Gamma}$ is invertible. Notice that $\widetilde{\Gamma}$ is a positive-entry matrix. If (7.55) holds, then $\widetilde{\Gamma}$ is strictly diagonal dominant. From Gershgorin's Theorem (Theorem A.7), it follows that $\widetilde{\Gamma}$ is invertible and a unique solution of (7.58) exists,

$$\mathbf{u}^*(\boldsymbol{\mu}) = \widetilde{\Gamma}^{-1} \cdot \widetilde{b}(\boldsymbol{\mu}),$$

which is an inner NE solution of $\mathcal{G}(\mathcal{M}, \Omega_i, \overline{L}_i)$. □

Next, based on the explicit solution (7.54) and on pricing coordination at the higher level, a recursive hierarchical algorithm is discussed. By Theorem 5.8 applied to the coupled OSNR game $\mathcal{G}(\mathcal{M}, \widehat{\Omega}_i, J_i)$ with costs J_i and coupled constraints (7.47), $(\mathbf{u}^*, \mu^*)$ is an optimal NE solution–Lagrange multiplier pair if and only if $\mathbf{u}^*$ is NG-feasible,

$$\sum_{i=1}^{m} u_i^*(\mu) \leq \widehat{P}, \quad u_i^* \in \Omega_i, \ i \in \mathcal{M} \tag{7.59}$$

$\mu^* \geq 0$, $\mu^*(\sum_{i=1}^{m} \mathbf{u}_i^* - \widehat{P}) = 0$ (complementary slackness condition) and the Lagrangian optimality condition

$$\mathbf{u}^* = \Big[\arg\min_{\mathbf{x} \in \Omega} \widetilde{L}\big(\mathbf{u}; \mathbf{x}; \mu^*\big)\Big]\Big|_{\mathbf{x}=\mathbf{u}} \tag{7.60}$$

holds. By Proposition 7.8 and (7.50), note that $\mathbf{u}^*(\mu)$ solving (7.60) can be found as an NE solution to the modified Nash game $\mathcal{G}(\mathcal{M}, \Omega_i, \overline{L}_i)$ with costs $\overline{L}_i$ (7.51), with no coupled constraints. For every given price μ, this NE solution $\mathbf{u}^*(\mu)$ is unique as in (7.54). Furthermore, from (7.54) it is seen that all components of $\mathbf{u}^*(\mu)$ decrease with μ. One can exploit the linear constraint and adjust the price μ to satisfy the slackness condition. Instead of maximizing $D(\mu)$, the optimal price μ^* can be obtained such that the slackness condition holds, i.e., as the point of interception between the curve representing total power, $u_T^*(\mu) = \sum_{i=1}^{m} u_i^*(\mu)$, with the level $\widehat{P}$ (Fig. 7.3). This method has the interpretation of a coordination mechanism. The link as the coordinator sets the price at the optimal value μ^*. The channels respond by adjusting their power levels to $u_i^*(\mu^*)$, which minimizes their own cost.

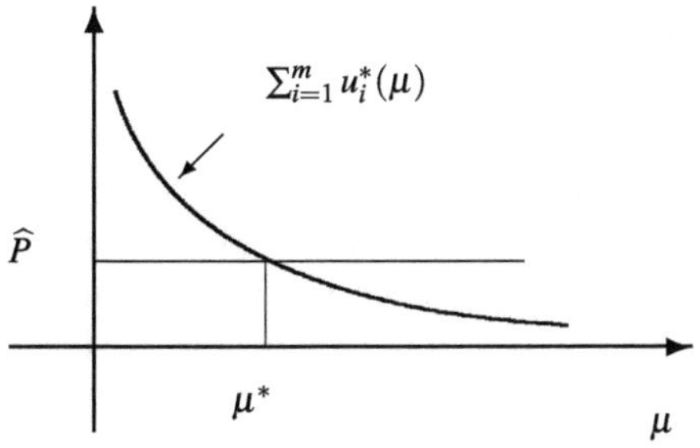

Fig. 7.3 Graphical plot of total power versus price

7.4.2 Iterative Algorithm

A hierarchical adjustment algorithm for both coordinating link price (higher-level) and channel powers (lower-level) is discussed.

Link Algorithm Every K iterations of the channel algorithm, the new link price μ is computed based on the received total power for all channels in the link $u_T(K) = \sum_{j=1} u_j(K)$ as

$$\mu(\bar{k}+1) = \left[\mu(\bar{k}) + \eta \quad \left(u_T(K) - \widehat{P}\right)\right]^+ \tag{7.61}$$

where $\bar{k}$ is the link iteration number, η is the step-size and $[z]^+ = \max\{z, 0\}$.

This simple price update based on Fig. 7.3 requires only measurement of total power. Moreover it corresponds to a gradient descent technique if link price is adjusted slower than channel powers. At the higher level, $\mu(\bar{k})$ acts as a coordination signal that aligns individual optimality with the system constraint, (7.47) or (7.59). This is the new price given to the channels, who repeat K iterations of the following algorithm.

Channel Algorithm Based on a pricing $\mu(\bar{k})$ from the link, the optimal channel power vector $\mathbf{u}^*(\mu(\bar{k}))$ can be found explicitly as in (7.54). This requires global centralized information. However, the following iterative update algorithm can be used:

$$u_i(n+1) = \frac{\beta_i}{\alpha_i + \mu(\bar{k})} - \frac{1}{a_i}\left(\frac{1}{y_i(n)} - \Gamma_{i,i}\right)u_i(n) \tag{7.62}$$

where n is the channel iteration number. This is again a decentralized algorithm, since the only information feedback is the individual channel y_i, which can be measured in real-time, and the channel "gain", $\Gamma_{i,i}$. For fixed μ this algorithm converges to the optimal NE solution (7.54). Convergence of the combined algorithms will be proved later on in the context of network topologies (see Chap. 8). Even if the optimal solution is coupled as in (7.54), it can be iteratively computed by using the decentralized algorithm (7.62). Thus individual channels do not have to coordinate with other channels at the lower-level game.

7.5 Notes

In this chapter, we presented a Nash game framework towards OSNR optimization in point to point links. We developed first the basic game formulation, without capacity constraints (Sect. 7.2). Then we considered indirectly the link capacity constraint (Sect. 6.4.3) by adding a penalty term to each channel cost function (Sect. 7.3). Conditions for the existence of a unique inner NE solution to this Nash game were given. The NE solution is analytically intractable and developing iterative algorithms is not immediate. We have studied the properties of the NE solution and presented two distributed algorithms: PUA and GA, towards finding the unique inner NE solution. The algorithms use only local measurements and the current load of the link.

In the second part, Sect. 7.4, we applied directly the results in Chap. 5 to a Nash game with coupled constraints in optical point-o-point links. The duality approach offered a natural way to hierarchically decompose the coupled OSNR Nash game into a lower-level Nash game, with no coupled constraints, and a higher-level link problem. Moreover, the lower-level Nash game is analytically tractable and its solution can be found iteratively, decentralized with respect to channels. This approach led to a solution that made full use of the input power resource. We discuss extension of these approaches to general network configurations in the next chapter.

Chapter 8
Games in Network Topologies

Abstract This chapter provides approaches on how to deal with games in more complicated network topologies, starting from the basic games in single point-to-point WDM fiber links studied in Chap. 7. The multi-link topologies studied are representative for selected paths extracted from a mesh configuration in which no closed loops are being formed by channel optical paths. In network configurations, coupled constraints are propagated along fiber links and constraint functions become complicated from the end-to-end point of view. The non-convexity introduces additional complexities for analysis. In this chapter, we present a partition approach. More precisely, we partition the general multi-link structure into stages each stage being a single link. Then we formulate a partitioned Nash game for general multi-link topologies composed of ladder-nested stage Nash games. We also show that convexity is ensured in single-sink topologies, so that a partition approach could be based on single-sink stages.

8.1 Introduction

We have seen that in their most complete form games in optical networks belong to the class of m-player games with coupled utilities and constraints. For single point-to-point link topologies these were studied in Chap. 7. The purpose of this chapter is to treat multi-link and mesh topologies. In fact end-to-end games *without coupled constraints* are direct extensions of the point-to-point games in Chap. 7, with the appropriate Γ network system matrix (as in Chap. 6) replacing the link system matrix. Because of this we shall not consider them here. Instead we shall focus on how to deal with games with coupled constraints in network topologies.

When coupled constraints are considered in network configurations, constraint functions become complicated from an end-to-end point of view. This is due to power propagation and constraint propagation along fiber links as shown in Example 6.12 in Chap. 6. The approach presented in Sect. 7.3 of Chap. 7 cannot easily be extended to such configurations. One reason is the difficulty in building an appropriate cost function that needs a complicated penalty pricing function term for the constraints. On the other hand the Lagrangian extension and decomposition results as developed in Sect. 7.4 of Chap. 7 provide a good path to follow. One thing to keep in mind is that the approach can be applied under convex constraints. Since convex-

L. Pavel, *Game Theory for Control of Optical Networks*,
Static & Dynamic Game Theory: Foundations & Applications,
DOI 10.1007/978-0-8176-8322-1_8,

ity may not be preserved across the network (see Example 6.12), this is unlike capacity constraints in flow control [21, 68]. This non-convexity introduces additional difficulties in directly applying the approach to network topologies from an end-to-end point of view. A possible approach is a partitioned game formulation and this is the approach we present. The main idea is that a general multi-link structure can be partitioned into stages whereby each stage is a γ-link or a single-sink structure, for which convexity is satisfied. A partitioned Nash game can be formulated for a general multi-link topology as being composed of o series of ladder-nested stage Nash games. Stage Nash games are played sequentially and solutions are interconnected. Based on Lagrangian extension and decomposition results, a hierarchical iterative algorithm towards computing a solution of the partitioned Nash game can be developed.

The remainder of this chapter is organized as follows. As a building block for the partitioned game approach, in Sect. 8.2 focuses on games in a γ-link topology. This is a simple network topology where power is adjustable not only at the transmitter sites but at all nodes (see Sect. 6.4.2). For simplicity we consider games with no coupled constraints in a γ-link so that the partitioned game approach is shown better. Then in Sect. 8.4 a partitioned game is presented for the case with constraints and general mesh and multi-link topologies. This follows the approach in Sect. 8.2 for the partitioned game where each stage game is a link game. Coupled capacity constraints are treated as in Sect. 7.4 for each stage game. Alternatively in Sect. 8.3 we show that for single-sink topologies convexity is maintained and thus approach in Sect. 7.4 can be directly applied to an end-to-end game. The material in this chapter is mostly based on [106, 112].

8.2 Games in γ-Link Topologies

As a stepping stone for solving games with coupled constraints in general network topologies in this section we consider a relatively simple topology of interconnection of L point-to-point links linked via network elements such as OXC or dynamic gain equalizers [126]. Thus parameters are adjustable at all nodes that have these elements (see Sect. 6.4.2). This allows us to develop a partitioned game approach into stages that will be used later on in multi-link and mesh topologies. For simplicity we treat only the case with no coupled constraints. We will consider coupled constraints from the following sections onwards (Sects. 8.3 and 8.4).

8.2.1 Partitioned Game Formulation

In this section we study γ-link topologies with a set of links $\mathcal{L} = \{1, \dots, L\}$ where channel powers are adjustable not only at Txs but also at some K bifurcation nodes via the adjustment factors γ (see Fig. 6.11, Sect. 6.4.2 and Fig. 8.1 below).

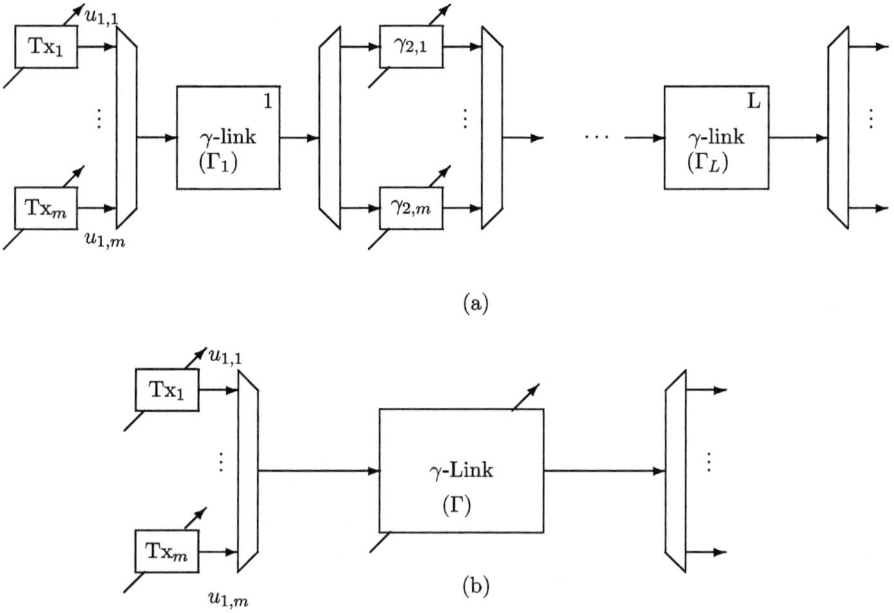

Fig. 8.1 A γ-link topology

We assume for simplicity that this is possible at each link so that $K = L$. We also assume for simplicity that the same set $\mathcal{M} = \{1, \dots, m\}$ of channels is transmitted over all links. We partition the multi-link structure into L stages depending on the nodes where channels are adjusted. The case where this is possible only at fewer nodes, i.e., $K < L$ is treated in [121]. Each stage has a point-to-point structure and stages are ladder-nested. A nested structure is an information structure in which each player has an access to the information acquired by all his precedents, i.e., players situated closer to the beginning of the decision process. If the difference between the information available to a player and his closer precedent involves only his actions, then the structure is ladder-nested [77]. The ladder-nested structure enables decomposition that in turn results in recursive procedure for its solution.

In effect we define a partitioned Nash game where we exploit the structure of each stage and the ladder-nested form of the game. The game is played in the following way. Channels transmitted over the same stage compete for input power resource. Actions of competition are taken in a fixed order among stages and solutions are interconnected. In the following we will formulate a partitioned Nash game composed of ladder-nested stage Nash games. We will also discuss iterative computation of equilibria and develop a three-level hierarchical algorithm.

First, we make some new notations for each stage. A multi-link structure of L links is partitioned into L stages, $\mathcal{L} = \{1, \dots, L\}$ and for each stage $l \in \mathcal{L}$, denote by $\mathbf{u}_l = [u_{i,l}]$ and $\mathbf{p}_l = [p_{i,l}]$, $i \in \mathcal{M} = \{1, \dots, m\}$ the input and output signal power vector, respectively. Similarly, denote by $\mathbf{n}_l^{\text{in}} = [n_{i,l}^{\text{in}}]$ and $\mathbf{n}_l^{\text{out}} = [n_{i,l}^{\text{out}}]$ the input and output noise power vector, respectively. Sometimes we write $\mathbf{u}_l$ as $\mathbf{u}_l = (\mathbf{u}_{l,-i}, u_{i,l})$

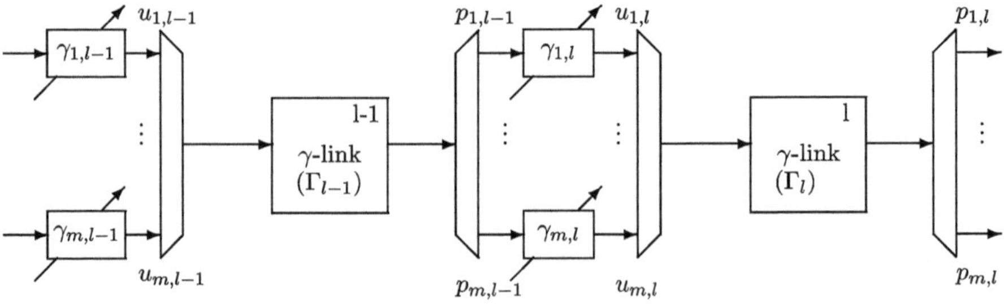

Fig. 8.2 A γ-link topology (recursive interconnection)

with the vector $\mathbf{u}_{l,-i}$ obtained by deleting the ith element from $\mathbf{u}_l$. Let $\gamma_l = [\gamma_{i,l}]$ denote the vector of channel adjustments at stage l such that

$$u_{i,l} = \gamma_{i,l} p_{i,l'}, \quad \forall i \in \mathcal{M}$$

where stage l' is the stage precedent to stage l for channel i, i.e., $l' = l - 1$ (Fig. 8.2). The adjustment parameter $\gamma_{i,l}$ is bounded within $[\gamma_{\min}, \gamma_{\max}]$. Each stage is composed of a link. Thus the OSNR for channel i, $i \in \mathcal{M}$, at the output of stage l, $l \in \mathcal{L}$, is defined as

$$y_{i,l} = \frac{p_{i,l}}{n_{i,l}^{\text{out}}}.$$

Based on this partition, the OSNR of channel i at the output of stage (link) l, can be obtained from Proposition 6.7 as

$$y_{i,l} = \frac{u_{i,l}}{n_{i,l}^{\text{in}} + \sum_{j \in \mathcal{M}} \Gamma_{l_{i,j}} u_{j,l}}, \tag{8.1}$$

where $\Gamma_l = [\Gamma_{l_{i,j}}]$ is the *link (stage) system matrix*. Based on (8.1), we define

$$\frac{1}{\delta Q_{i,l}} := \frac{1}{y_{i,l}} - \frac{1}{y_{i,l'}} = \sum_{j \in \mathcal{M}} \Gamma_{l_{i,j}} \frac{u_{j,l}}{u_{i,l}} \tag{8.2}$$

where stage l' is the link precedent to stage l for channel i. In this case of γ-link topologies this is simply $l' = l - 1$. It follows that $\frac{1}{\delta Q_{i,l}}$ measures the OSNR degradation of channel i from stage l' to stage l. Thus, instead of maximization of OSNR from Tx to Rx, one can consider minimization of OSNR degradation between stages.

We formulate a partitioned Nash game, composed of lower-level ladder-nested stage Nash games. On stage l, a Nash game is played with each channel attempting to minimize its individual cost over m channels. Games on stages are played in sequence and solutions are interconnected as will be shown below.

We first specify channel individual cost function $J_{i,l}$ in the lower-level stage Nash game. Channel individual cost function $J_{i,l}$ has a similar form as J_i (7.3) (Chap. 7), defined as a difference between a pricing function $\alpha_{i,l} u_{i,l}$ and a utility function.

Unlike the games in Chap. 7, herein the utility function reflects the associated channel's OSNR degradation. Based on the OSNR model (8.1) and the new measure variable $\frac{1}{\delta Q_{i,l}}$ (8.2) for stage l, let us define the cost function $J_{i,l}$ as

$$J_{i,l} = \alpha_{i,l} u_{i,l} - \mathcal{U}_{i,l}, \quad \forall i \in \mathcal{M} \tag{8.3}$$

where in this case

$$\mathcal{U}_{i,l} = \beta_{i,l} \ln\left(1 + \frac{a_{i,l}}{\frac{1}{\delta Q_{i,l}} - \Gamma_{l_{i,i}}}\right) \tag{8.4}$$

That is, the utility function $\mathcal{U}_{i,l}$ reflects a preference for lower OSNR degradation of channel i on stage l. Substituting (8.2) into (8.3) yields

$$J_{i,l}(\mathbf{u}_{l,-i}, u_{i,l}) = \alpha_{i,l} u_{i,l} - \beta_{i,l} \ln\left(1 + a_{i,l} \frac{u_{i,l}}{\tilde{X}_{l,-i}}\right), \tag{8.5}$$

where

$$\tilde{X}_{l,-i} = \sum_{j \in \mathcal{M}, j \neq i} \Gamma_{l_{i,j}} u_{j,l} \tag{8.6}$$

It follows that $J_{i,l}(\mathbf{u}_{l,-i}, u_{i,l})$ is continuously differentiable in its arguments and convex in $u_{i,l}$. Let us denote the lower-level stage Nash game by $\mathcal{G}(\mathcal{M}, \Omega_i, J_{i,l})$. Recall that the overall action space Ω is convex, then from Proposition 4.5 $\mathcal{G}(\mathcal{M}, \Omega_i, J_{i,l})$ admits an NE solution.

We exploit the partitioned Nash game composed of L lower-level stage Nash games, given by $\mathcal{G}(\mathcal{M}, \Omega_i, J_{i,l})$. Each $\mathcal{G}(\mathcal{M}, \Omega_i, J_{i,l})$ is played by m channels such that $\frac{1}{\delta Q_{i,l}}$ is minimized for each channel i on stage l. The lower-level stage Nash games are played sequentially (in a precedence order) with the interpretation that across all L stages, $\sum_{l \in \hat{\mathcal{R}}_i} J_{i,l}$ is related to the overall OSNR degradation for channel i. Solutions of all $\mathcal{G}(\mathcal{M}, \Omega_i, J_{i,l})$ are interconnected, explained as follows.

Let $\mathbf{u}_l^* = [u_{i,l}^*]$ be an NE solution of $\mathcal{G}(\mathcal{M}, \Omega_i, J_{i,l})$. Recall that channel powers are adjustable at bifurcation nodes and $\gamma_{i,l}$ is the adjustable parameter for channel i on stage l. Given the precedent actions of channel i, i.e., $\mathbf{u}_{i,l'}^*$ and $\mathbf{p}_{i,l'}^*$, the adjustment for channel i on stage l is obtained as

$$\gamma_{i,l}^* = \frac{\mathbf{u}_{i,l}^*}{\mathbf{p}_{i,l'}^*}$$

In a vector form, $\gamma_l^* = [\gamma_{i,l}^*]$, $i \in \mathcal{M}$, satisfies

$$\mathbf{u}_l^* = \operatorname{diag}(\gamma_l^*) \cdot \mathbf{p}_{-l}^*,$$

where $\mathbf{p}_{-l}$ consists of the corresponding channel signal powers from the output of other stages for all $i \in \mathcal{M}$. Recall that each adjustment parameter is bounded within

$[\gamma_{\min}, \gamma_{\max}]$. The partitioned Nash game admits a solution if each $\gamma_{i,l}^* \in [\gamma_{\min}, \gamma_{\max}]$. Alternatively, γ_l^* can be written as

$$\gamma_l^* = F_l(\mathbf{u}_l^*, \mathbf{p}_{-l}^*),$$

where F_l is a one-to-one mapping. The reader is referred to [117, 121] for a detailed proof (Theorem 1).

Remark 8.1 Note that using a system-like cost interpretation the overall $L \times m$ game between γ-links and channels has a cost function

$$J_t = \sum_{l=1}^{L} \sum_{i=1}^{m} J_{i,l} \tag{8.7}$$

This could be interpreted as the sum of net utilities over all γ-links and over channels. After changing the summation order this gives

$$J_t = \sum_{i=1}^{m} \tilde{J}_i, \quad \text{with } \tilde{J}_i = \sum_{l=1}^{L} J_{i,l} \tag{8.8}$$

Thus J_t can be interpreted as the sum of net utilities over channels, each channel with cost $\tilde{J}_i$

$$\tilde{J}_i = \sum_{l=1}^{L} \left[\alpha_{i,l} u_{i,l} - \beta_{i,l} \ln\left(1 + \frac{a_{i,l}}{\frac{1}{\delta Q_{i,l}} - \Gamma_{l_{i,i}}}\right)\right]$$

An interesting parallel can be made between $\tilde{J}_i$ and the cost J_i (7.3) (Chap. 7), used in the end-to-end game. This relation is more evident if one considers the case of identical γ-links. In this case the parameters $\alpha_{i,l}$, $\beta_{i,l}$ can be taken to be the same for all γ-links, $\alpha_{i,l} = \alpha_i$, $\beta_{i,l} = \beta_i$, $a_{i,l} = a_i$ and $\tilde{J}_i$ can be rewritten as

$$\tilde{J}_i = \alpha_i \sum_{l=1}^{L} u_{i,l} - \beta_i \sum_{l=1}^{L} \ln\left(1 + \frac{a_i}{\frac{1}{\delta Q_{i,l}} - \Gamma_{i,i}}\right) \tag{8.9}$$

Comparing this with J_i (7.3) (Chap. 7), it can be seen that $\tilde{J}_i$, (8.9), is similarly expressed as the difference between a linear pricing term and a logarithmic utility term. In fact $\tilde{J}_i$, (8.9), generalizes J_i since it captures in pricing and utility the contribution of each γ-span from Tx to Rx.

Remark 8.2 We started the formulation from a game with action space in R^{mL} that was further decomposed. This formulation has a computational advantage when compared to an overall link m-player game with action space in R^L, and a possible cost $\tilde{J}_i$. For the latter game formulation one cannot find an explicit, analytically tractable NE solution, due to the coupling between channel powers at one γ-link and all previous spans. Alternatively, the game formulation thus developed corresponds to a ladder-nested structure: in the game between the γ-spans the lth player decision is taken after the $(l-1)$th player's action. The overall NE solution is still coupled but it has decoupled necessary existence conditions. Moreover, the specific triangular structure of these conditions enables a recursive computation of the unique NE solution. For more details the reader is referred to [121], where a similar partitioned Nash game formulation was proposed for the number of stages $K < L$. As in [121] herein the number of channels at each γ-link (stage) was identical for simplicity. In the following two sections this formulation is extended to multi-link and mesh topologies which are more complex. Firstly, at each stage, the number of channels at different stages (links) can be variable and, secondly, more importantly the coupled link capacity constraint has to be considered.

Next we discuss an iterative algorithm. On each stage l, the following distributed iterative channel algorithm is used:

$$u_{i,l}(n+1) = \frac{\beta_{i,l}}{\alpha_{i,l}} - \left(\frac{1}{y_{i,l}(n)} - \frac{1}{y_{i,l'}} - \Gamma_{l_{i,i}} \right) \frac{u_{i,l}(n)}{a_{i,l}} \tag{8.10}$$

where n is the iteration number, stage $l' = l - 1$ is the precedent of stage l and $y_{i,l'}$ is invariable during the channel iteration on stage l. The algorithm (8.10) is distributed in the sense that each channel i on stage l updates its signal power $u_{i,l}$ based on the feedback information, i.e., its OSNR at the output of stage l, $y_{i,l}$, and fixed parameters, $\boldsymbol{\mu}_{l,l}(t)$ and $y_{i,l'}$. It is worth mentioning that the algorithm operates synchronously. It is possible to consider asynchronous updates and extend the analysis in that case.

Theorem 8.3 *If for all $i \in \mathcal{M}$, $a_{i,l}$ are selected such that* (8.11)

$$\sum_{j \neq i} \Gamma_{l_{i,j}} < a_{i,l}, \quad \forall i \in \mathcal{M} \tag{8.11}$$

then channel algorithm (8.10) *converges to the inner NE solution* $\mathbf{u}_l^*$.

Proof By Theorem 7.1 (point-to-point link) applied to link l, if on stage (link) l, $a_{i,l}$ is selected satisfying (8.11), then an inner unique NE solution $\mathbf{u}_l^*$ exists given as in (7.9). Component-wise, (7.9) is written for $i \in \mathcal{M}$ as

$$u_{i,l}^* = \frac{\beta_{i,l}}{\alpha_{i,l}} - \frac{1}{a_{i,l}} \sum_{j \in \mathcal{M}, j \neq i} \Gamma_{l_{i,j}} u_{j,l}^* \tag{8.12}$$

The rest of the proof follows directly as the proof of Lemma 4 in [121], stated here for completeness. Let

$$e_{i,l}(n) := u_{i,l}(n) - u_{i,l}^*(\boldsymbol{\mu}_l)$$

The corresponding vector form is $e_l(n) = [\ldots, e_{i,l}(n), \ldots]^T$. Let us also define

$$\left\| e_l(n) \right\|_\infty := \max_{i \in \mathcal{M}} \left| e_{i,l}(n) \right|$$

Assuming that $y_{i,l'}$ is stationary it follows that

$$e_{i,l}(n+1) = -\frac{1}{a_{i,l}} \sum_{j \in \mathcal{M}, j \neq i} \Gamma_{l_{i,j}} e_{j,l}(n)$$

Using this it can be shown that under condition (8.11), $\|e_l(n+1)\|_\infty \leq C_0 \|e_l(n)\|_\infty$, where $0 \leq C_0 < 1$ and $\|e_l(n)\|_\infty \leq C_0^n \|e_l(0)\|_\infty$, such that the sequence $\{e_l(n)\}$ converges to 0. Therefore channel algorithm (8.10) converges to the inner NE solution $\mathbf{u}_l^*$. □

Note that for a given γ-link matrix Γ_l, the selected $a_{i,l}$ factors can be set such that the diagonal dominance condition holds. If Γ_l is itself diagonal dominant, then a possible choice is $a_{i,l} = \Gamma_{l_{i,i}}$. If all γ-links are identical then the same $a_{i,l} = a_i$ can be used.

8.3 Games in Single-Sink Topologies

In the simple network topology in the previous section, i.e., γ-link topology, we did not consider coupled constraints. Towards treating the full case of network topologies with coupled constraints, we now turn to games with coupled constraints but we restrict the network topology to another simple case, called single-sink topology. As we will see this particular topology has the feature of maintaining convexity of the coupled constraints.

In a single-sink topology with a set $\mathcal{L} = \{1, \ldots, L\}$ of links and a set $\mathcal{M} = \{1, \ldots, m\}$ of channels, channels are added at different links, but dropped off at the output of the same link (e.g. Fig. 8.3).

Compared to the point-to-point topology here we can have different number channels over different links. We order the m channels in the following way: a set $\mathcal{M}_1^a = \{1, \ldots, m_1\}$ of m_1 channels are added into the network at the first link, a set

$$\mathcal{M}_l^a = \left\{ \sum_{j=1}^{l-1} m_j + 1, \ldots, \sum_{j=1}^{l} m_j \right\}$$

of m_l channels are added at link l and

$$\sum_{l=1}^{L} m_l = m$$

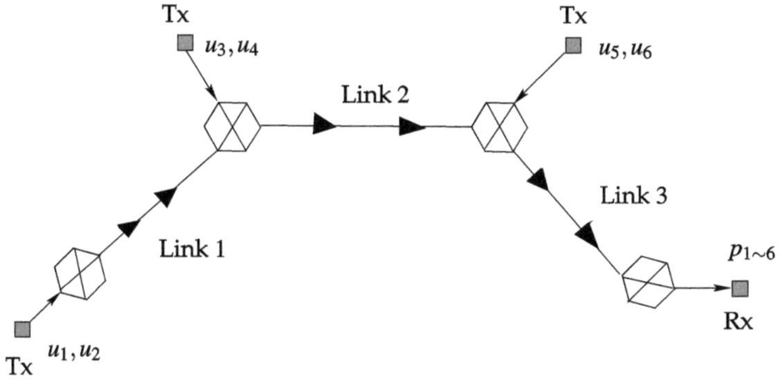

Fig. 8.3 A single-sink topology

As before the input power of each channel $i \in \mathcal{M}$ at Tx is denoted by $u_i \in \Omega_i = [u_{\min}, u_{\max}]$ with $u_{\max} > u_{\min} \geq 0$. For each channel $i \in \mathcal{M}$, let $u_{i,l}$ and $p_{i,l}$ denote the signal power of channel i at the input and output of link l, respectively. Each link l has a total power target $\widehat{P}_l$, thus overall there are L coupled constraints.

8.3.1 Convexity Analysis (Coupled Constraints)

In this section we analyze the convexity of constraints in a class of multi-link topologies where only one sink exists.

Example 8.4 Consider an example shown in Fig. 8.3, where three channels ($m = 3$) are transmitted over a single-sink three-link network ($L = 3$).

Similar to Example 6.12 in Chap. 6, the link capacity constraints $g_l(\mathbf{u})$ can be written as

$$g_1(\mathbf{u}) = \sum_{i=1}^{2} u_{1,i} - P_1^0 = u_1 + u_2 - P_1^0 \leq 0$$

$$g_2(\mathbf{u}) = \sum_{i=1}^{4} u_{2,i} - P_2^0 = \sum_{i=1}^{2} p_{1,i} + u_3 + u_4 - P_2^0 = u_3 + u_4 - \left(P_2^0 - P_1^0\right) \leq 0$$

$$g_3(\mathbf{u}) = \sum_{i=1}^{6} u_{3,i} - P_3^0 = \sum_{i=1}^{4} p_{2,i} + u_5 + u_6 - P_3^0 = u_5 + u_6 - \left(P_3^0 - P_2^0\right) \leq 0,$$

where the system constraint (6.29) was used twice. Thus the coupled constraint set for link $l \in \mathcal{L}$, $\overline{\Omega}_l = \{\mathbf{u} \in \Omega \mid g_l(\mathbf{u}) \leq 0\}$ is convex and $\overline{\Omega} = \bigcap_{l \in \mathcal{L}} \overline{\Omega}_l$ is also convex.

This result can be generalized for single-sink multi-link topologies with m channels, presented in the following lemma.

Lemma 8.5 *Consider a single-sink topology with a set of links $\mathcal{L}$. For any link $l \in \mathcal{L}$, the coupled constraint set $\overline{\Omega}_l$ is convex. The overall coupled action space $\overline{\Omega} = \bigcap_{l\in\mathcal{L}} \overline{\Omega}_l$ is a convex set.*

Proof Let us denote $\bar{m}_{l-1} = \sum_{j=1}^{l-1} m_j$ and $\bar{m}_l = \sum_{j=1}^{l} m_j$. Then

$$\mathcal{M}_l^a = \{\bar{m}_{l-1} + 1, \dots, \bar{m}_{l-1} + m_l\}$$

Thus the link capacity constraint (6.39) can be rewritten as

$$g_l(\mathbf{u}) = \sum_{j=1}^{m_l} u_{\bar{m}_{l-1}+j} - \triangle\widehat{P}_l \leq 0, \tag{8.13}$$

where $\triangle\widehat{P}_l$ is a modified power constraint related to the coupled propagated constraints. In fact $\triangle\widehat{P}_l$ is the remaining total power constraint after the propagated optical power of channels is considered. In Example 8.4, $\triangle P_2^0 = P_2^0 - P_1^0$. One can rewrite $g_l(\mathbf{u})$ as

$$g_l(\mathbf{u}) = \sum_{j=\bar{m}_{l-1}+1}^{\bar{m}_l} u_j - \triangle\widehat{P}_l \tag{8.14}$$

or

$$g_l(\mathbf{u}) = \mathbf{e}_{m_l}^T \mathbf{u} - \triangle\widehat{P}_l \tag{8.15}$$

where

$$e_{m_l}^T = \left[\mathbf{0}_{\bar{m}_{l-1}}^T, \mathbf{1}_{m_l}^T, \mathbf{0}^T\right] \tag{8.16}$$

is an $1 \times m$ vector with $\mathbf{1}_{m_l}$, an $m_l \times 1$ all ones vector, and $\mathbf{0}$, a vector of all 0 components with appropriate dimension. Thus $\overline{\Omega}_l$ is convex for all l, and so is $\overline{\Omega}$. □

Remark 8.6 From (8.14), one can see that in the inequality constraint for link l, $g_l(\mathbf{u}) \leq 0$, only channels in the set $\mathcal{M}_l^a$ are coupled, i.e., channels that are added into the network at link l are coupled. This is advantageous for a hierarchical decomposition, as will be shown later.

8.3.2 Hierarchical Decomposition

Based on the convexity result in Lemma 8.5, let us consider now an m-player Nash game in a single-sink topology where there are L coupled constraints. The action space is

$$\overline{\Omega} = \bigcap_{l\in\mathcal{L}} \{\mathbf{u} \in \Omega \mid g_l(\mathbf{u}) \leq 0\},$$

where Ω is the Cartesian product of Ω_i, $i \in \mathcal{M}$. The action set of channel $i \in \mathcal{M}$ is the projection of $\overline{\Omega}$ on channel i's direction, namely,

$$\widehat{\Omega}_i(\mathbf{u}_{-i}) = \{\xi \in \Omega_i \mid g_l(\mathbf{u}_{-i}, \xi) \leq 0, \ \forall l \in \mathcal{L}\}.$$

As before in the Nash game formulation, each channel $i \in \mathcal{M}$ minimizes its individual cost function $J_i : \overline{\Omega} \to \mathbb{R}$,

$$J_i(\mathbf{u}_{-i}, u_i) = \alpha_i u_i - \mathcal{U}_i(\mathbf{u}_{-i}, u_i), \tag{8.17}$$

with a utility $\mathcal{U}_i$ indicating preference for better OSNR:

$$\mathcal{U}_i(\mathbf{u}_{-i}, u_i) = \beta_i \ln\left(1 + \frac{a_i}{\frac{1}{yi} - \Gamma_{i,i}}\right) = \ln\left(1 + \frac{a_i u_i}{n_i^0 + \sum_{j \neq i} \Gamma_{i,j} u_j}\right),$$

where $\alpha_i, \beta_i > 0$ are weighting factors, $a_i > 0$ is a channel specific parameter and $\Gamma = [\Gamma_{i,j}]$ is the network system matrix being defined in (6.36).

We denote the game by $\mathcal{G}(\mathcal{M}, \widehat{\Omega}_i, J_i)$, which is in the class of m-player Nash games with coupled utilities and coupled constraints. It follows from (8.17) that J_i is continuously differentiable in its arguments and convex in u_i. Note that the overall coupled action space $\overline{\Omega}$ is compact. By Lemma 8.5, $\overline{\Omega}$ is convex as well. Then from Proposition 5.3, $\mathcal{G}(\mathcal{M}, \widehat{\Omega}_i, J_i)$ admits an NE solution. In the following, we use Lagrangian extension and decomposition results presented in Chap. 5 to compute an NE solution.

An NE solution of $\mathcal{G}(\mathcal{M}, \widehat{\Omega}_i, J_i)$ can be computed by using the Lagrangian extension to the game-theoretic framework. We continue using the notations defined in Chap. 5 for the separable NG cost function $\widetilde{J}(\mathbf{u}; \mathbf{x})$, (4.7),

$$\widetilde{J}(\mathbf{u}; \mathbf{x}) = \sum_{i=1}^{m} J_i(\mathbf{u}_{-i}, x_i),$$

the separable augmented constraint $\widetilde{g}(\mathbf{u}; \mathbf{x})$, (5.8),

$$\widetilde{g}(\mathbf{u}; \mathbf{x}) = \sum_{i=1}^{m} g(\mathbf{u}_{-i}, x_i),$$

the augmented Lagrangian function $\widetilde{L}(\mathbf{u}; \mathbf{x}; \boldsymbol{\mu})$, (5.11),

$$\widetilde{L}(\mathbf{u}; \mathbf{x}; \boldsymbol{\mu}) = \widetilde{J}(\mathbf{u}; \mathbf{x}) + \boldsymbol{\mu}^T \widetilde{g}(\mathbf{u}; \mathbf{x})$$

and the dual cost function $D(\boldsymbol{\mu})$, (5.17),

$$D(\boldsymbol{\mu}) = \widetilde{L}(\mathbf{u}^*; \mathbf{u}^*; \boldsymbol{\mu}), \tag{8.18}$$

where $\mathbf{u}^*$ is such that $\mathbf{u} = \mathbf{u}^*$ satisfies $\mathbf{u} = \arg\min_{\mathbf{x} \in \Omega} \widetilde{L}(\mathbf{u}; \mathbf{x}; \boldsymbol{\mu})$. In a fixed-point notation,

$$D(\boldsymbol{\mu}) = \left[\min_{\mathbf{x} \in \Omega} \widetilde{L}(\mathbf{u}; \mathbf{x}; \boldsymbol{\mu})\right]\Big|_{\arg\min_{\mathbf{x} \in \Omega} \widetilde{L} = \mathbf{u}},$$

where $\widetilde{g}(u; u) \leq 0$.

Then $\mathcal{G}(\mathcal{M}, \widehat{\Omega}_i, J_i)$ is related to a constrained minimization of $\widetilde{J}$, (4.7), (5.8), with respect to the second argument x, that admits a fixed-point solution. Individual components of a solution $\mathbf{u}^*$ to this constrained minimization constitute an NE solution to $\mathcal{G}(\mathcal{M}, \widehat{\Omega}_i, J_i)$ in the sense of Definition 5.2. From Remark 5.6, we know that $\mathbf{u}^*$ is obtained by first minimizing $\widetilde{L}(\mathbf{u}; \mathbf{x}; \boldsymbol{\mu})$ with respect to x. The next step involves finding a fixed-point solution which depends on $\boldsymbol{\mu}$, $\mathbf{u}^*(\boldsymbol{\mu})$. An NE solution–Lagrange multiplier pair $(\mathbf{u}^*(\boldsymbol{\mu}^*), \boldsymbol{\mu}^*)$ is obtained if ${\boldsymbol{\mu}^*}^T g(\mathbf{u}^*(\boldsymbol{\mu}^*)) = 0$ and $\boldsymbol{\mu}^* \geq 0$.

The dual cost function $D(\boldsymbol{\mu})$ is related to the minimization of the associated Lagrangian function $\widetilde{L}(\mathbf{u}; \mathbf{x}; \boldsymbol{\mu})$. For $\mathcal{G}(\mathcal{M}, \widehat{\Omega}_i, J_i)$ with convex coupled constraints, Proposition 5.9 yields a decomposition into a lower-level Nash game with no coupled constraints and a higher-level problem for pricing. Particularly, Remark 8.6 states that only channels that are added into the network at link l, are coupled. This benefits the hierarchical decomposition for $D(\boldsymbol{\mu})$ by Proposition 5.9.

Theorem 8.7 *Consider $\mathcal{G}(\mathcal{M}, \widehat{\Omega}_i, J_i)$ on a single-sink multi-link topology with L links (L coupled constraints). A set, $\mathcal{M}_l^a$, of channels are added into network at link $l \in \mathcal{L}$, $\mathcal{L} = \{1, \ldots, L\}$ and $\mathcal{M} = \bigcup_{l \in \mathcal{L}} \mathcal{M}_l^a$. The dual cost function $D(\boldsymbol{\mu})$, (8.18), can be decomposed as*

$$D(\boldsymbol{\mu}) = \sum_{i=1}^{m} \overline{L}_i\big(\mathbf{u}_{-i}^*(\boldsymbol{\mu}), u_i^*(\boldsymbol{\mu}), \mu_{r(i)}\big) + \sum_{i=1}^{m} \sum_{l=1}^{L} \mu_l \big(\mathbf{e}_{m_l,-i}^T \mathbf{u}_{-i}^*(\boldsymbol{\mu}) - \Delta \widehat{P}_l\big) \quad (8.19)$$

where $r(i)$ refers to link r where channel i is added, $\mathbf{e}_{m_l,-i}$ is an $(m-1)$ vector obtained by deleting the ith element of the m vector $\mathbf{e}_{m_l}$, and $\mathbf{u}^*(\boldsymbol{\mu})$ is an NE solution to $\mathcal{G}(\mathcal{M}, \Omega_i, \overline{L}_i)$ where

$$\overline{L}_i(\mathbf{u}_{-i}, x_i, \mu_{r(i)}) = J_i(\mathbf{u}_{-i}, x_i) + \mu_{r(i)} x_i. \quad (8.20)$$

This is similar to the point-to-point case (see Chap. 7) but herein each of the L constraints is accounted for.

Proof Let us order the L links and m channels in the same way as in the proof of Lemma 8.5. Then the constraint function (8.15) can be written in a two-argument form

$$g_l(\mathbf{u}_{-i}, x_i) = \mathbf{e}_{m_l}^T \cdot \mathbf{u} - \Delta \widehat{P}_l = \begin{cases} \mathbf{e}_{m_l,-i}^T \mathbf{u}_{-i} + x_i - \Delta \widehat{P}_l, & l = r(i) \\ \mathbf{e}_{m_l,-i}^T \mathbf{u}_{-i} - \Delta \widehat{P}_l, & l \neq r(i) \end{cases} \quad (8.21)$$

Since J_i and g_l are continuously differentiable and convex functions, by using Proposition 5.9, the dual cost function $D(\boldsymbol{\mu})$ can be decomposed as

$$D(\boldsymbol{\mu}) = \sum_{i=1}^{m} L_i\big(\mathbf{u}_{-i}^*(\boldsymbol{\mu}), \mathbf{u}_i^*(\boldsymbol{\mu}), \boldsymbol{\mu}\big)$$

where

$$L_i(\mathbf{u}_{-i}, x_i, \boldsymbol{\mu}) = J_i(\mathbf{u}_{-i}, x_i) + \boldsymbol{\mu}^T g(\mathbf{u}_{-i}, x_i)$$

Using (8.21) one finds that, $\forall i \in \mathcal{M}$, L_i is given as

$$\begin{aligned} L_i &= J_i(\mathbf{u}_{-i}, x_i) + \mu_{r(i)} g_{r(i)}(\mathbf{u}_{-i}, x_i) + \sum_{l \neq r(i)} \mu_l g_l(\mathbf{u}_{-i}, x_i) \\ &= J_i(\mathbf{u}_{-i}, x_i) + \mu_{r(i)} x_i + \sum_l \mu_l \big(\mathbf{e}_{m_l,-i}^T \mathbf{u}_{-i} - \triangle \widehat{P}_l\big) \end{aligned} \tag{8.22}$$

The following arguments are similar to those in Corollary 1 in [120], outlined here. Let us minimize RHS of (8.22) with respect to x_i and then solve for a fixed-point solution $x_i = u_i$. Since only the first two terms on RHS of (8.22) depend on x_i, isolating the terms without x_i yields

$$\begin{aligned} D(\boldsymbol{\mu}) &= \sum_{i=1}^{m} \Big[\min_{x_i \in \Omega_i} \overline{L}_i(\mathbf{u}_{-i}, x_i, \mu_{r(i)}) \Big] \Big|_{\arg\min_{x_i \in \Omega_i} \overline{L}_i = u_i} \\ &\quad + \sum_{i=1}^{m} \sum_{l=1}^{L} \mu_l \big(\mathbf{e}_{m_l,-i}^T \mathbf{u}_{-i} - \triangle \widehat{P}_l\big), \end{aligned} \tag{8.23}$$

with $\overline{L}_i$ as in (8.20). After minimization in each subproblem on RHS of (8.23), x_i^* is obtained as a function of $\mathbf{u}_{-i}$, namely,

$$x_i^* = x_i(\mathbf{u}_{-i})$$

A fixed-point solution denoted by $x_i = u_i$ can be obtained by setting $x_i(\mathbf{u}_{-i}) = u_i$ and solving simultaneously the m equations for a vector solution denoted by $\mathbf{u}^*$, or $\mathbf{u}^*(\boldsymbol{\mu})$ since it depends on $\boldsymbol{\mu}$. Substituting $\mathbf{u}^*(\boldsymbol{\mu})$ on RHS of (8.23) yields (8.19). □

Theorem 8.7 leads to a hierarchical decomposition. An NE solution–Lagrange multiplier pair $(\mathbf{u}^*(\boldsymbol{\mu}^*), \boldsymbol{\mu}^*)$ can be obtained by solving a lower-level uncoupled Nash game $\mathcal{G}(\mathcal{M}, \Omega_i, \overline{L}_i)$ and a higher-level problem for coordination (i.e., setting price $\boldsymbol{\mu}$). The hierarchical game decomposition offers computational advantages. For a given price $\boldsymbol{\mu} \geq 0$, the lower-level game with no coupled constraints admits an inner closed-form explicit NE solution. The following result gives conditions for the existence of the NE solution.

Proposition 8.8 *For each given* $\boldsymbol{\mu} \geq 0$, $\mathcal{G}(\mathcal{M}, \Omega_i, \overline{L}_i)$ *admits an NE solution if* a_i *satisfies*

$$\sum_{j \neq i} \Gamma_{i,j} < a_i, \quad \forall i \in \mathcal{M} \tag{8.24}$$

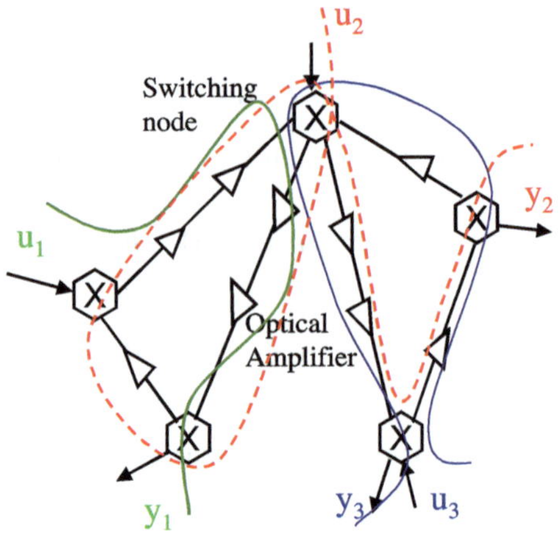

Fig. 8.4 A mesh topology with three channel optical paths

The inner NE solution $\mathbf{u}^*(\boldsymbol{\mu})$ *is unique, given as*

$$\mathbf{u}^*(\boldsymbol{\mu}) = \widetilde{\Gamma}^{-1} \cdot \widetilde{b}(\boldsymbol{\mu}),$$

where $\widetilde{\Gamma} = [\widetilde{\Gamma}_{i,j}]$ *and* $\widetilde{b}(\boldsymbol{\mu}) = [\widetilde{b}_i(\boldsymbol{\mu})]$ *are defined as*

$$\widetilde{\Gamma}_{i,j} = \begin{cases} a_i & j = i, \\ \Gamma_{i,j} & j \neq i, \end{cases} \qquad \widetilde{b}_i(\boldsymbol{\mu}) = \frac{a_i \beta_i}{\alpha_i + \boldsymbol{\mu}_{r(i)}} - n_i^0, \tag{8.25}$$

with $\Gamma_{i,j}$ *defined in* (6.36).

The proof follows similar to the proof of Proposition 7.9 for the point-to-point topology except that it involves the network matrix and $\mu_{r(i)}$.

Based on this single-sink case and the method in Sect. 8.2, a partitioning method for general multi-link topologies was proposed in [101, 112]. Such a topology is partitioned into stages where each stage has a single-sink structure and algorithms developed based on the extragradient method [70, 71]. However, this partition does not fully exploit the power adjustment flexibility, is not completely scalable and cannot be easily extended to treat mesh topologies. In the next section we present an approach that addresses this most general case, of how to treat mesh topologies.

8.4 Games with Coupled Constraints: Mesh Topologies

After these preliminary simple network topologies (γ-link in Sect. 8.2 and single-sink topologies in Sect. 8.3), we now consider games with coupled constraints in general mesh and multi-link network topologies. A possible case with three channel optical paths is shown in Fig. 8.4, where each channel is represented by a different color (wavelength).

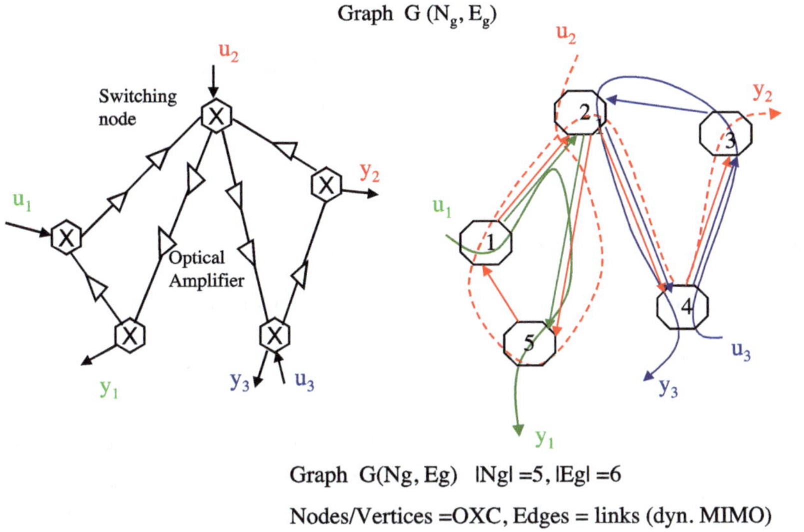

Fig. 8.5 Graph with adjustable nodes for Fig. 8.4

For such general mesh or multi-link topologies, an NE solution of the end-to-end game with coupled constraints is intractable because of the non-convexity of propagated constraints. Thus the results in Sect. 8.3 cannot be directly extended to general multi-link or mesh topologies. We will use an approach that builds on the simpler results in Sect. 8.2. In particular we will use a partition technique as in Sect. 8.2 together with a hierarchical decomposition (for coupled constraints) as in Chap. 7 or Sect. 8.3. The same notations as in Sects. 8.2 and 8.3 are used.

Recall that the starting point for setting an end-to-end game was the fact that channel powers are assumed to be independently adjustable only at Txs. In fact, this restriction is relaxed. In optical networks, optical switches not only perform the function of selectively dropping some of wavelengths or adding selected wavelengths while letting others pass through, but also provide the flexibility of optical power adjustment for each channel [2]. For the example in Fig. 8.4 these five nodes are shown in Fig. 8.5.

The simplest example we consider is the one obtained by extracting from Fig. 8.4 a single quasi-ring. Thus consider the quasi-ring topology shown in Fig. 8.6 as the simplest example. In Fig. 8.6, we use $\mathrm{Tx}_i/\mathrm{Rx}_i$ to indicate the Tx/Rx that channel i uses. We also use l_i to refer link l. The quasi-ring topology is a ring-type topology with partially closed loops being formed by channel optical paths [2]. In Fig. 8.6, the optical paths of channels 1 and 3 are $l_1 \to l_2 \to l_3$ and $l_3 \to l_1 \to l_2$, respectively. Note that each of the links can be either the intermediate or the end link on a specific channel optical path. Thus how to define a single-sink structure that potentially can be partitioned is not immediate. Another issue of concern is the interconnection among stages. However, the simplest way is to consider just a link by link partitioning. This alternative approach is scalable and easily applicable, and moreover fully

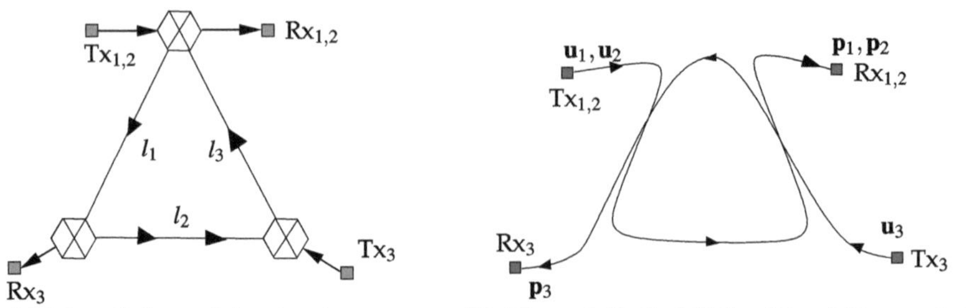

(a) Three-link quasi-ring topology (b) Channel Optical Paths (Quasi-Ring routes)

Fig. 8.6 Quasi-ring topology

uses the flexibility of power adjustment at each node. Thus the network is partitioned as in Sect. 8.2 into stages composed only of single optical links. By breaking the loop and selecting one stage as the start, stages can be placed sequentially in a ladder-nested form. One benefit of such a partition is that the convexity of coupled constraints propagated along links on each stage is automatically satisfied.

In the partitioned Nash game, each link Nash game is played towards minimizing channel OSNR degradation. Based on such a partition, the hierarchical decomposition is applicable to each link game, leading to a lower-level uncoupled game for channels on each link and a higher-level problem for link pricing. Computation of equilibria is based on a three-level hierarchical algorithm. Such a partition simplifies the structure of each stage, makes it regular and scalable, and benefits the development of a link pricing algorithm. This approach can be thus used for general mesh topologies.

We can apply Lemma 6.8 and use the recursive OSNR model for the network. Consider link l in Fig. 6.10. The OSNR of channel i at the output of link l is given as

$$\frac{1}{y_{i,l}} = \frac{1}{y_{i,l'}} + \sum_{j \in \mathcal{M}} A_{j,l} \Gamma_{l_{i,j}} \frac{u_{j,l}}{u_{i,l}}, \quad i \in \mathcal{M}_l, \tag{8.26}$$

where link l' is the link precedent to link l for channel i. Using (8.26), let us define

$$\frac{1}{\delta Q_{i,l}} = \frac{1}{y_{i,l}} - \frac{1}{y_{i,l'}} = \sum_{j \in \mathcal{M}} A_{j,l} \Gamma_{l_{i,j}} \frac{u_{j,l}}{u_{i,l}} \tag{8.27}$$

It follows that $\frac{1}{\delta Q_{i,l}}$ measures the OSNR degradation of channel i from link l' to link l. Thus instead of maximization of yi from Tx to Rx, consider minimization of each $\frac{1}{\delta Q_{i,l}}$ between links, i.e., minimization of individual OSNR degradation. In the next section, by using the similar methodology as in Sect. 8.2, a partitioned Nash game composed of stage Nash games is presented. Each stage is a link and the stage Nash game is formulated to minimize the OSNR degradation, and moreover the coupled constraints are considered.

8.4.1 Partitioned Game Formulation

In this section we discuss a partitioned game framework with link capacity constraints in mesh topologies. As mentioned above the network is partitioned into stages composed of single links. By partitioning, each stage game is defined for a link, thus a link game, as in Sect. 8.2. Each link Nash game is played towards minimizing channel OSNR degradation. Then a partitioned Nash game composed of link games is defined. The set of OSNR degradation minimizations on stages/links is related to OSNR maximization from Tx to Rx (see Remark 8.1 in Sect. 8.2). This setup is described below.

On each link $l \in \mathcal{L}$, the set of channels is $\mathcal{M}_l = \{1, \dots, m_l\}$. Let us consider the link capacity constraint on link l:

$$g_l(\mathbf{u}_l) = \sum_{i \in \mathcal{M}} A_{i,l} u_{i,l} - \widehat{P}_l \le 0, \quad \forall l \in \mathcal{L}, \tag{8.28}$$

where $A_{i,l} = 1$ if channel i is transmitted on link l and otherwise, $A_{i,l} = 0$. Let $\overline{\Omega}_l$ denote the action space on link l,

$$\overline{\Omega}_l = \bigl\{\mathbf{u}_l \in \Omega \mid g_l(\mathbf{u}_l) \le 0\bigr\}, \tag{8.29}$$

where Ω denotes the Cartesian product of Ω_i. The action set of individual channel $i \in \mathcal{M}_l$ is defined as the projection of $\overline{\Omega}_l$ on channel i's direction, namely,

$$\widehat{\Omega}_{i,l}(u_{l,-i}) = \bigl\{\xi \in \Omega_i \mid g_l(u_{l,-i}, \xi) \le 0\bigr\},$$

where $u_{l,-i}$ is obtained by deleting $u_{i,l}$ from vector $\mathbf{u}_l$. It can be seen that both the action space $\overline{\Omega}_l$ and the individual action set $\widehat{\Omega}_{i,l}(u_{l,-i})$ are compact and convex.

The link Nash game is played with each channel attempting to minimize its individual cost with respect to its OSNR degradation. We specify a channel cost function $J_{i,l}$ for channel i on each link l. Similar to the channel cost function defined in Sect. 8.2, $J_{i,l}$ is defined as a difference between a pricing function $A_{i,l}\alpha_{i,l}u_{i,l}$ and a utility function $\mathcal{U}_{i,l}$ which reflects the associated channel's OSNR degradation, namely,

$$J_{i,l} = A_{i,l}\alpha_{i,l}u_{i,l} - \mathcal{U}_{i,l}, \quad \forall i \in \mathcal{M}, \tag{8.30}$$

with

$$\mathcal{U}_{i,l} = \beta_{i,l} \ln\left(1 + \frac{A_{i,l} a_{i,l}}{\frac{1}{\delta Q_{i,l}} - \Gamma_{l_{i,i}}}\right), \tag{8.31}$$

where $\beta_{i,l} > 0$ indicates the strength of the channel's desire to minimize its OSNR degradation and $a_{i,l} > 0$ is for scalability. Substituting (8.27) into (8.30) yields

$$J_{i,l} = A_{i,l}\alpha_{i,l}u_{i,l} - \beta_{i,l} \ln\left(1 + a_{i,l}\frac{A_{i,l}u_{i,l}}{X_{l,-i}}\right), \quad \forall i \in \mathcal{M}, \tag{8.32}$$

where $X_{l,-i} = \sum_{j \neq i, j \in \mathcal{M}} A_{j,l} \Gamma_{l_{i,j}} u_{j,l}$. It follows that $J_{i,l}$ is continuously differentiable in its arguments and convex in $u_{i,l}$.

We denote such a link Nash game by $\mathcal{G}(\mathcal{M}, \widehat{\Omega}_{i,l}, J_{i,l})$. Note that the individual cost function $J_{i,l}$ is generally defined for each channel $i \in \mathcal{M}$. If channel i is not transmitted on link l, i.e., $i \notin \mathcal{M}_l$, then $A_{i,l} = 0$ and $J_{i,l} = 0$, which means that the decision of channel i, $i \notin \mathcal{M}_l$, does not affect the decisions made by other channels $j \in \mathcal{M}_l$. Thus $\mathcal{G}(\mathcal{M}, \widehat{\Omega}_{i,l}, J_{i,l})$ is equivalently to a reduced Nash game played among m_l channels. Furthermore, the existence of an NE solution of $\mathcal{G}(\mathcal{M}, \widehat{\Omega}_{i,l}, J_{i,l})$ is guaranteed by Proposition 5.3.

We exploit the partitioned Nash game. By partitioning and selecting one stage as the start, stages can be sorted sequentially with the interpretation that across all stages, $\sum_{l \in \mathcal{R}_i} J_{i,l}$ is related to the overall OSNR degradation for channel i. Solutions of all $\mathcal{G}(\mathcal{M}, \widehat{\Omega}_{i,l}, J_{i,l})$, $l \in \mathcal{L}$ are interconnected similar to Sect. 8.2. The explanation is given as follows.

Recall that channel powers are adjustable at optical switches and $\gamma_{i,l}$ is the adjustable parameter for channel i on stage l. The vector form is $\gamma_l = [\gamma_{i,l}]$. Let $\mathbf{u}_l^* = [u_{i,l}^*]$ be an NE solution of $\mathcal{G}(\mathcal{M}, \widehat{\Omega}_{i,l}, J_{i,l})$. The corresponding signal power vector at the output of link l is $\mathbf{p}_l^* = [p_{i,l}^*]$ and the corresponding augmented output power vector defined in Lemma 6.10 is $\mathbf{p}^*$. Note that for those channel $i \notin \mathcal{M}_l$, values of $u_{i,l}^*$ and $p_{i,l}^*$ are randomly set. By using (6.37) in Lemma 6.10, optimal $\gamma_{i,l}$ can be obtained by solving the corresponding component-wise equation in

$$\mathbf{u}_l^* = \operatorname{diag}(\gamma_l^*) \cdot \operatorname{diag}(V_{i,l}^T) \cdot \mathbf{p}^*.$$

Finally, the partitioned Nash game admits a solution if each $\gamma_{i,l}^* \in [\gamma_{\min}, \gamma_{\max}]$.

Compared to Sect. 8.2 the interconnection matrix $\operatorname{diag}(V_{i,l})$ for the full network topology appears as well as the coupled constraints.

8.4.2 Hierarchical Decomposition

In this section, we first use Lagrangian extension presented in Chap. 5 to obtain an NE solution. Then we discuss an iterative hierarchical algorithm for computation of equilibria of the partitioned Nash game. Recall that if channel i is not transmitted on link l, its decision does not affect the decisions made by other channels $j \in \mathcal{M}_l$. Thus $\mathcal{G}(\mathcal{M}, \widehat{\Omega}_{i,l}, J_{i,l})$ is equivalently to a reduced Nash game played among m_l channels. The computation of equilibria is based on this fact and the algorithm is developed for the equivalent reduced Nash game, i.e., $\mathcal{G}(\mathcal{M}_l, \widehat{\Omega}_{i,l}, J_{i,l})$. We use the mark "$^-$" to indicate the associated reduced vector. For example, $\bar{\mathbf{u}}_l$ is the reduced vector obtained by removing those elements $u_{i,l}$, $i \notin \mathcal{M}_l$, from $\mathbf{u}_l$. Sometimes we write $\bar{\mathbf{u}}_l$ as $\bar{\mathbf{u}}_l = (\bar{\mathbf{u}}_{l,-i}, u_{i,l})$ with the vector $\bar{\mathbf{u}}_{l,-i}$ obtained by deleting the ith element from the reduced vector $\bar{\mathbf{u}}_l$.

On each link $l \in \mathcal{L}$, $\mathcal{G}(\mathcal{M}_l, \widehat{\Omega}_{i,l}, J_{i,l})$ admits an NE solution From Proposition 5.3. An NE solution can be computed by using the Lagrangian extension to the game-theoretic framework presented in Chap. 5. We first define several augmented functions for $\mathcal{G}(\mathcal{M}_l, \widehat{\Omega}_{i,l}, J_{i,l})$. The augmented NG cost function $\widetilde{J}_l(\bar{\mathbf{u}}_l; \bar{\mathbf{x}}_l)$ is defined as

$$\widetilde{J}_l(\bar{\mathbf{u}}_l; \bar{\mathbf{x}}_l) = \sum_{i \in \mathcal{M}_l} J_{i,l}(\bar{\mathbf{u}}_{l,-i}, \bar{x}_{i,l}),$$

where $J_{i,l}$ is defined in (8.32) with $A_{i,l} = 1$. The two-argument constraints are given as

$$\widetilde{g}_l(\bar{\mathbf{u}}_l; \bar{\mathbf{x}}_l) = \sum_{i \in \mathcal{M}_l} g_l(\bar{\mathbf{u}}_{l,-i}, \bar{x}_{i,l})$$

Both $\widetilde{J}_l(\bar{\mathbf{u}}_l; \bar{\mathbf{x}}_l)$ and $\widetilde{g}_l(\bar{\mathbf{u}}_l; \bar{\mathbf{x}}_l)$ are separable in the second argument $\bar{\mathbf{x}}_l$. The associated two-argument Lagrangian function $\widetilde{L}_l$ is defined as

$$\widetilde{L}_l(\bar{\mathbf{u}}_l; \bar{\mathbf{x}}_l; \boldsymbol{\mu}_l) = \widetilde{J}_l(\bar{\mathbf{u}}_l; \bar{\mathbf{x}}_l) + \boldsymbol{\mu}_l \widetilde{g}_l(\bar{\mathbf{u}}_l; \bar{\mathbf{x}}_l), \tag{8.33}$$

where the scalar $\boldsymbol{\mu}_l$ is the Lagrange multiplier. The associated dual cost function $D_l(\boldsymbol{\mu}_l)$ is defined as

$$D_l(\boldsymbol{\mu}_l) = \widetilde{L}_l\big(\bar{\mathbf{u}}_l^*; \bar{\mathbf{u}}_l^*; \boldsymbol{\mu}_l\big), \tag{8.34}$$

where $\bar{\mathbf{u}}_l^*$ is such that $\bar{\mathbf{u}}_l = \bar{\mathbf{u}}_l^*$ satisfies

$$\bar{\mathbf{u}}_l = \arg\min_{\bar{\mathbf{x}}_l \in \overline{\Omega}} \widetilde{L}(\bar{\mathbf{u}}_l; \bar{\mathbf{x}}_l; \boldsymbol{\mu}_l)$$

where $\overline{\Omega}$ denotes the Cartesian product of Ω_i, $i \in \mathcal{M}_l$.

The $\mathcal{G}(\mathcal{M}_l, \widehat{\Omega}_{i,l}, J_{i,l})$ is related to a constrained minimization of $\widetilde{J}_l(\bar{\mathbf{u}}_l; \bar{\mathbf{x}}_l)$ with respect to $\bar{\mathbf{x}}_l$, which admits a fixed-point solution $\bar{\mathbf{u}}_l^*$. Individual components of $\bar{\mathbf{u}}_l^*$ constitute an NE solution to $\mathcal{G}(\mathcal{M}_l, \widehat{\Omega}_{i,l}, J_{i,l})$. In the Lagrangian optimality condition, $\bar{\mathbf{u}}_l^*$ is obtained by first minimizing $\widetilde{L}_l(\bar{\mathbf{u}}_l; \bar{\mathbf{x}}_l; \boldsymbol{\mu}_l)$ with respect to $\bar{\mathbf{x}}_l$. The next step involves finding a fixed-point solution $\bar{\mathbf{u}}_l^*$ by setting $\bar{\mathbf{x}}_l = \bar{\mathbf{u}}_l$. Since $\bar{\mathbf{u}}_l^*$ depends on $\boldsymbol{\mu}_l$, an NE solution–Lagrange multiplier pair $(\bar{\mathbf{u}}_l^*(\boldsymbol{\mu}_l^*), \boldsymbol{\mu}_l^*)$ is obtained if

$$\boldsymbol{\mu}_l^* g_l\big(\bar{\mathbf{u}}_l^*\big) = 0 \quad \text{and} \quad \boldsymbol{\mu}_l^* \geq 0.$$

The minimization of $\widetilde{L}_l(\bar{\mathbf{u}}_l; \bar{\mathbf{x}}_l; \boldsymbol{\mu}_l)$ can be decomposed by Proposition 5.9. Then for each link l applying Proposition 7.8 in Chap. 7 yields a decomposition in terms of $\overline{L}_{i,l}$. Alternatively since each stage l is a single optical link, which is the simplest single-sink structure, the decomposition results (Theorem 8.7) for single-sink topologies can be directly used. Thus $\mathcal{G}(\mathcal{M}_l, \widehat{\Omega}_{i,l}, J_{i,l})$ can be naturally decomposed into a lower-level Nash game with no coupled constraints and a higher-level link problem for pricing (Lagrange multiplier). By using Theorem 8.7, after decomposition, the lower-level Nash game is obtained with the following individual cost function:

$$\overline{L}_{i,l}(\bar{\mathbf{u}}_{l,-i}, \bar{x}_{i,l}, \boldsymbol{\mu}_l) = J_{i,l}(\bar{\mathbf{u}}_{l,-i}, \bar{x}_{i,l}) + \boldsymbol{\mu}_l \bar{x}_{i,l}, \quad \forall i \in \mathcal{M}_l, \tag{8.35}$$

where $J_{i,l}$ is defined in (8.32) and an uncoupled action set Ω_i. We denote this game by $\mathcal{G}(\mathcal{M}_l, \Omega_i, \overline{L}_{i,l})$. Notice that $\overline{L}_{i,l}$ (8.35) has a similar form with $\overline{L}_i$ (7.51) in Chap. 7. By using Proposition 7.9 we obtain the following result characterizing the NE solution of $\mathcal{G}(\mathcal{M}_l, \Omega_i, \overline{L}_{i,l})$.

Corollary 8.9 *For each given $\boldsymbol{\mu}_l \geq 0$, $\mathcal{G}(\mathcal{M}_l, \Omega_i, \overline{L}_{i,l})$ admits an NE solution if $a_{i,l}$ in $\overline{L}_{i,l}$ (8.35) satisfies*

$$\sum_{j\neq i, j\in\mathcal{M}_l} \Gamma_{l_{i,j}} < a_{i,l}, \quad \forall i \in \mathcal{M}_l \tag{8.36}$$

The inner NE solution $\bar{\mathbf{u}}_l^*(\boldsymbol{\mu}_l)$ is unique, given as

$$\bar{\mathbf{u}}_l^*(\boldsymbol{\mu}_l) = \widetilde{\Gamma}_l^{-1} \cdot \bar{b}_l(\boldsymbol{\mu}_l) \tag{8.37}$$

where $\widetilde{\Gamma}_l = [\widetilde{\Gamma}_{l_{i,j}}]$ with

$$\widetilde{\Gamma}_{l_{i,j}} = \begin{cases} a_{i,l}, & j = i \\ \Gamma_{l_{i,j}}, & j \neq i \end{cases} \quad i, j \in \mathcal{M}_l,$$

where $\bar{b}_l(\boldsymbol{\mu}_l) = [b_{i,l}(\boldsymbol{\mu}_i)]$ with

$$b_{i,l}(\boldsymbol{\mu}_l) = \frac{a_{i,l}\beta_{i,l}}{\alpha_{i,l} + \boldsymbol{\mu}_l}, \quad i \in \mathcal{M}_l$$

It follows from (8.37) that $\sum_{i\in\mathcal{M}_l} u_{i,l}^*(\boldsymbol{\mu}_l)$ decreases when $\boldsymbol{\mu}_l$ increases. This monotonicity property will be used in the development of an iterative hierarchical algorithm.

8.4.3 Primal–Dual (Channel-Link) Iterative Algorithm

We discuss an iterative hierarchical algorithm based on the NE solution (8.37) of the lower-level Nash game and coordination at the higher level.

For given $\boldsymbol{\mu}_l \geq 0$, $\bar{u}_l^*(\boldsymbol{\mu}_l)$ in Corollary 8.9 is a solution to solve the Lagrangian optimality condition. Recall that $(\bar{\mathbf{u}}_l^*(\boldsymbol{\mu}_l^*), \boldsymbol{\mu}_l^*)$ is an NE solution–Lagrange multiplier pair if

$$\boldsymbol{\mu}_l^* g_l(\bar{\mathbf{u}}_l^*) = 0 \quad \text{and} \quad \boldsymbol{\mu}_l^* \geq 0.$$

By using (8.28), $\boldsymbol{\mu}_l^* \geq 0$ can be obtained by solving

$$\boldsymbol{\mu}_l^* \left(\sum_{i\in\mathcal{M}_l} \mathbf{u}_{i,l}^*(\boldsymbol{\mu}_l^*) - \widehat{P}_l \right) = 0 \tag{8.38}$$

Based on these, an iterative hierarchical algorithm can be developed for both link pricing and channel power adjustment.

Channel Algorithm Based on the similarity between $\mathcal{G}(\mathcal{M}_l, \Omega_i, \overline{L}_{i,l})$ and $\mathcal{G}(\mathcal{M}, \Omega_i, \overline{J}_{i,l})$ developed for γ-point topologies in Sect. 8.2, one can adapt the channel algorithm:

$$u_{i,l}(n+1) = \frac{\beta_{i,l}}{\alpha_{i,l} + \boldsymbol{\mu}_l(\bar{k})} - \left(\frac{1}{y_{i,l}(n)} - \frac{1}{y_{i,l'}} - \Gamma_{l_{i,i}}\right)\frac{u_{i,l}(n)}{a_{i,l}}, \quad \forall i \in \mathcal{M}_l \tag{8.39}$$

where link l' is the precedent of link l. Note that both $y_{i,l'}$ and $\boldsymbol{\mu}_l(\bar{k})$ are invariable during the channel iteration on link l. Each channel i on link l updates its signal power $u_{i,l}$ based on the feedback information, i.e., its OSNR at the output of link l, $y_{i,l}$, and fixed parameters, $\boldsymbol{\mu}_l(t)$ and $y_{i,l'}$. Note that compared to (8.10) in Sect. 8.2 the term $\boldsymbol{\mu}_l$ appears due to coupled constraint, while compared to (7.62) in Chap. 7) the term $y_{i,l'}$ appears due to partitioning in link stages and interconnection between link l' and link l. Then applying Theorem 8.3 in Sect. 8.2 one can obtain the following convergence result.

Corollary 8.10 *If for all $i \in \mathcal{M}_l$, $a_{i,l}$ in $\overline{L}_{i,l}$* (8.35) *are selected such that* (8.36) *is satisfied. Then for each given $\boldsymbol{\mu}_l \geq 0$, channel algorithm* (8.39) *converges to the inner NE solution $\bar{\mathbf{u}}_l^*(\boldsymbol{\mu}_l)$.*

Link Algorithm The link algorithm is a gradient projection algorithm [24, 81], developed based on (8.38). On each link l, after every K iterations of the channel algorithm (8.39), the new link price $\boldsymbol{\mu}_l$ is generated at each iteration time $\bar{k}$, according to the following link algorithm:

$$\boldsymbol{\mu}_l(\bar{k}+1) = \left[\boldsymbol{\mu}_l(\bar{k}) + \eta\left(\sum_{i \in \mathcal{M}_l} u_{i,l}\big(\boldsymbol{\mu}_l(\bar{k})\big) - \widehat{P}_l\right)\right]^+ \tag{8.40}$$

where $\eta > 0$ is a step-size and $[z]^+ = \max\{z, 0\}$. Practically, K is sufficiently large such that each channel power converges to its solution. Thus given the total power of all channels on the link l, the link algorithm (8.40) is completely distributed and can be implemented by individual links using only local information. A proof for the convergence of this link algorithm is provided in [111].

The link algorithm above is a simple projection algorithm. Theorem 1 in [111] provides an explicit upper-bound for the step-size η related to the NE solution. It can be mentioned that an extragradient method can been used on each stage for the link pricing [112]. The extragradient method is a modified projection algorithm, and sufficient conditions for its convergence depend on a Lipschitz constant which is intractable.

Remark 8.11 For any given $\boldsymbol{\mu}_l \geq 0$, Theorem 8.3 shows that the channel algorithm converges to the NE solution which is unique in the sense of inner-ness. Though uniqueness of a solution is not guaranteed, $\boldsymbol{\mu}_l^*$, $(\mathbf{u}_l^*, \boldsymbol{\mu}_l^*)$ is an optimal NE solution–Lagrange multiplier pair and the algorithm converges to one such pair.

8.5 Notes

For mesh and multi-link network topologies we treated several cases in increasing order of complexity. First we treated simple γ-link topologies and we formulated a partitioned Nash game composed of link Nash games, where for simplicity no coupled constraints were considered. In the partitioned Nash game, each link Nash game is played towards minimizing channel OSNR degradation. Based on such a partition, the hierarchical decomposition is applicable to each link Nash game.

Next we considered games with coupled utility and constraints in multi-link topologies, extending the approach in Chap. 7 where only single point-to-point fiber links were treated. In multi-link topologies constraints propagate along link and convexity may not be preserved. The general multi-link topology was dealt with by formulating a partitioned Nash game composed of stage Nash games, where each stage has a single-link structure, i.e., a γ-link structure. Connections between stages are realized by exploiting the flexibility of extra adjustments at bifurcation points. By selecting a starting link, the game on each link can be played sequentially and the hierarchical decomposition leads to a lower-level Nash game for channels with no coupled constraints and a higher-level problem for link pricing. Computation of equilibria is based on a hierarchical algorithm. An alternative partitioned game could be based on single-sink structure (which has convexity ensured), but the single-link partition has a more regular and scalable structure. A possible implementation is described later on in Chap. 10.

In Chaps. 7 and 8 game-theoretic approaches have been used to solve channel OSNR optimization with link capacity constraints in optical networks. In particular, while we formulated a game-theoretic framework towards optimizing channel OSNR, however, we have not taken desired channel OSNR targets into account yet. In a Nash game in the presence of player selfishness, Nash equilibria may not optimize overall system performance. These efficiency issues as well as robustness of network control algorithms derived from game-theoretic formulations in the presence of time-delay will be addressed in the following two chapters.

Chapter 9
Nash Equilibria Efficiency and Numerical Studies

Abstract This chapter provides an alternative constrained OSNR optimization approach. This framework can be used to investigate the effects of parameters in the game-theoretic approach, i.e., the efficiency of Nash equilibria. A system optimization problem is formulated towards achieving an OSNR target for each channel while satisfying the link capacity constraint. In the game case we show that OSNR targets can be achieved and efficiency can be possibly improved by appropriate selection of game parameters.

9.1 Introduction

In previous chapters we showed how a game-theoretic framework can be formulated towards optimizing channel OSNR, but we have not taken desired channel OSNR targets into account yet. In a Nash game in the presence of player selfishness, Nash equilibria may not optimize overall system performance. These efficiency issues will be addressed next, focusing on a single point-to-point link game. The material in this chapter is mostly from [102–104]. A combination approach can be found in [168].

We first consider a system optimization framework towards achieving a target OSNR level for each channel, while minimizing the interference and hence improving the overall performance. Link capacity constraint is also considered. This framework is used to investigate the effects of parameters in a game-theoretic approach.

Efficiency of Nash equilibria has been studied extensively [1, 38, 64, 122, 129, 134]. The efficiency loss as compared to an socially optimal system performance is quantified by a ratio called "Price of Anarchy" [73]. Work on how to quantify the efficiency loss under separable and under non-separable costs can be found in [64] and [122], respectively. Results suggest that, provided a pricing mechanism is chosen properly, the selfish behavior of players in a Nash game may not degrade the system performance arbitrarily [64, 134]. Such approaches will be reviewed in Chap. 12 for routing and path coloring via game approaches in optical networks. In this chapter, instead of looking from the view point of degree of efficiency (Price of Anarchy), we study the efficiency of an OSNR game by investigating the effects of parameters in individual game cost functions. In particular, we use the system

L. Pavel, *Game Theory for Control of Optical Networks*,
Static & Dynamic Game Theory: Foundations & Applications,
DOI 10.1007/978-0-8176-8322-1_9,

optimization framework to measure the efficiency of a game as in Chap. 7. This can be motivated by the fact that the aggregate cost function in a game-theoretic formulation is not always strictly convex; hence the optimal solution of the associated constrained optimization problem is not available. Nevertheless, we show that a possible cost function in the system optimization has an approximate interpretation as the game cost function in the game-theoretic formulation. We compare the numerical results based on this and show the effects of pricing mechanisms.

9.2 A System Optimization Formulation

Consider a single point-to-point WDM link with a set of $\mathcal{M} = \{1, \dots, m\}$ channels transmitted over the link and two types of constraint: the link capacity constraint as before

$$\sum_{i \in \mathcal{M}} u_i - P^0 \leq 0,$$

where P^0 is the link total power target, and the target OSNR constraint

$$OSNR_i \geq \widehat{\gamma}_i, \quad \forall i \in \mathcal{M}, \tag{9.1}$$

where $\widehat{\gamma}_i$ is the target OSNR for channel i. As before the signal optical power at Tx u_i is bounded in $\Omega_i = [0, u_{\max}]$ with constant $u_{\max} > P^0$, for all $i \in \mathcal{M}$.

Consider an alternative OSNR optimization setup from the perspective of system performance, as opposed to a game-theoretic approach as developed in previous chapters. Specifically consider the following cost:

$$\begin{array}{ll} \min & C(\mathbf{u}) \\ \text{subject to} & u_i \in \Omega_i, \quad \forall i \in \mathcal{M}, \\ & OSNR_i \geq \widehat{\gamma}_i, \quad \forall i \in \mathcal{M}, \\ & \sum_{j \in \mathcal{M}} u_j \leq P^0, \end{array}$$

where $C(\mathbf{u})$ is the system cost function, defined as the sum of all individual costs, $C_i(u_i)$,

$$C(\mathbf{u}) = \sum_{i \in \mathcal{M}} C_i(u_i),$$

hence it is separable. Each individual cost function $C_i(u_i)$ is a generic cost function that satisfies the following assumption:

Assumption 9.1 *Each $C_i(u_i)$ is strictly convex, continuously differentiable and*

$$\lim_{u_i \to 0} C_i(u_i) = +\infty \tag{9.2}$$

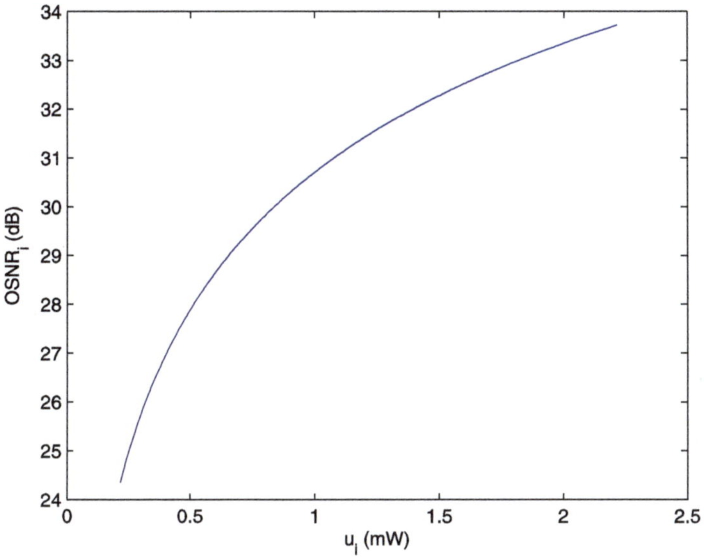

Fig. 9.1 $OSNR_i$ vs. u_i

This cost function can be defined similarly to the form of (7.3), $C_i = P_i - U_i$. The pricing function P_i is a linear function of u_i. The utility function U_i is a logarithmic function of u_i, which quantifies approximately the link's demand or the channel's willingness to pay for a certain level of $OSNR_i$ based on the relationship between $OSNR_i$ and u_i. This relationship is illustrated approximately in Fig. 9.1.

By using the OSNR model (6.16), (9.1) can be rewritten as

$$\frac{u_i}{n_i^0 + \sum_{j \in \mathcal{M}} \Gamma_{i,j} u_j} \geq \widehat{\gamma}_i,$$

or

$$u_i + \sum_{j \in \mathcal{M}} (-\widehat{\gamma}_i \Gamma_{i,j}) u_j \geq n_i^0 \widehat{\gamma}_i,$$

or in vector form

$$T\mathbf{u} \geq b, \tag{9.3}$$

where

$$T = \begin{bmatrix} 1 - \widehat{\gamma}_1 \Gamma_{1,1} & -\widehat{\gamma}_1 \Gamma_{1,2} & \cdots & -\widehat{\gamma}_1 \Gamma_{1,m} \\ -\widehat{\gamma}_2 \Gamma_{2,1} & 1 - \widehat{\gamma}_2 \Gamma_{2,2} & \cdots & -\widehat{\gamma}_2 \Gamma_{2,m} \\ \vdots & \vdots & \ddots & \vdots \\ -\widehat{\gamma}_m \Gamma_{m,1} & -\widehat{\gamma}_m \Gamma_{m,2} & \cdots & 1 - \widehat{\gamma}_m \Gamma_{m,m} \end{bmatrix}, \qquad b = \begin{bmatrix} n_1^0 \widehat{\gamma}_1 \\ n_2^0 \widehat{\gamma}_2 \\ \vdots \\ n_m^0 \widehat{\gamma}_m \end{bmatrix}$$

All T's off-diagonal entries $-\widehat{\gamma}_i \Gamma_{i,j}$ are less than zero. By Definition A.9, T is a Z-matrix.

From the total power constraint and $u_i \geq 0$, $\forall i \in \mathcal{M}$, we have $u_i \leq P^0$. Recalling that u_i is bounded in $\Omega_i = [0, u_{\max}]$ and $u_{\max} > P^0$, we can deduce that the conditions $\sum_{j\in\mathcal{M}} u_j \leq P^0$ and $u_i \in \Omega_i$, $\forall i \in \mathcal{M}$ are equivalent to $\mathbf{1}^T\mathbf{u} \leq P^0$ and $\mathbf{u} \geq 0$, where $\mathbf{1}$ is the $m \times 1$ all ones vector. Therefore, the constraint set of the system optimization problem is $\Omega := \{\mathbf{u} \in \mathbb{R}^m \mid T\mathbf{u} \geq b, \mathbf{1}^T\mathbf{u} \leq P^0 \text{ and } \mathbf{u} \geq 0\}$. This constrained system optimization problem can be formulated as

$$\begin{array}{ll} \min & C(\mathbf{u}) \\ \text{subject to} & \mathbf{u} \in \Omega \end{array} \tag{9.4}$$

and denote this problem by $OPT(\Omega, C)$. Condition (9.2) in Assumption 9.1 ensures that the solution to $OPT(\Omega, C)$ does not hit $u_i = 0$, $\forall i \in \mathcal{M}$. Let $\widehat{\gamma} = [\widehat{\gamma}_1, \dots, \widehat{\gamma}_m]^T$. The following result characterizes the unique solution of $OPT(\Omega, C)$.

Theorem 9.1 *If the following conditions on* $\widehat{\boldsymbol{\gamma}}$ *hold*:

$$\widehat{\gamma}_i < \frac{1}{\sum_{j\in\mathcal{M}} \Gamma_{i,j}}, \quad \forall i \in \mathcal{M}, \tag{9.5}$$

where $\Gamma = [\Gamma_{i,j}]$ *is the system matrix defined in* (6.17) *and*

$$\mathbf{1}^T \cdot \widetilde{T}(\widehat{\boldsymbol{\gamma}}) \cdot b(\widehat{\boldsymbol{\gamma}}) \leq P^0, \tag{9.6}$$

with $b(\widehat{\boldsymbol{\gamma}}) = [n_1^0\widehat{\gamma}_1, \dots, n_m^0\widehat{\gamma}_m]^T$ *and* $\widetilde{T}(\widehat{\boldsymbol{\gamma}}) = T^{-1}(\widehat{\boldsymbol{\gamma}})$, *then the constraint set* Ω *is non-empty and* $OPT(\Omega, C)$ *has a unique positive solution* u^{opt}.

Proof We first show that the constraint set Ω is non-empty. Note that in the link OSNR model, the system matrix Γ (6.17) is a positive matrix, so if (9.5) is satisfied, we have

$$1 - \widehat{\gamma}_i\Gamma_{i,i} > \widehat{\gamma}_i \sum_{j\in\mathcal{M}, j\neq i} \Gamma_{i,j} > 0, \quad \forall i \in \mathcal{M},$$

or equivalently,

$$1 - \widehat{\gamma}_i\Gamma_{i,i} > \sum_{j\in\mathcal{M}, j\neq i} |-\widehat{\gamma}_i\Gamma_{i,j}|, \quad \forall i \in \mathcal{M},$$

which implies that the Z-matrix T has positive diagonal entries and by Definition A.6, T is strictly diagonally dominant. According to Theorem A.7, each eigenvalue of T has a positive real part. Then it follows from Theorem A.10 that T is an M-matrix. So it has the following properties: $T\mathbf{u} \geq b > 0$ implies $\mathbf{u} \geq 0$, and T^{-1} is non-negative. Thus

$$\mathbf{u} \geq T^{-1}b := \widetilde{T}b \tag{9.7}$$

and then we have $\mathbf{1}^T\mathbf{u} \geq \mathbf{1}^T \cdot \widetilde{T} \cdot b$. Note that both $\widetilde{T}$ and b depend on $\widehat{\boldsymbol{\gamma}}$, i.e., $\widetilde{T} = \widetilde{T}(\widehat{\boldsymbol{\gamma}})$ and $b = b(\widehat{\boldsymbol{\gamma}})$. So

$$\mathbf{1}^T \cdot \mathbf{u} \geq \mathbf{1}^T \cdot \widetilde{T}(\widehat{\boldsymbol{\gamma}}) \cdot b(\widehat{\boldsymbol{\gamma}})$$

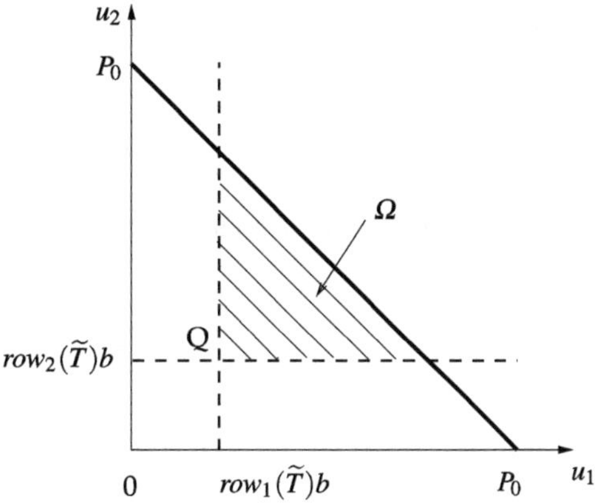

Fig. 9.2 Illustration of the constraint set Ω

Let $\widetilde{u} = \widetilde{T}b$. Then $T\widetilde{u} = b$. Also we have $\mathbf{1}^T \cdot \widetilde{u} = \mathbf{1}^T \cdot \widetilde{T}b$. By (9.6), $\mathbf{1}^T \cdot \widetilde{u} \le P^0$. It follows that

$$\widetilde{u} \in \{\mathbf{u} \in \mathbb{R}^m \mid T\mathbf{u} \ge b, \mathbf{1}^T\mathbf{u} \le P^0\}$$

Thus the above set is non-empty if both (9.5) and (9.6) are satisfied. Since $T\mathbf{u} \ge b > 0$ implies $\mathbf{u} \ge 0$, we have proved that if $\widehat{\boldsymbol{\gamma}}$ is selected such that (9.5) and (9.6) are satisfied, the constraint set Ω is non-empty.

Moreover, the constraint set Ω is convex and we have $0 \le u_i \le P_0$, $\forall i \in \mathcal{M}$. So Ω is bounded. In addition, it is also closed since it consists of the intersection of half-spaces. Thus this system optimization problem is a strictly convex optimization problem on a convex compact constraint set [30], which, following Proposition A.2, always admits a unique globe minimum, u^{opt}. □

Example 9.2 To illustrate the constraint set Ω and the conditions (9.5) and (9.6) consider a simple example in Fig. 9.2 where $m = 2$. From (9.7), we have

$$\begin{bmatrix} u_1 \\ u_2 \end{bmatrix} \ge \begin{bmatrix} \text{row}_1(\widetilde{T}) \cdot b \\ \text{row}_2(\widetilde{T}) \cdot b \end{bmatrix}$$

where $\text{row}_i(A)$ is defined as the ith row of the matrix A. From Fig. 9.2, it is readily seen that if $\sum_{i=1}^{2} \text{row}_i(\widetilde{T})b \le P^0$, the intersection point Q lies in the set of the total input power constraint, i.e.,

$$\left(\text{row}_1(\widetilde{T}) \cdot b, \text{row}_2(\widetilde{T}) \cdot b\right) \in \{\mathbf{u} \in \mathbb{R}^2 \mid u_1 + u_2 \le P_0\}$$

Therefore the constraint set Ω is non-empty.

Remark 9.3 Recall that n^0 denotes the input noise power at Tx and may include external noise, such as thermal noise. If the input noise is neglected, n^0 includes only external noise, which is negligible [45]. So $b = [n_1^0\widehat{\gamma}_1, \ldots, n_m^0\widehat{\gamma}_m]^T \approx 0$ and $P^0 \ge \mathbf{1}^T \cdot \widetilde{T} \cdot b \approx 0$. This means that the constraint set is non-empty under the first condition (9.5). Thus the OSNR target $\widehat{\gamma}_i$ can be selected in a distributed way based on the first condition (9.5).

Remark 9.4 Let us take a close look at the second condition (9.6). Recall that $T = I - \text{diag}(\widehat{\boldsymbol{\gamma}})\Gamma$, where I is an identity matrix. We know from the proof of Theorem 9.1 that T is an M-matrix. By Theorem A.11, $\rho(\text{diag}(\widehat{\boldsymbol{\gamma}})\Gamma) < 1$ and

$$T^{-1} = \big(I - \text{diag}(\widehat{\boldsymbol{\gamma}})\Gamma\big)^{-1} = \sum_{k=0}^{\infty} \text{diag}\big(\widehat{\gamma}^k\big)\Gamma^k$$

exists, which is positive component-wise. We can rewrite (9.6) as

$$\mathbf{1}^T \cdot \sum_{k=0}^{\infty} \text{diag}\big(\widehat{\gamma}^k\big)\Gamma^k \cdot \text{diag}(\widehat{\boldsymbol{\gamma}}) \cdot n^0 \leq P^0 \tag{9.8}$$

If $\widehat{\gamma}_i$ increases (given $\widehat{\gamma}_j$, $j \neq i$), LHS of (9.8) will increase. We can find a maximum OSNR target $\widehat{\gamma}_{\max}$ by solving the following equation:

$$\widehat{\gamma}_{\max} \cdot \mathbf{1}^T \cdot (I - \widehat{\gamma}_{\max}\Gamma)^{-1} \cdot n^0 = P^0 \tag{9.9}$$

Based on the link OSNR model, we know that the performance for each channel is interference limited. In addition, (9.9) shows that the OSNR targets significantly affect the capacity of a link: each link decides the OSNR threshold $\widehat{\gamma}_{\max}$ by using (9.9). Any new channel with a required OSNR target no more than $\widehat{\gamma}_{\max}$ will be admitted to transfer over the link. This idea can be used for links to develop channel admission control schemes. Similar algorithms exist for wireless networks [65, 66].

9.3 Distributed Algorithm

Since $OPT(\Omega, C)$ is a constrained convex optimization problem, there are several computational methods for solving it [22]. A possible approach is to use a barrier (or penalty) function to relax the constrained optimization problem. Barrier functions set a barrier against leaving the feasible region: If the optimal solution occurs at the boundary of the feasible region, the procedure moves from the interior to the boundary [22, 143]. By an appropriate choice of barrier functions, the solution of the relaxed system problem can arbitrarily approximate the one of the original problem $OPT(\Omega, C)$.

To see this, note that, from the proof of Theorem 9.1 that $T\mathbf{u} \geq b > 0$ implies $\mathbf{u} \geq 0$, $OPT(\Omega, C)$ can be rewritten succinctly as

$$\begin{array}{ll} \min & C(\mathbf{u}) \\ \text{subject to} & \widehat{T}\mathbf{u} \geq \widehat{b}, \end{array} \tag{9.10}$$

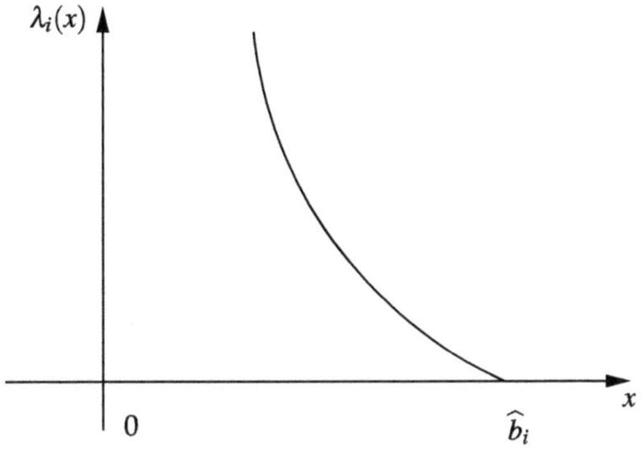

Fig. 9.3 A barrier function example

where

$$\widehat{T} = \begin{bmatrix} T \\ -\mathbf{1}^T \end{bmatrix} \quad \text{and} \quad \widehat{b} = \begin{bmatrix} b \\ -P^0 \end{bmatrix}$$

Now consider a barrier function $\lambda_i : \mathbb{R} \to \mathbb{R}$ with the following properties:

(P.1) $\forall i \in \mathcal{M}$, $\lambda_i(x)$ is non-increasing, continuous and

$$\lim_{u_i \to \infty} \int_{\widehat{b}_i}^{y_i(\mathbf{u})} \lambda_i(x)\, dx \to -\infty \tag{9.11}$$

where

$$y_i(\mathbf{u}) := \text{row}_i(\widehat{T})u. \tag{9.12}$$

(P.2) $\lambda_i(x)$ attains the value 0 if $x > \widehat{b}_i$, where $\widehat{b}_i$ is defined in (9.10).

Such an example is shown in Fig. 9.3.

With this barrier function λ_i, which has properties (P.1) and (P.2), construct a function

$$V_p(\mathbf{u}) = \sum_{i \in \mathcal{M}} C_i(u_i) - \sum_{i \in \mathcal{M}} \int_{\widehat{b}_i}^{y_i(\mathbf{u})} \lambda_i(x)\, dx \tag{9.13}$$

and define a relaxed system problem:

$$\min_{\mathbf{u} \geq 0} V_p(\mathbf{u}) \tag{9.14}$$

Based on (P.1), i.e., the non-increasing property of the barrier function, and from Assumption 9.1, i.e., that $C_i(u_i)$ is strictly convex, it follows that $V_p(\mathbf{u})$ is strictly convex and hence has a unique internal minimum value such that

$$\frac{\partial V_p(\mathbf{u})}{\partial u_i} = C_i'(u_i) - \text{row}_i\big(\widehat{T}^T\big)\lambda\big(y(\mathbf{u})\big) = 0, \quad \forall i \in \mathcal{M}$$

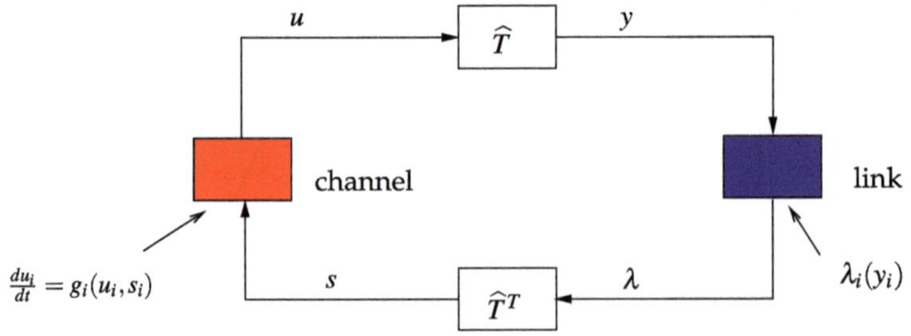

Fig. 9.4 Primal algorithm: block representation

Thus solving the set of above equations we obtain the unique solution $\bar{u}^{\text{opt}}$ of (9.14):

$$\bar{u}_i^{\text{opt}} = C_i'^{-1}\big(\text{row}_i\big(\widehat{T}^T\big)\lambda\big(y\big(\bar{u}^{\text{opt}}\big)\big)\big), \quad \forall i \in \mathcal{M}$$

The barrier function $\lambda_i(\cdot)$ can be selected such that the unique solution of (9.14) may arbitrarily closely approximate the optimal solution of $OPT(\Omega, C)$. For example, a barrier function can be defined as [68]

$$\lambda_i(x) = \frac{[\widehat{b}_i - x + \varepsilon]^+}{\varepsilon^2},$$

where $[x]^+ = \max\{x, 0\}$.

For the relaxed system problem a primal algorithm can be defined as

$$\dot{u}_i(t) = g_i(u_i, s_i) = -k_i \frac{\partial V_p(\mathbf{u})}{\partial u_i} = -k_i\big(C_i'\big(u_i(t)\big) - s_i(t)\big), \quad \forall i \in \mathcal{M}, \tag{9.15}$$

where the coefficient $k_i > 0$ and $s_i(t)$ is defined as

$$s_i(t) = \text{row}_i\big(\widehat{T}^T\big)\lambda\big(y\big(u(t)\big)\big), \tag{9.16}$$

and $\lambda(\cdot) = [\lambda_1(\cdot), \ldots, \lambda_{m+1}(\cdot)]^T$, with each $\lambda_i(\cdot)$ satisfying (P.1) and (P.2).

Algorithm (9.15) is a gradient algorithm and can be implemented in a distributed way. Each channel varies its input power u_i gradually as in (9.15), while the link (network system) calculates the vector $s(t) = [s_1(t), \ldots, s_m(t)]^T$ based on the received input powers, OSNR preference and link constraint, and then feeds this updated information back to each channel. This algorithm is represented in Fig. 9.4.

The following theorem states that the unique equilibrium of the algorithm (9.15) corresponds to the unique solution of (9.14), $\bar{u}^{\text{opt}}$. Moreover, the solution is globally asymptotically stable.

Theorem 9.5 (See [102]) *The unique solution $\bar{u}^{\text{opt}}$ to the relaxed system optimization problem* (9.14) *is globally asymptotically stable for the system* (9.15).

Proof Since $\bar{u}^{\text{opt}}$ is the unique solution to the equations $\frac{\partial V_p(\mathbf{u})}{\partial u_i} = 0$, $\forall i \in \mathcal{M}$, it follows that it is the unique equilibrium point of the system (9.15). Next, from $V_p(\mathbf{u})$

being strictly convex it follows that $\bar{u}^{\text{opt}}$ is the global minimum point of the function $V_p(\mathbf{u})$. Let $C = V_p(\bar{u}^{\text{opt}})$. Then $V_p(\mathbf{u}) > C$ for all $u \neq \bar{u}^{\text{opt}}$, and we construct a Lyapunov function for the system (9.15):

$$V(\mathbf{u}) = V_p(\mathbf{u}) - C$$

It follows that $V(\mathbf{u}) = 0$ when $u = \bar{u}^{\text{opt}}$, and that $V(\mathbf{u}) > 0$ when $u \neq \bar{u}^{\text{opt}}$. That is, the function $V(\mathbf{u})$ is positive definite with respect to the equilibrium point $u = \bar{u}^{\text{opt}}$. Taking the derivative of $V(\mathbf{u})$ along the trajectory of the system gives

$$\dot{V}(\mathbf{u}) = \sum_{i \in \mathcal{M}} \left(\frac{\partial}{\partial u_i} V_p(\mathbf{u}) \cdot \dot{u}_i \right) = \sum_{i \in \mathcal{M}} k_i \left(\frac{\partial}{\partial u_i} V_p(\mathbf{u}) \right)^2$$

Since $\dot{V}(\mathbf{u}) = 0$ when $u = \bar{u}^{\text{opt}}$, and $\dot{V}(\mathbf{u}) < 0$ when $u \neq \bar{u}^{\text{opt}}$, it follows that $\dot{V}(\mathbf{u})$ is negative definite with respect to the equilibrium point $u = \bar{u}^{\text{opt}}$. The conclusion follows by Lyapunov stability theory. □

Note that the unique solution of the relaxed system problem (9.14) may arbitrarily closely approximate the optimal solution of the original system problem (9.10) with an approximate selection of the barrier function. For details the reader s referred to [101].

9.4 Numerical Study of Efficiency in Nash Game

Next we show how the system optimization framework can be used to investigate the effects of parameters for the game-theoretic framework. We motivate such an approach by first showing that the aggregate cost function in a game-theoretic formulation is not always strictly convex. An individual cost function in the system optimization formulation has an approximate interpretation a game cost function. Based on this a numerical study of the effects of parameters in a game setup can be performed [103, 104].

Recall that for a game $\mathcal{G}(\mathcal{M}, \hat{\Omega}_i, J_i)$ as formulated in Chap. 7, the cost function is composed of a pricing function and a utility function. The utility function is defined so as to indicate each channel's preference for a better OSNR. We are interested to know how much cost is added or utility is lost due to each player's selfish behavior in a Nash game, i.e., we consider the loss of efficiency versus the socially optimal cost. Among many possible social welfare functions, the aggregate cost function can be considered as a candidate

$$J(\mathbf{u}) := \sum_{i \in \mathcal{M}} J_i(\mathbf{u})$$

Note that while the individual costs are convex in each player's action u_i, in general the aggregate game cost function is not necessarily convex in $\mathbf{u}$. The following simple example shows this. For simplicity we omit the penalty term and noise in the OSNR model.

Example 9.6 Consider a Nash game with three players ($m = 3$) with individual costs,

$$J_i(\mathbf{u}) = u_i - \ln \frac{u_i}{\sum_{j \neq i} u_j}, \qquad i = 1, 2, 3$$

It follows that

$$\frac{\partial^2 J_1}{\partial u_1^2} = \frac{u_2 + u_3}{u_1} > 0$$

$$\frac{\partial^2 J_2}{\partial u_1^2} = -\frac{1}{(u_1 + u_3)^2} < 0$$

$$\frac{\partial^2 J_3}{\partial u_1^2} = -\frac{1}{(u_1 + u_2)^2} < 0$$

Therefore,

$$\frac{\partial^2 J}{\partial u_1^2} = \frac{u_2 + u_3}{u_1} - \frac{1}{(u_1 + u_3)^2} - \frac{1}{(u_1 + u_2)^2}$$

Note that $\frac{\partial^2 J}{\partial u_1^2}$ has not a definite sign, so that $J(\mathbf{u})$ is not always convex with respect to u_1, even though $J_1(\mathbf{u})$ is strictly convex with respect to u_1.

Thus using a standard approach of setting a constrained minimization problem associated with this aggregate cost function is not immediate. Alternatively, recall the system optimization problem $OPT(\Omega, C)$ and consider a cost function $C_i(u_i)$ for channel i defined as

$$C_i(u_i) = \alpha_i u_i - \beta_i \ln u_i, \tag{9.17}$$

where $\alpha_i > 0$ and $\beta_i > 0$. This cost function $C_i(u_i)$ is obviously strictly convex and continuously differentiable. Moreover $C_i(u_i) \to +\infty$ as $u_i \to 0$. Comparing to J_i in the above example it can be seen that $C_i(u_i)$ in $OPT(\Omega, C)$ has an interpretation similar to the one of the cost function $J_i(\mathbf{u})$ in $\mathcal{G}(\mathcal{M}, \hat{\Omega}_i, J_i)$, and moreover C_i is uncoupled in u for a given set of other power u_{-i}. Based on the relation between these two formulations, i.e., $OPT(\Omega, C)$ and $\mathcal{G}(\mathcal{M}, \hat{\Omega}_i, J_i)$, we can use the central cost function in $OPT(\Omega, C)$ as a welfare function of $\mathcal{G}(\mathcal{M}, \hat{\Omega}_i, J_i)$.

This is shown next, based on the system optimization being used to measure the efficiency of the NE solution numerically. We compare by simulation the two approaches: system optimization approach and the game-theoretic approach, respectively. Consider a link with six channels ($m = 6$) and total power target is

$P^0 = 2.5$ mW (3.98 dBm). Then the system matrix Γ is obtained as

$$\Gamma = \begin{bmatrix} 7.463 & 7.378 & 7.293 & 7.210 & 7.127 & 6.965 \\ 7.451 & 7.365 & 7.281 & 7.198 & 7.115 & 6.953 \\ 7.438 & 7.353 & 7.269 & 7.186 & 7.103 & 6.942 \\ 7.427 & 7.342 & 7.258 & 7.175 & 7.093 & 6.931 \\ 7.409 & 7.324 & 7.240 & 7.157 & 7.075 & 6.914 \\ 7.387 & 7.303 & 7.219 & 7.136 & 7.055 & 6.894 \end{bmatrix} \times 10^{-5}.$$

Within the set of six channels, there are two levels of OSNR target, a 26 dB level desired on the first three channels and a 22 dB OSNR level on the next three channels. Conditions (9.5) and (9.6) on the target OSNR are satisfied. So the feasible constraint set is non-empty. The coefficients in (9.17) are selected as $\alpha_i = 1, i = 1, \ldots, 6$, and $\beta = [0.5, 0.51, 0.52, 0.3, 0.31, 0.32]$. Recalling the relationship between channel OSNR and channel input power shown in Fig. 9.1, the values of β_i implicitly indicate channel OSNR preferences. The coefficient k_i is fixed for each channel with $k_i = 0.01, i = 1, \ldots, 6$ and we select a barrier function

$$\lambda_i(x_i) = 1000\big(\max\{0, \hat{b}_i - x_i\}\big)^6, \tag{9.18}$$

where $x_i(\mathbf{u}) = \text{row}_i(\widehat{T})u$. Notice that $\lambda_i(u_i)$ is zero when the constraints are satisfied. So there is a penalty with any violation of the constraints.

Let the NE solution of $\mathcal{G}(\mathcal{M}, \hat{\Omega}_i, J_i)$ be denoted by u^*. Algorithm GA 7.36 (see Chap. 7) is used, i.e.,

$$\dot{u}_i(t) = -\mu\left(\alpha_i + \frac{1}{(P^0 - \sum_{j\in\mathcal{M}} u_j(t))^2} - \frac{\beta_i a_i / u_i(t)}{(\frac{1}{OSNR_i(t)} + a_i - \Gamma_{i,i})}\right),$$

where the step-size is selected as $\mu = 0.01$.

Remark 9.7 Theorem 7.7 states that GA converges to the NE solution if (7.40) and (7.41) are satisfied, where $u_{\min} > 0$ is a positive lower bound on each u_i. A lower bound on β_i is set by (7.41). Since each channel attempts to select larger β_i for the purpose of higher OSNR, the lower bound does not affect the results of the efficiency study.

In $OPT(\Omega, C)$, instead of using a generic $C_i(u_i)$, (9.17) is used as the cost function for channel i, based on the primal algorithm (9.15) and the barrier function as in (9.18). Thus the equilibrium point of (9.15) closely approximates the solution of $OPT(\Omega, C)$. Initial channel powers are selected for both cases as

$$u(0) = [0.216\ 0.221\ 0.226\ 0.231\ 0.236\ 0.241]\ (\text{mW})$$

The approximate optimal solution of $OPT(\Omega, C)$ is achieved as

$$u^{\text{opt}} = [0.5\ 0.51\ 0.52\ 0.3\ 0.31\ 0.32]\ (\text{mW})$$

so that system cost with respect to u^{opt} is $C(u^{\text{opt}}) = 4.5789$.

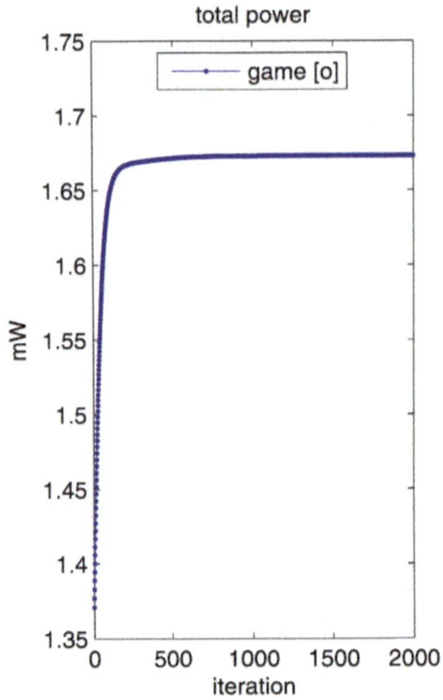

Fig. 9.5 u_T and $OSNR_i$ in game with $\alpha_i = 0.001$

We first present three cases in which the parameter selection strategy is not used as a guideline (thus it is possible that the game settles down at an NE solution where channels do not reach their OSNR targets). In all cases, the user-defined parameters β_i in $\mathcal{G}(\mathcal{M}, \hat{\Omega}_i, J_i)$ are chosen as same as β_i in $OPT(\Omega, C)$. A fixed α_i is used of 0.001, 1, and 20, respectively. With these pricing mechanisms, the total power (u_T) vs. iteration and channel OSNR vs. channel number are shown in Figs. 9.5, 9.6 and 9.7 for three cases.

Thus without proper pricing mechanism, OSNR targets may not be achieved for some channels or all channels, while the link capacity constraint is satisfied in all cases. Furthermore, we notice that the penalty term

$$\frac{1}{P^0 - \sum_{j \in \mathcal{M}} u_j}$$

in $\mathcal{G}(\mathcal{M}, \hat{\Omega}_i, J_i)$ plays a key role with small α_i. In other words, with larger α_i (say, $\alpha_i = 20$ in the third case), total power is smaller than the link capacity constraint. While with smaller α_i in the first two cases, total power approaches the constraint and higher channel OSNR is possibly achieved.

Channel powers, u^{opt} and u^*, in three games versus channel number are shown in Fig. 9.8. System cost $C(\mathbf{u}) = \sum_{i \in \mathcal{M}} C_i(u_i)$ is evaluated via u^{opt} and u^*, respectively and is shown in Table 9.1. Results imply that larger α_i degrades system performance and even violates the system constraints.

Next three other cases are shown in which proper pricing mechanisms are chosen such that OSNR targets for all channels are achieved. In the game-theoretic frame-

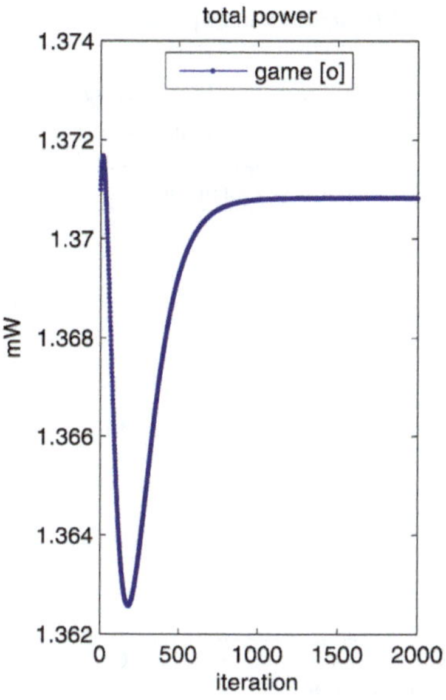

Fig. 9.6 u_T and $OSNR_i$ in game with $\alpha_i = 1$

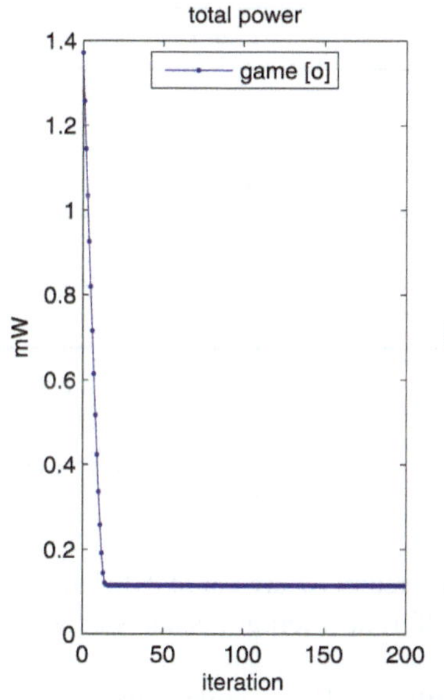

Fig. 9.7 u_T and $OSNR_i$ in game with $\alpha_i = 20$

work $\mathcal{G}(\mathcal{M}, \hat{\Omega}_i, J_i)$, the parameter selection strategy (7.28) is used such that proper pricing mechanisms are chosen and OSNR targets for all channels are achieved. Although the parameter selection strategy acts as a guideline for the selection of

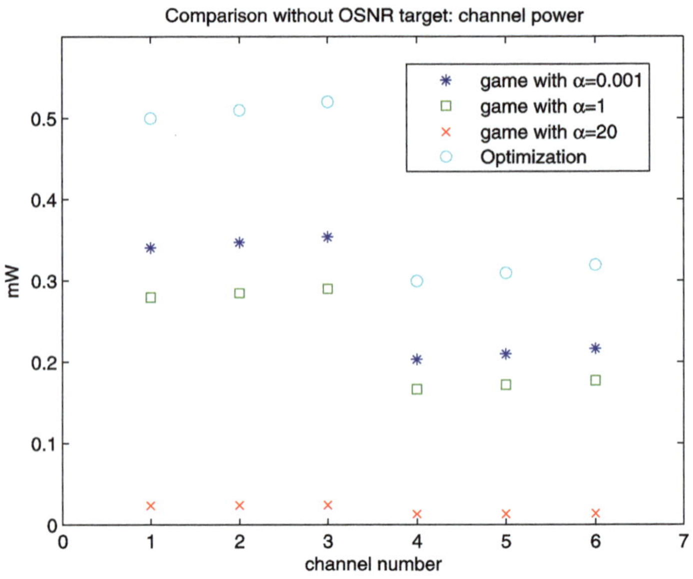

Fig. 9.8 Comparison: channel power in games

Table 9.1 System cost values with different α_i

	$C(u^*)$	$C(u^{\text{opt}})$
$\alpha_i = 0.001$	4.7403	4.5789
$\alpha_i = 1$	4.9282	4.5789
$\alpha_i = 20$	9.7804	4.5789

Table 9.2 Parameters: β_i

	β_i
Game $[a]$	[3.8 4.8 5.8 2.6 3.0 3.5]
Game $[b]$	[5.5 7.0 9.4 4.0 4.5 5.0]
Game $[c]$	[10 12 15 8.4 8.5 8.3]

each β_i, it is practically intractable. Parameters α_i are set at 1 for all cases and β_i are selected as in Table 9.2 such that different pricing mechanisms are chosen for $\mathcal{G}(\mathcal{M}, \hat{\Omega}_i, J_i)$.

Since we do not use Monte Carlo method [44] to simulate, we select β_i in three games by using the following rules. Firstly, β_i increases for each channel, i.e., Game [c] has the largest β_i compared to Game [a] and Game [b]. Secondly, Game [b] has the largest ratio of β_i to $\beta_{\min}$.

The efficiency of these two solutions u^* and u^{opt}, is compared by evaluating the system cost $C(\mathbf{u})$. The corresponding system cost values are obtained and shown in Table 9.3.

Table 9.3 System cost values with different β_i

	$C(u^*)$	$C(u^{\text{opt}})$
Game [a]	4.6171	4.5789
Game [b]	4.6216	4.5789
Game [c]	4.6057	4.5789

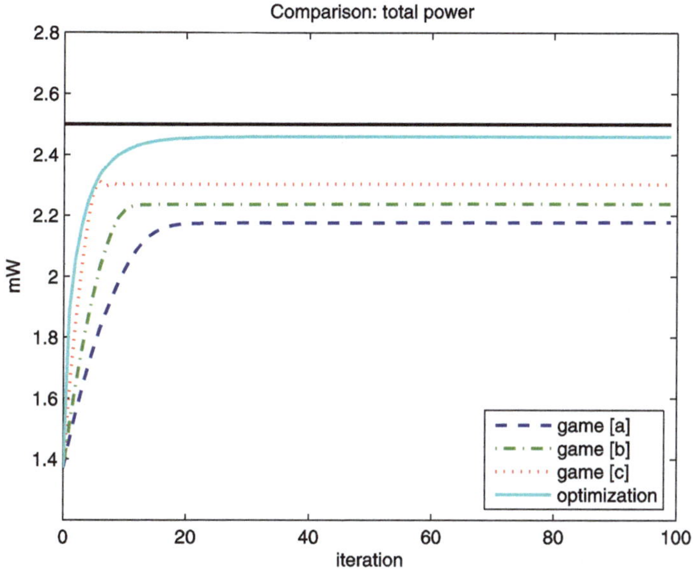

Fig. 9.9 Comparison: total power

The results in Table 9.3 (compared with Table 9.1) show that the efficiency in the solution of the Nash game (user optimization) can be improved by a proper pricing mechanism. It is a well-known that no full efficiency can be obtained from results in economics [38], transportation [129] and network resource allocation [64]. However, the Nash game solution gets very close to the optimal solution for system optimization (see Table 9.3).

Figure 9.9 shows total power versus iteration. Channel power and channel OSNR versus channel number are shown Figs. 9.10 and 9.11, respectively. Both constraints (link capacity constraint and channel OSNR target) are satisfied in all cases. From Fig. 9.11 it is seen that among the three cases, channel final OSNR values in Game [c] get closest to the optimal solution of $OPT(\Omega, C)$.

We recall that the parameters β_i in the Nash game are upper-bounded by the condition (7.23) in Theorem 7.4, or in other words, the ratio of β_i to $\beta_{\min}$ is upper bounded. This condition restricts each channel asking unilaterally for a much higher OSNR target than others. This phenomenon is also reflected in the selections of β_i in three cases. We take Game [b] for an example. In this case, $\beta_3 = 9.4$ which is greatly larger than other β_i, indicating that channel 3 asks for a highest OSNR level.

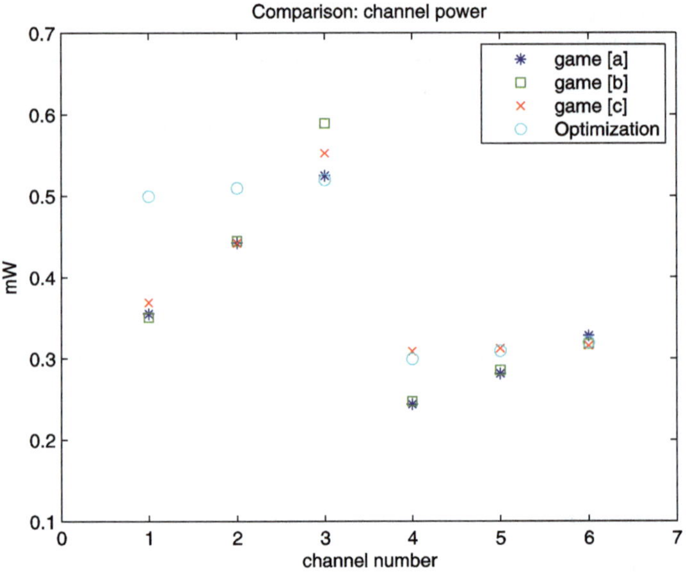

Fig. 9.10 Comparison: channel power

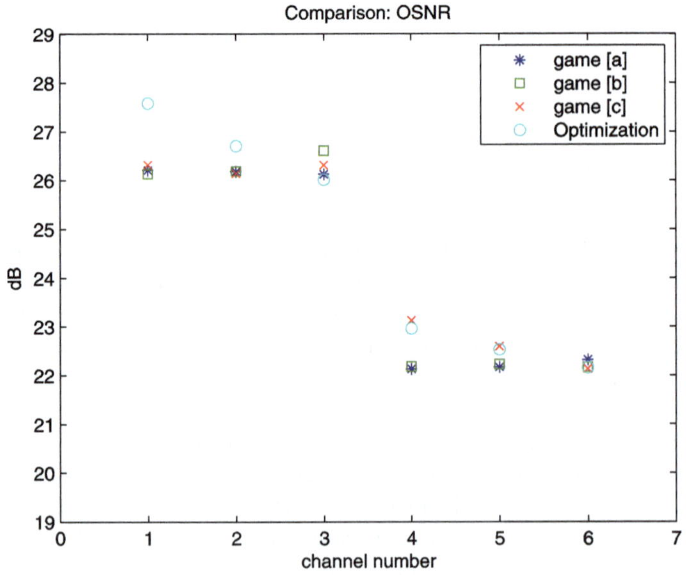

Fig. 9.11 Comparison: channel OSNR

Meanwhile, channel 4 has the smallest $\beta_4 = 3.9$. Thus the largest ratio of β_i to $\beta_{\min}$ in Game [b] is around 2.41, which is the largest ratio among the three cases (2.23, 2.41, 1.81, in Game [a], [b], [c], respectively). Recall that Game [b] results in the largest system cost value $C(u^*)$, 4.6216, which implies a loss of efficiency of the Nash equilibrium in the case when β_i deviates from the average.

9.5 Notes

In this chapter we discussed and alternative constrained OSNR optimization from the perspective of system performance. This system optimization framework can be used to measure the efficiency of Nash equilibria of the Nash game presented in Chap. 7. A numerical approach can be used to study the effects of parameters in individual game cost functions. Results show that the OSNR target in the game-theoretic framework can be achieved and the efficiency can be possibly improved by appropriate selection of parameters.

Chapter 10
Simulations and Experimental Studies

Abstract This chapter provides simulation and experimental results for various algorithms studied in previous chapters. The first section describes the physical setup. This is followed by simulations and experimental results based on implementing iterative algorithms in Chap. 7 for a Nash game with two, three, and five channels, respectively, in a point-to-point link topology. The last section presents results for partitioned Nash game framework in a multi-link topology and a quasi-ring topology, based on implementing the hierarchal algorithms studied in Chap. 8. Full use of the flexibility of channel power adjustment at each optical switch is assumed and the multi-link is partitioned into stages with single links. Simulation and experimental results are given for each type of network topology.

10.1 Introduction

In the following we present simulation and experimental results for the various algorithms studied in some of the previous chapters. Simulations are done in MATLAB while the experimental results are performed on a physical setup described below.

An optical network test system (ONTS) composed of the following optical devices has been used: stabilized light sources (LS), variable optical attenuators (VOA), optical spectrum analyzer (OSA), optical Erbium-Doped Fiber Amplifier (OA), tunable bandpass fiber optic filters, fiber optical couplers and an ASE broadband source. Parts of the ONTS are shown in Fig. 10.1.

Any transmitter (Tx) is composed of a light source (of a particular wavelength) and a variable optical attenuator. Each channel input power is adjustable by setting the value of the corresponding VOA. An OSA provides accurate and comprehensive measurement capabilities for spectral analysis, and can be used in ONTS to measure channel optical powers and OSNRs. An OA amplifies optical signals and noise simultaneously. Tunable filters are used to adjust the center wavelength of a narrow passband so that it separates optical signals with different wavelengths. Fiber optical couplers are used to combine or distribute optical powers from single (or multiple) input to single (or multiple) output. Couplers are designed bi-directionally and thus can be used as a coupler or a splitter. Input optical noise in ONTS is obtained by using the ASE broadband source.

L. Pavel, *Game Theory for Control of Optical Networks*,
Static & Dynamic Game Theory: Foundations & Applications,
DOI 10.1007/978-0-8176-8322-1_10,

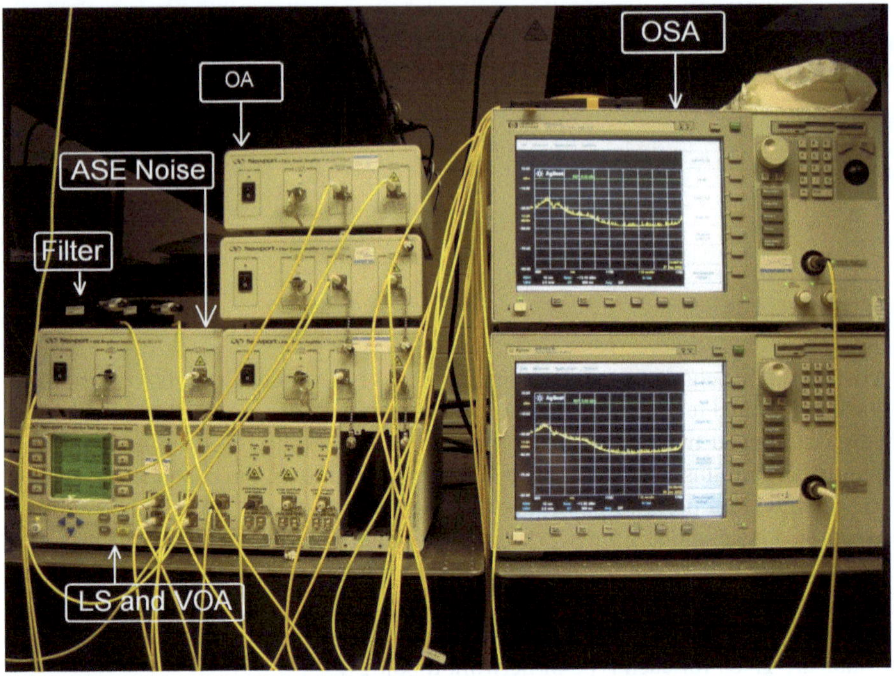

Fig. 10.1 ONTS: devices

Results have been implemented in LabVIEW[1] is used for the purpose of communication and control. Communication between instruments and computers is realized via GPIB (General Purpose Interface Bus, also referred to as IEEE-488). LabVIEW first communicates with the light sources to initialize them (e.g., wavelength selection and light source power setting). This is followed by communication with VOAs to get and set the attenuation configurations and then with OSAs to measure the OSNR level and output power of each channel. Distributed optimization algorithms are embedded in the control block by using the MathScript Node in LabVIEW.

The ONTS with a single optical link is setup shown in Fig 10.2.

10.2 Point-to-Point Link Topologies

The first section presents results for implementation of the iterative algorithms in Chap. 7 for point-to-point links, either simulations in MATLAB or experimental results on the ONTS physical setup.

[1]LabVIEW (short for Laboratory Virtual Instrumentation Engineering Workbench). LabVIEW, as a platform and development environment for a visual programming language from National Instruments®.

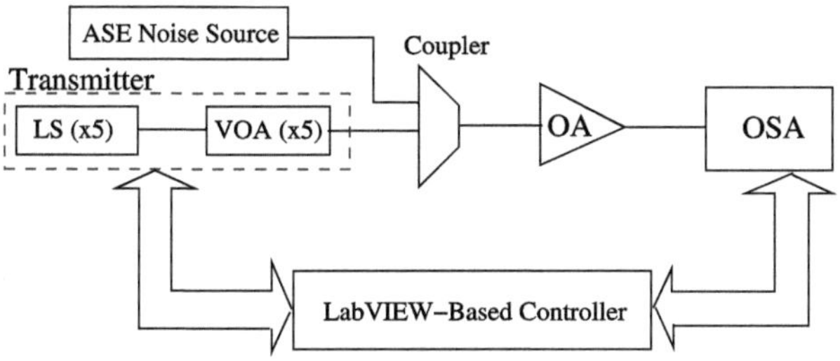

Fig. 10.2 ONTS configuration setup: single link

10.2.1 Simulations Results

Consider multiple channels transmitted over a single optical link with a total power capacity constraint $P^0 = 2.5$ mW (3.98 dBm). The associated system matrix Γ for $m = 3$ is obtained as

$$\Gamma = \begin{bmatrix} 0.7463 & 0.7378 & 0.7293 \\ 0.7451 & 0.7365 & 0.7281 \\ 0.7438 & 0.7353 & 0.7269 \end{bmatrix} \times 10^{-4}$$

For the iterative algorithms, the channel parameters in cost functions are selected satisfying both the sufficient conditions for the existence of a unique NE solution, (7.22)–(7.24), and the conditions for stability of the update scheme, (7.40)–(7.41). Simulations are repeated with the following selected parameters: $\alpha = [0.001, 0.001, 0.001]$, $\beta = [1, 3, 2]$ and $a = [1, 1, 1]$.

Two distributed algorithms are implemented in simulation. At each iteration time $(n + 1)$, $u_i(n + 1)$ is obtained via PUA, i.e., via solving (7.33),

$$\alpha_i + \frac{1}{(P^0 - \sum_{j \in \mathcal{M}, j \neq i} u_j(n) - u_i(n+1))^2} = \frac{\beta_i a_i}{(\frac{1}{OSNR_i(n)} - \Gamma_{i,i} + a_i)u_i(n+1)}$$

A discretized version of GA (7.36) is used to compute $u_i(n + 1)$:

$$u_i(n+1) = u_i(n) - \mu\left(\alpha_i + \frac{1}{(P^0 - \sum_{j \in \mathcal{M}} u_j(n))^2} - \frac{\beta_i a_i}{(\frac{1}{OSNR_i(n)} + a_i - \Gamma_{i,i})u_i(n)}\right),$$

where $\mu = 0.01$. In addition, the initial channel power value in GA are set as $u(0) =$ [0.116 0.121 0.126] (mW).

A Nash game with two channels is considered first. Two channels compete for the power resources and the game settles down at the NE solution via PUA and GA. Figures 10.3, 10.4 and 10.5 show the evolutions of channel input power, total power and OSNR. It can be observed that wide fluctuations are obtained when running PUA. These fluctuations are largely avoided when using GA. Moreover, during the iterative process, the total power constraint is violated when using PUA, but not when using GA.

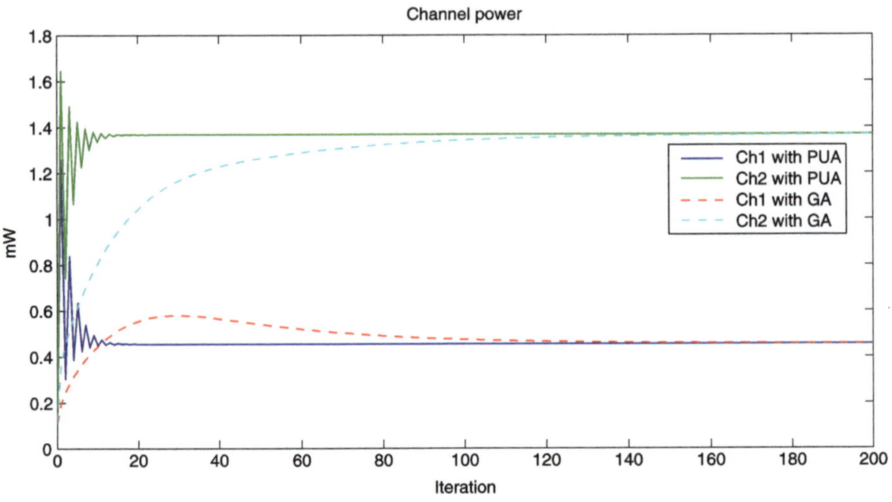

Fig. 10.3 PUA and GA: channel power

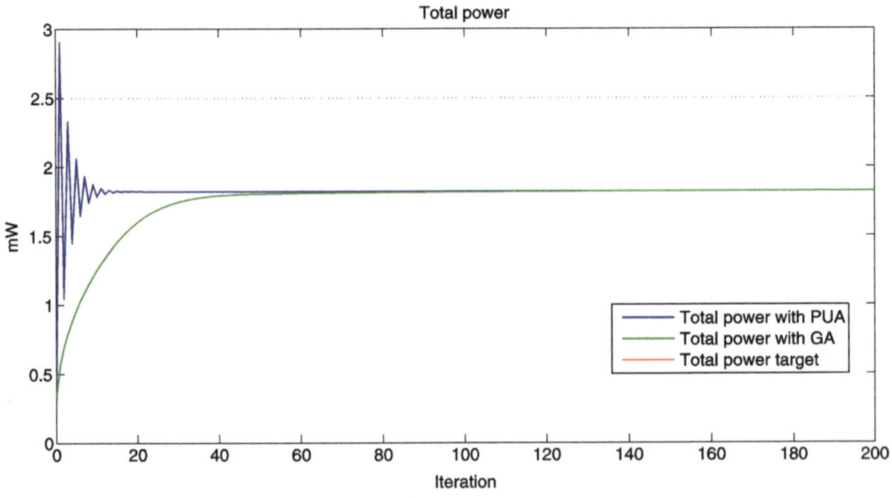

Fig. 10.4 PUA and GA: total power

Next we consider a noncooperative game with three channels ($m = 3$). Input power, total input power and OSNR vs. iteration time are plotted in Figs. 10.6, 10.7 and 10.8, respectively. Throughout the adjustments, total input power does not exceed the link capacity constraint, in accordance with the fact that the trajectory lies in $\bar{\Omega}$ when the initial state is in $\bar{\Omega}_\delta$. Moreover, total input power reaches the equilibrium point more quickly than the input powers and is below the power capacity constraint, P^0. These facts partly reflect that the extra term gives only a small effect on the cost rather than distorting the cost, however, it prevents channels from

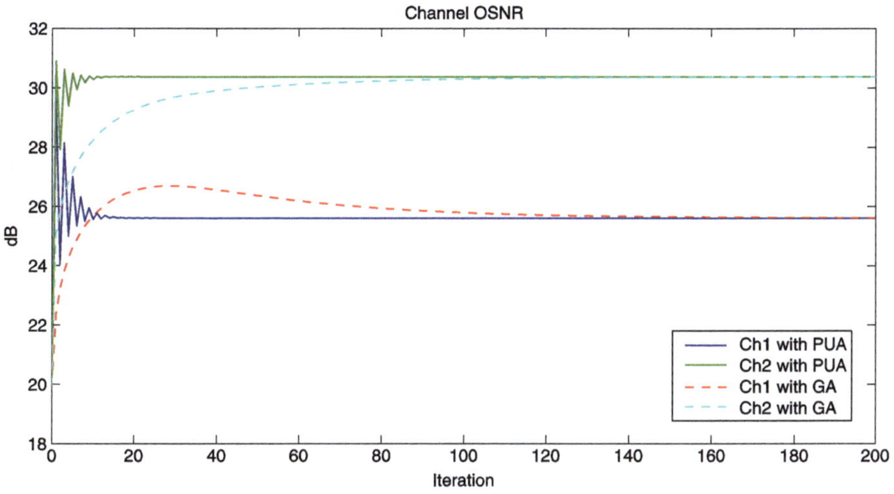

Fig. 10.5 PUA and GA: channel OSNR

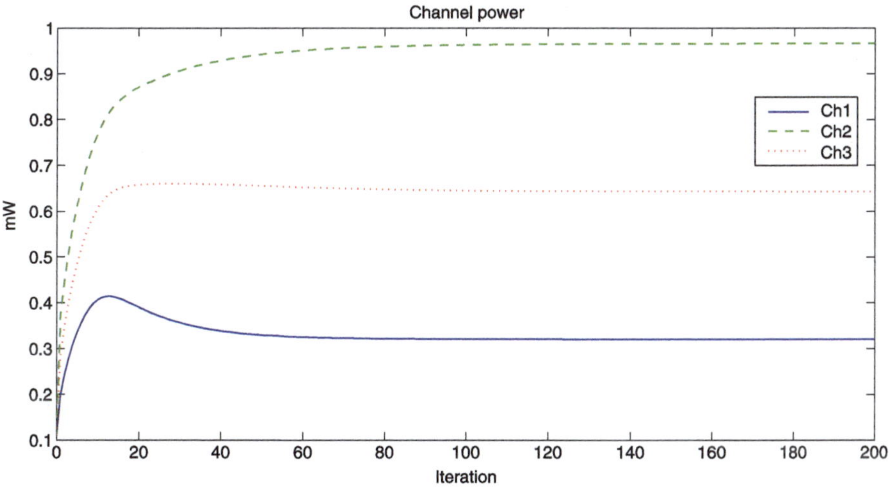

Fig. 10.6 GA: channel power

making full use of input power resource. Simulation results showed above prove the performance of the algorithms and validate the analytic results.

10.2.2 Experimental Results

Consider the ONTS with a single optical link as shown in Fig. 10.2, with one OA. Multiple channels are transmitted with the following wavelengths: 1533.47 nm,

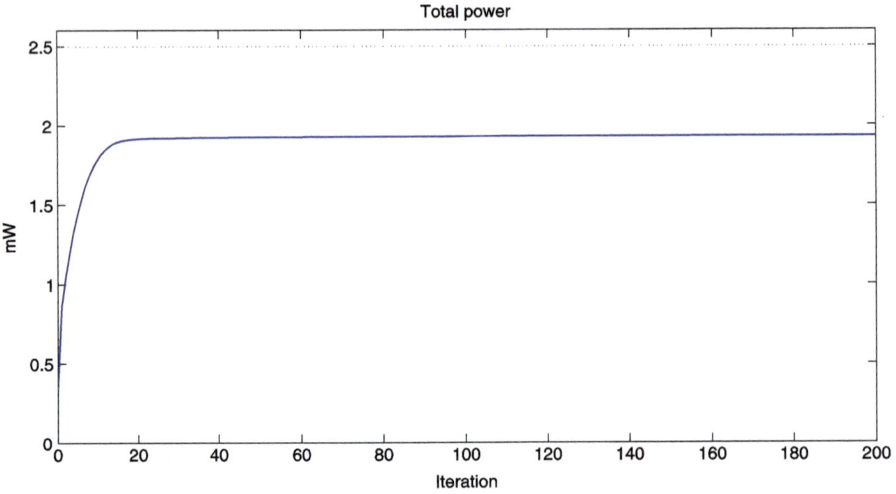

Fig. 10.7 GA: total power

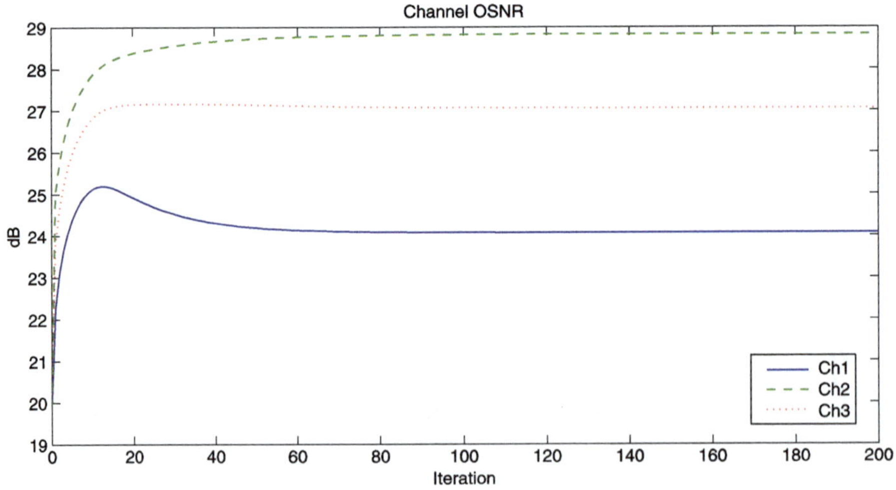

Fig. 10.8 GA: channel OSNR

1535.04 nm, 1537.40 nm, 1555.75 nm and 1558.17 nm. We continue using the notations defined in Chap. 7. Since only one OA is used, i.e., $N = 1$ in (6.17), diagonal elements of the system matrix Γ are derived from (6.17), i.e.,

$$\Gamma_{i,i} = \frac{ASE_i}{P^0},$$

where $P^0 = 1.5$ mW is the constant total power target of this link and ASE_i is the ASE noise, defined as

$$ASE_i = 2n_{sp}(G_i - 1)h\nu_i B,$$

where

n_{sp} amplifier excess noise factor, $n_{sp} = 1$
h Planck's constant, $h = 6.626$
B optical bandwidth, $B = 10^{10}$
ν_i optical frequency of channel i
G_i gain of OA at the wavelength of channel i

The two algorithms, PUA and GA, are applied. In PUA, each $u_i(n+1)$ is obtained by solving the following equation:

$$\alpha_i + \frac{1}{(P^0 - \sum_{j\in\mathcal{M}, j\neq i} u_j(n) - u_i(n+1))^2} = \frac{\beta_i a_i}{(\frac{1}{OSNR_i(n)} - \Gamma_{i,i} + a_i)u_i(n+1)}$$

In GA,

$$u_i(n+1) = u_i(n) - \mu\left(\alpha_i + \frac{1}{(P^0 - \sum_{j\in\mathcal{M}} u_j(n))^2} - \frac{\beta_i a_i}{(\frac{1}{OSNR_i(n)} + a_i - \Gamma_{i,i})u_i(n)}\right),$$

where the step-size μ is selected as $\mu = 0.0025$ for $m = 2$ and $\mu = 0.0002$ for $m = 5$. We set α_i at 0.01 and a_i at 1, respectively, for each channel. The following values are chosen for the channel (player) specific parameters: β_i, $\beta = [10, 20]$ for $m = 2$ and $\beta = [5, 10, 20, 25, 30]$ for $m = 5$.

Using the results in Chap. 7, PUA is applied to a Nash game with two channels ($m = 2$). GA is applied to two Nash games: one with two channels ($m = 2$) and one with five channels ($m = 5$), respectively.

Firstly, a Nash game with two channels ($m = 2$) is considered: two channels compete for the power resources via PUA and GA, respectively. Experimental data are plotted in MATLAB. Figures 10.9 and 10.10 show the evolutions of channel input power and total power, respectively.

Experiment results show that fluctuations in PUA are largely avoided in GA. We note some difference between results of GA and PUA in Fig. 10.9, which does not exist in simulation results. This is justified as follows: in ONTS, when running PUA, we solve the first-order condition equation by using the measured OSNR value and input power such that the new input power is achieved. A calibration for the insertion (power) loss of equipment has to be included in the control block. However, the measured data are not accurate even with calibration. Unlike this, when using GA this issue is not encountered. A snapshot of final results on OSA after PUA settling down is shown in Fig. 10.11.

Next the Nash game with five channels ($m = 5$) is considered. The evolutions of channel input power and total power are shown in Figs. 10.12 and 10.13, respectively.

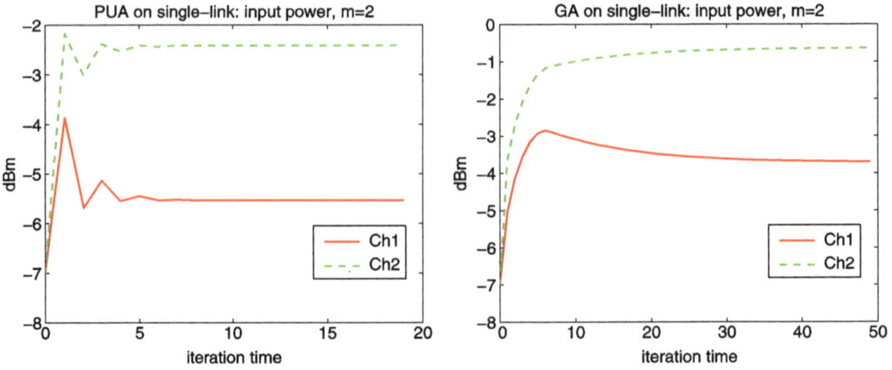

(a) PUA (b) GA

Fig. 10.9 Evolution of input power: $m = 2$

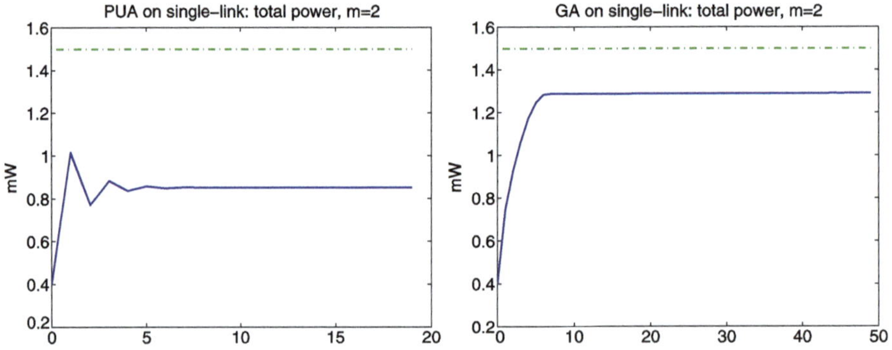

(a) PUA (b) GA

Fig. 10.10 Evolution of total power: $m = 2$

Since all channels are under same conditions, for example, same noise profile and same network price α_i, they compete for the power resource via the value of β_i: the larger the β_i is, the higher the OSNR and input power values are obtained. Snapshots of initial states and final results on OSA are shown in Figs. 10.14 and 10.15, respectively.

10.3 Multi-link and Mesh Topologies

In this section, we consider the partitioned Nash game framework to study several types of network topology and implement the hierarchical algorithm to compute its solution. The first two types of network topology are the multi-link topology and the quasi-ring topology, [18]. Both of them are representative for selected paths extracted from a mesh configuration. A multi-link topology was studied in Chap. 8;

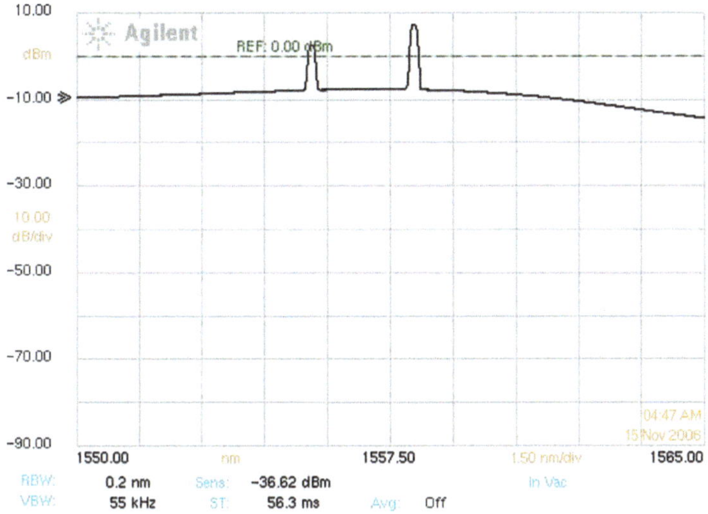

Fig. 10.11 Snapshot on OSA: results of PUA

Fig. 10.12 Evolution of input power with GA, $m = 5$

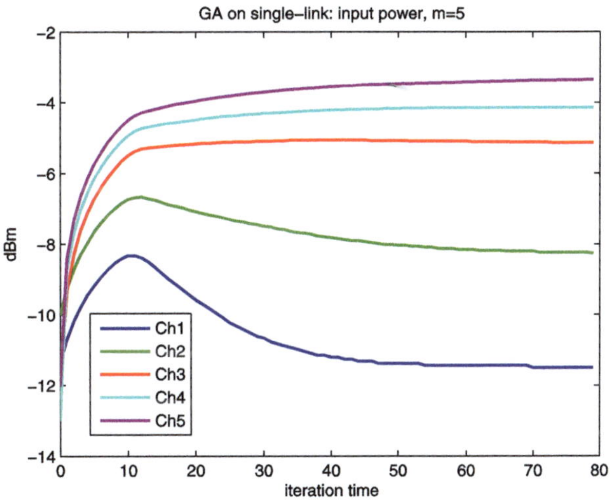

herein we also fully use the flexibility of channel power adjustment at each optical switch and partition the multi-link structure into stages with single links. A mesh network topology is studied next. Simulation results are given for each type of network topology.

Some assumptions and parameters defined in simulation for all types of network topology are presented first. As before, in all topologies, we assume that each link is with same number of amplified spans and all optical amplifiers deployed along links have the same gain spectral shape. The dynamic adjustment parameter $\gamma_{l,i}$ is bounded within $[\gamma_{\min}, \gamma_{\max}] = [0, 10]$ for all $l \in \mathcal{L}$ and for all $i \in \mathcal{M}$. In partitioned

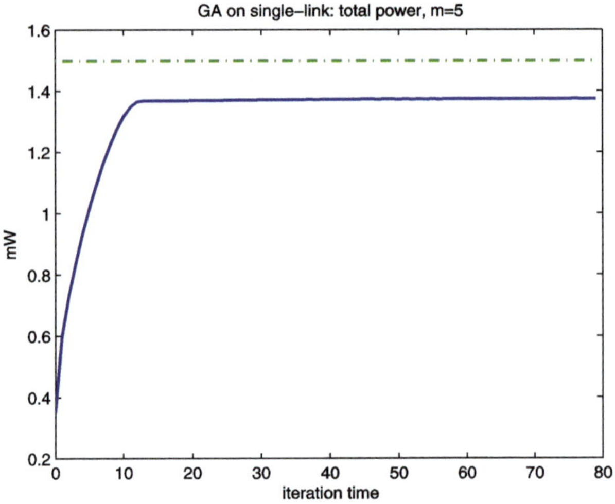

Fig. 10.13 Evolution of total power with GA, $m = 5$

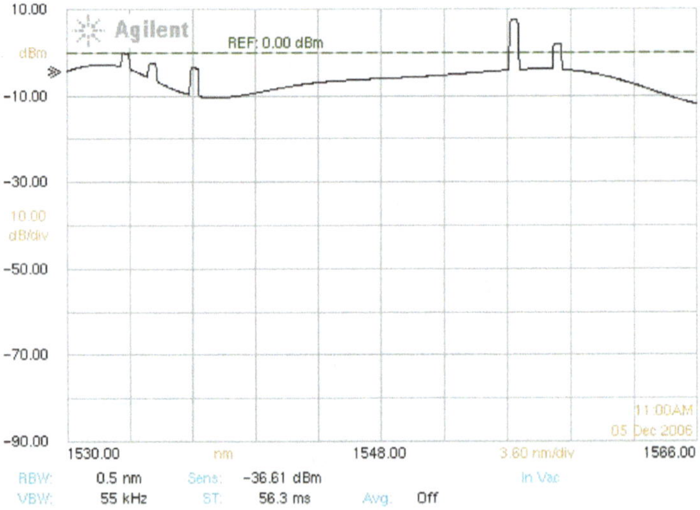

Fig. 10.14 Snapshot on OSA: results of GA, $m = 5$

Nash games, the individual channel cost function $J_{l,i}$ is with the following parameter selection:

$$\begin{aligned}\alpha_{l,i} &= 10 \times \Gamma_{l_{i,i}},\\ \beta_{l,i} &= 1 + 0.1 \times i,\\ a_{l,i} &= 50 \times l \times \Gamma_{l_{i,i}}, \quad i \in \mathcal{M}_l, l \in \mathcal{L},\end{aligned}$$

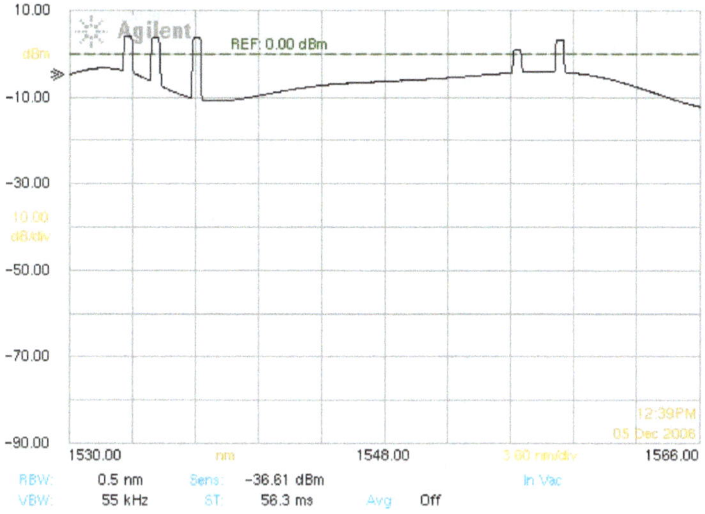

Fig. 10.15 Snapshot on OSA: results of GA, $m = 5$

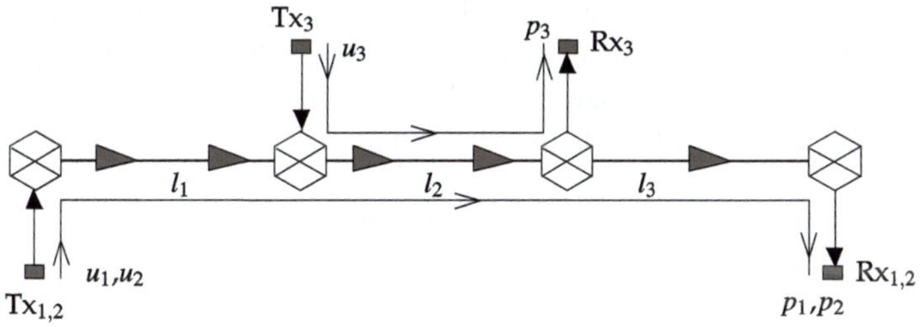

Fig. 10.16 Three-player game on a three-link topology

where $\Gamma_{l_{i,i}}$ is the diagonal element in the link system matrix Γ_l. Values of $\Gamma_{l_{i,i}}$ are obtained in each individual network topology. Note that the condition (8.36) is satisfied.

10.3.1 Multi-link Topologies

We implement the partition approach to a simple multi-link topology with three links and three channels shown in Fig. 10.16. Initially channels 1 and 2 are added on link 1 and channel 3 is added on link 2.

Constrained game problems in multi-link topologies were studied in Chap. 8 by introducing a partitioned Nash game with stages. We consider the case in which channel powers are adjustable at each optical switching node. A partition method is

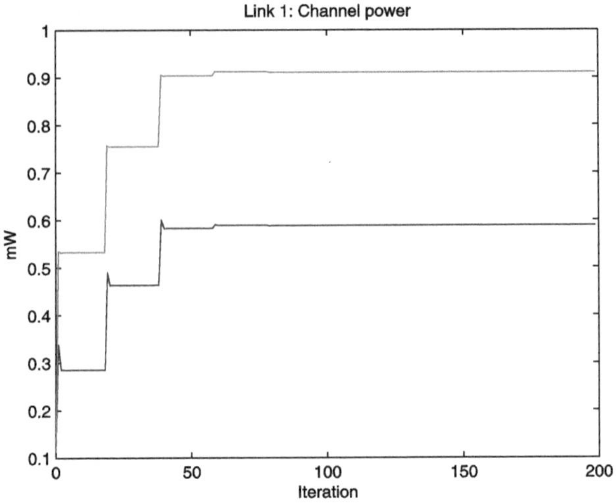

Fig. 10.17 Multi-link case: channel input power on link 1

used in which each link is a stage. This partition simplifies the partitioned structure and the convexity condition is naturally satisfied, as we discussed before.

We present MATLAB simulation results for the topology shown in Fig. 10.16 by applying the iterative hierarchical algorithm (8.39), (8.40). Link total power targets are $P_1^0 = 1.5$ mW, $P_2^0 = 2.5$ mW and $P_3^0 = 2.0$ mW. Diagonal elements of each link system matrix Γ_l, $l = 1, 2, 3$, are obtained as

$$\begin{aligned} \Gamma_{1_{1,1}} &= 1.139 \times 10^{-4}, & \Gamma_{1_{2,2}} &= 3.604 \times 10^{-4}, & & \\ \Gamma_{2_{1,1}} &= 6.804 \times 10^{-5}, & \Gamma_{2_{2,2}} &= 2.162 \times 10^{-4}, & \Gamma_{2_{3,3}} &= 6.839 \times 10^{-4}, \\ \Gamma_{3_{1,1}} &= 8.505 \times 10^{-5}, & \Gamma_{3_{2,2}} &= 2.703 \times 10^{-4}, & & \end{aligned}$$

while the step-size in the link algorithm is $\eta = 0.8$.

After partitioning, the stage Nash game on link 1 is played first. Figure 10.17 shows the evolution vs. iteration of channel input power on link 1.

For every 20 iterations, the link adjusts its price via the link algorithm and then channels readjust their powers. Figure 10.18 show the evolutions of total power and link price. After the Nash game on link 1 settles down, the stage Nash game on link 2 starts to play, followed by the game on link 3. These evolutions are shown in Figs. 10.19, 10.20, 10.21 and 10.22.

The following are the final achieved values for the adjustable parameters:

$$\gamma^* = \begin{bmatrix} 0 & 0 & 0 \\ 3.6429 & 0.6240 & 0 \\ 5.1592 & 1.7449 & 0 \end{bmatrix}$$

For channels added directly from Tx, we set the γ value to 0, since channel power is adjusted at Tx. Thus γ^* is feasible with respect to the predefined range $[\gamma_{\min}, \gamma_{\max}]$.

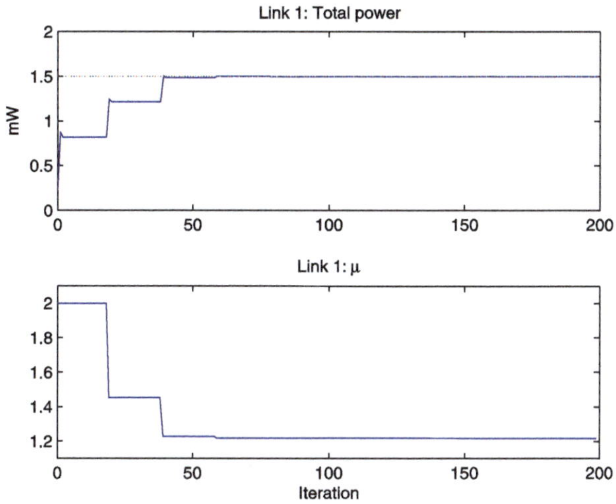

Fig. 10.18 Multi-link case: total power and link price on link 1

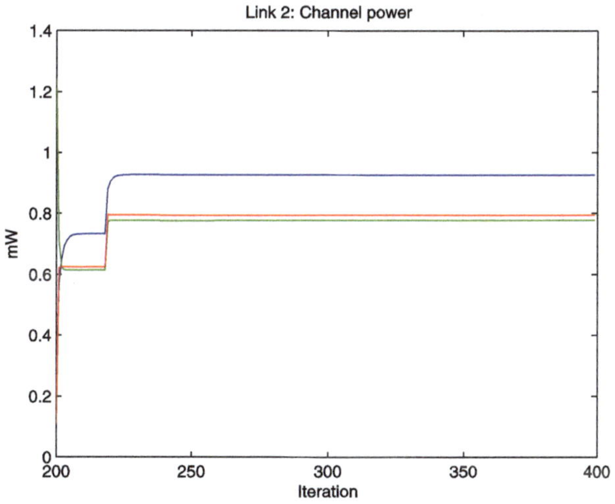

Fig. 10.19 Multi-link case: channel input power on link 2

10.3.2 Quasi-ring Topologies

In multi-link topologies, links are inter-connected in a ladder-nested manner and the L link Nash games can be automatically played in a precedence or parallel order. Situations are different in quasi-ring topologies. We take the quasi-ring topology in Fig. 8.6 for an example. The optical paths of channels 1 and 3 are $l_1 \to l_2 \to l_3$ and $l_3 \to l_1 \to l_2$, respectively. Each of the links is the intermediate or the end of some channel optical path. We partition this structure into three stages and each stage is a link. A partitioned Nash game is formulated composed of three-link Nash games. On each link Nash game, the convexity of link capacity constraints is automatically satisfied. By breaking the closed loop and selecting one link as the start, link Nash games can be played sequentially.

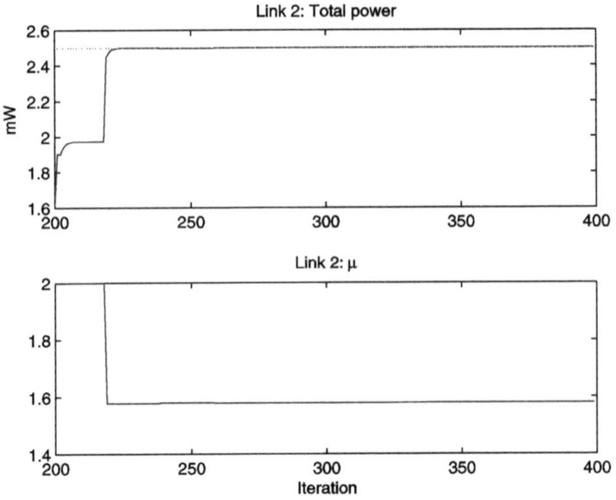

Fig. 10.20 Multi-link case: total power and link price on link 2

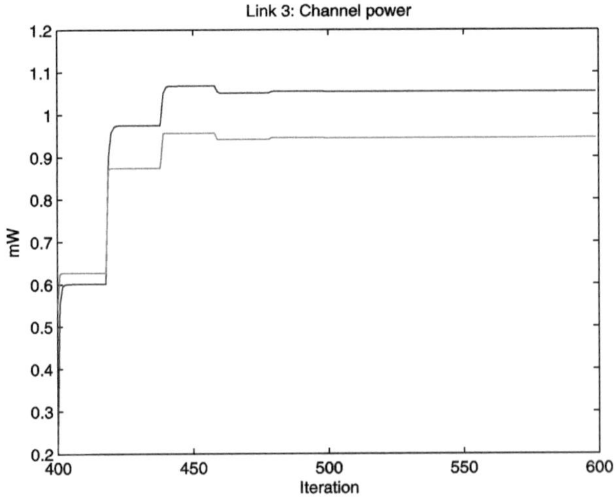

Fig. 10.21 Multi-link case: channel input power on link 3

Let us take a closer look at the simple quasi-ring topology in Fig. 8.6(a) and three channels whose optical paths are shown in Fig. 8.6(b). We break the closed loop and select link 3 as the starting link. The unfolded configuration is shown in Fig. 10.23.

The overall recursive process is such that stage Nash games on links are played sequentially: $l_3 \to l_1 \to l_2$. On link 3, the adjustable parameters for channels 1 and 2 are initially set as $\gamma_{3,1}^1$ and $\gamma_{3,2}^1$, respectively, where the superscript ^{1} indicates the number of iteration of the game among links. The game on link 3 settles down at $u_3^*(\mu_3^*)$ with the corresponding channel output power $p_3^*(\mu_3^*)$. Sequentially, the game on link 1 is played with an NE solution, $u_1^*(\mu_1^*)$. Then channel output power is $p_1^*(\mu_1^*)$. Given $p_3^*(\mu_3^*)$, and the adjustable parameter on link 1, $\gamma_{1,3}^*$, is determined. The game on link 2 is played after that. The NE solution of this game is $u_2^*(\mu_2^*)$ and the channel output power is $p_2^*(\mu_2^*)$. Then the adjustable parameters on

Fig. 10.22 Multi-link case: total power and link price on link 3

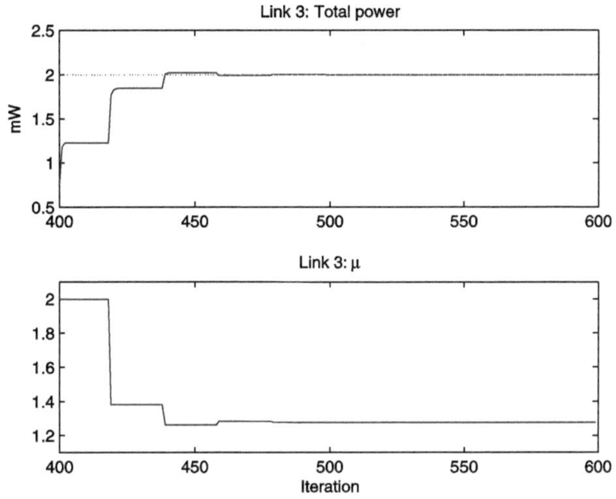

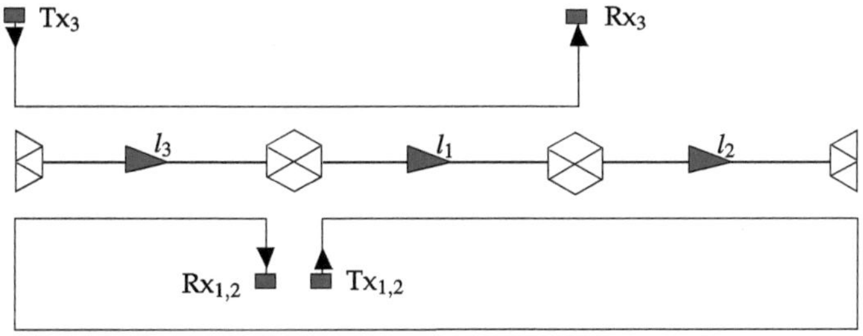

Fig. 10.23 Unfolded quasi-ring topology with the starting link, link 3

link 2, $\gamma_{2,i}^*$, $i = 1, 2$, are determined. With the given $p_2^*(\mu_2^*)$, link 3 determines its adjustable parameters by $\gamma_{3,i}^* = \frac{u_{3,i}^*(\mu_3^*)}{p_{2,i}^*(\mu_2^*)}$, $i = 1, 2$.

We present MATLAB simulation results for the topology shown in Fig. 8.6. Link total power targets are $P_1^0 = 1.5$ mW, $P_2^0 = 2.5$ mW and $P_3^0 = 2.0$ mW. Diagonal elements of each link system matrix Γ_l, $l = 1, 2, 3$, are obtained as

$$\Gamma_{1_{1,1}} = 2.166 \times 10^{-4}, \qquad \Gamma_{1_{2,2}} = 6.852 \times 10^{-4}, \qquad \Gamma_{1_{3,3}} = 2.2 \times 10^{-3},$$

$$\Gamma_{2_{1,1}} = 3.611 \times 10^{-4}, \qquad \Gamma_{2_{2,2}} = 1.100 \times 10^{-3},$$

$$\Gamma_{3_{1,1}} = 2.708 \times 10^{-4}, \qquad \Gamma_{3_{2,2}} = 8.565 \times 10^{-4}, \qquad \Gamma_{3_{3,3}} = 2.7 \times 10^{-3}$$

A step-size of $\eta = 0.1$ is used in the link algorithm. The partitioned Nash game is played as described above. For every 20 iterations, the link adjusts its price via the link algorithm and then channels readjust their powers. Evolutions in time of channel input power, total power and link price on each link l are shown in Figs. 10.24–10.29, respectively.

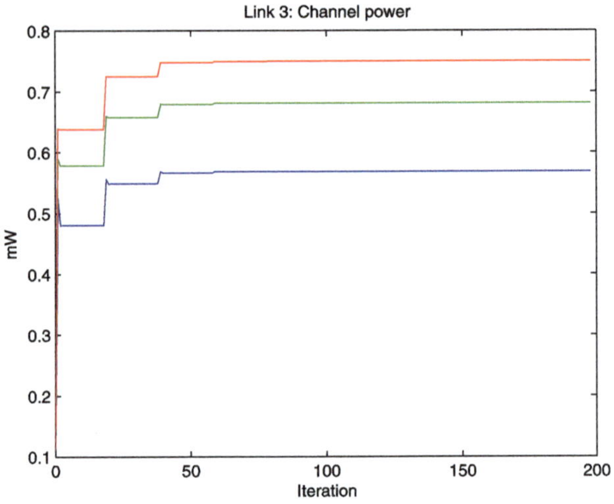

Fig. 10.24 Quasi-ring case: channel input power on link 3

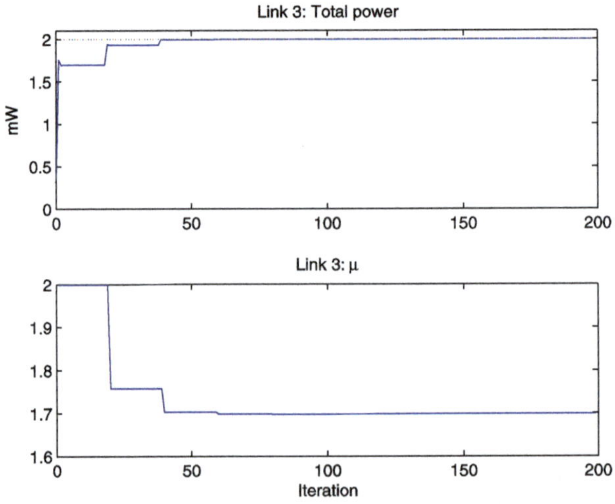

Fig. 10.25 Quasi-ring case: total power and link price on link 3

The adjustable parameters for three links and three channels are obtained as in the following 3 × 3 matrix:

$$\gamma^* = \begin{bmatrix} 0 & 0 & 0.7208 \\ 7.5983 & 1.4868 & 0 \\ 1.5277 & 0.6042 & 0 \end{bmatrix}$$

The overall game settles down since γ^* is feasible. Note that a different starting link can be selected, say, link 1, such that games on links are played sequentially: $l_1 \to l_2 \to l_3$. Typically we select the starting link where channels are added directly from Txs.

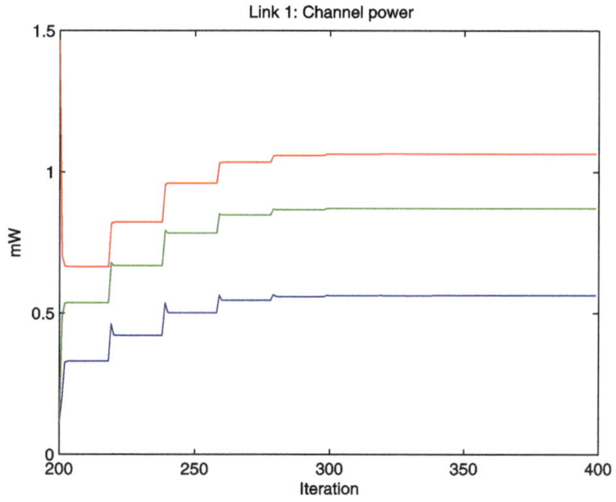

Fig. 10.26 Quasi-ring case: channel input power on link 1

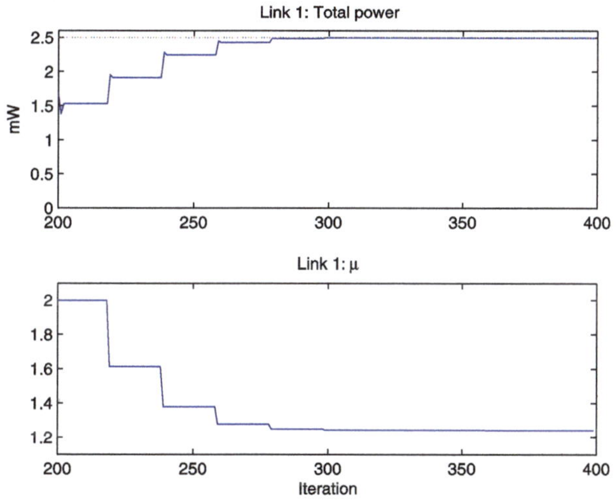

Fig. 10.27 Quasi-ring case: total power and link price on link 1

10.3.3 Mesh Topologies

We study a mesh network topology as shown in Fig. 10.30(a), where eight channels are transmitted over six links. The channel routes are shown in Fig. 10.30(b).

It can be seen from Fig. 10.30(b) that there exists a closed loop among links 1, 4, and 6, which is formed by the optical paths of channels 3, 4, 5, and 6. We break the closed loop and select link 1 as the starting link. This unfolded configuration is shown in Fig. 10.31.

The overall recursive play process is described as follows. Games on links 1, 6, and 4 are played in a precedence order: $l_1 \rightarrow l_6 \rightarrow l_4$. Since the closed loop among links 1, 4, and 6 is unfolded, on link 1 the adjustable parameters for channels 5

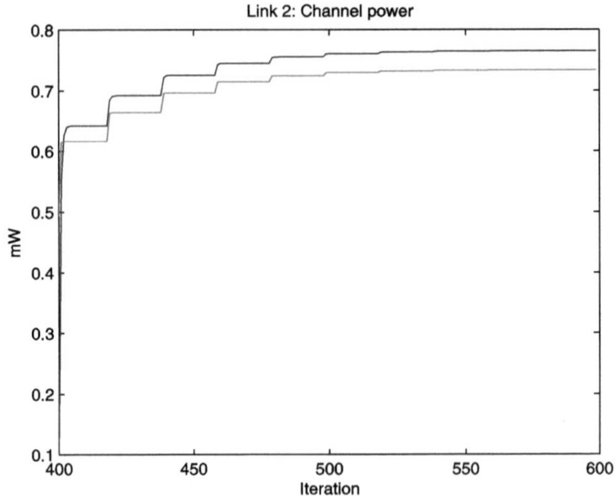

Fig. 10.28 Quasi-ring case: channel input power on link 2

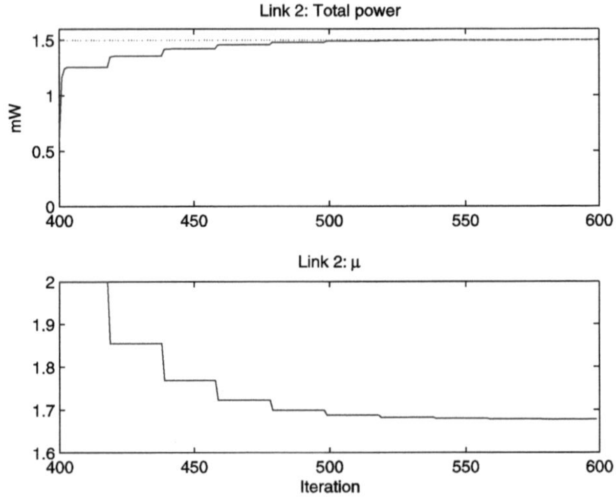

Fig. 10.29 Quasi-ring case: total power and link price on link 2

and 6 are initially set as $\gamma_{1,5}^{1}$ and $\gamma_{1,6}^{1}$, respectively. The games on links 1, 2, and 5 can be played in a parallel order, while the game on link 3 is played after games on links 2 and 5 settle down. The last game to play is the one on link 4. After all the games settle down on $\mathbf{u}_l^*(\boldsymbol{\mu}_l^*)$ with corresponding channel output powers $\mathbf{p}_l^*(\boldsymbol{\mu}_l^*)$ and adjustable parameters γ_l^*, link 1 re-determines its adjustable parameter γ_1^* according to $p_4^*(\boldsymbol{\mu}_4^*)$ because of the closed loop among links 1, 4, and 6. Again we note that link 1 is not the only choice for the starting link. Typically we select the link where channels are added directly from transmitters as the starting link.

A user-defined step-size $\eta = 0.6$ is used in the simulation. Link total power targets are $P^0 = [2.0\ 1.5\ 2.0\ 2.5\ 1.5\ 1.5]$ mW. Diagonal elements of each link system matrix Γ_l, $l = 1, \ldots, 6$, are obtained as $\Gamma_{1_{3,3}} = 0.0014$, $\Gamma_{1_{4,4}} = 0.0027$, $\Gamma_{1_{5,5}} = 0.0027$, $\Gamma_{1_{6,6}} = 0.0013$, $\Gamma_{2_{1,1}} = 7.212 \times 10^{-4}$, $\Gamma_{2_{2,2}} = 0.001$, $\Gamma_{3_{1,1}} =$

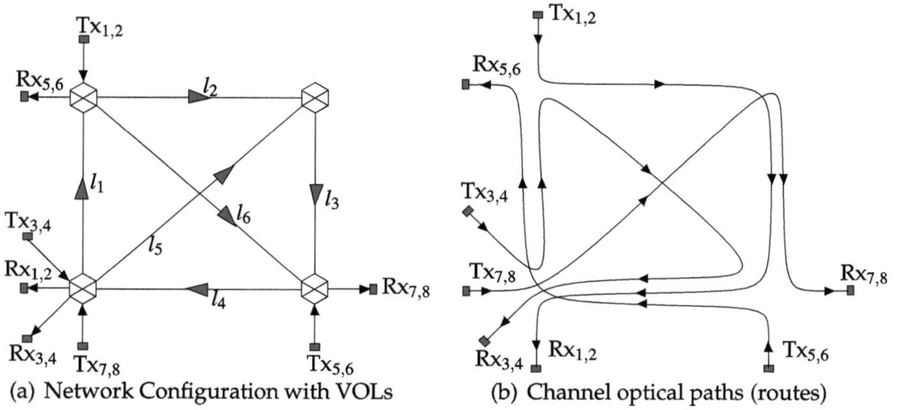

Fig. 10.30 Mesh network topology

Fig. 10.31 Unfolded mesh topology with the starting link, link 1

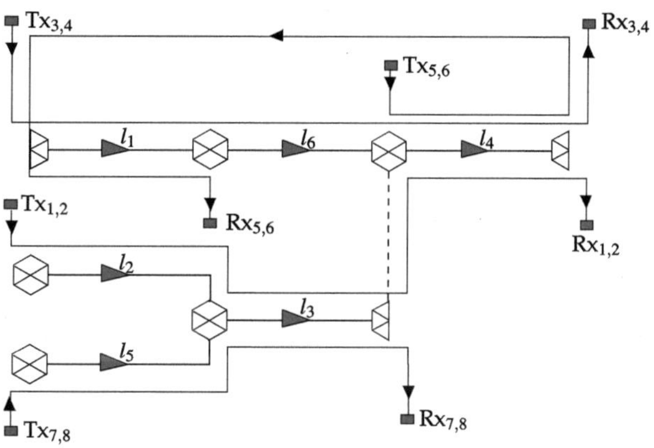

5.409×10^{-4}, $\Gamma_{3_{2,2}} = 8.564 \times 10^{-4}$, $\Gamma_{3_{7,7}} = 8.483 \times 10^{-4}$, $\Gamma_{3_{8,8}} = 5.34 \times 10^{-4}$, $\Gamma_{4_{1,1}} = 4.327 \times 10^{-4}$, $\Gamma_{4_{2,2}} = 6.852 \times 10^{-4}$, $\Gamma_{4_{3,3}} = 0.0011$, $\Gamma_{4_{4,4}} = 0.0022$, $\Gamma_{4_{5,5}} = 0.0022$, $\Gamma_{4_{6,6}} = 0.0011$, $\Gamma_{5_{7,7}} = 0.0011$, $\Gamma_{5_{8,8}} = 7.12 \times 10^{-4}$, $\Gamma_{6_{3,3}} = 0.0018$ and $\Gamma_{6_{4,4}} = 0.0036$. We show the evolutions of channel input power, total power and link price on links 1, 2, and 4 in Figs. 10.32–10.37, respectively. The final adjustable parameter values for six links and eight channels are obtained as in the 6×8 matrix below:

$$\gamma^* = \begin{bmatrix} 0 & 0 & 0 & 0 & 0.7067 & 1.3548 & 0 & 0 \\ 0 & 0 & 0 & 0 & 0 & 0 & 0 & 0 \\ 0.6893 & 0.4308 & 0 & 0 & 0 & 0 & 0.6535 & 1.0415 \\ 1.1210 & 0.7271 & 0.8486 & 0.4335 & 0 & 0 & 0 & 0 \\ 0 & 0 & 0 & 0 & 0 & 0 & 0 & 0 \\ 0 & 0 & 2.5839 & 1.1778 & 0 & 0 & 0 & 0 \end{bmatrix}$$

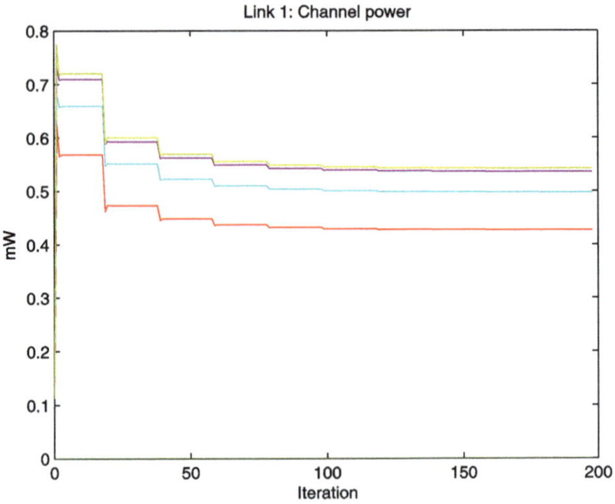

Fig. 10.32 Mesh case: channel input power on link 1

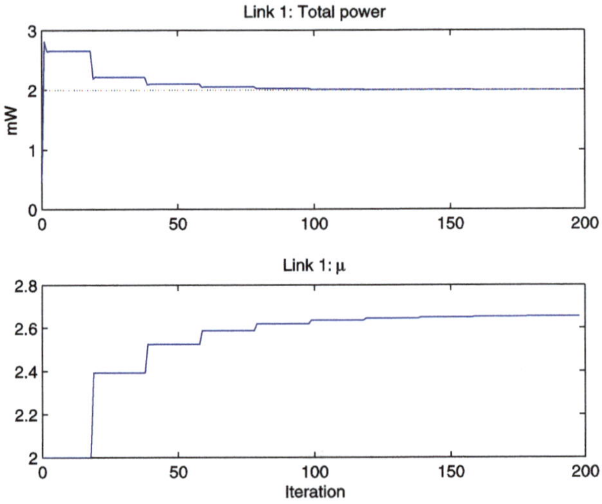

Fig. 10.33 Mesh case: total power and link price on link 1

10.3.4 Experimental Results

Next experimental results are presented on the same ONTS where this time three links are setup as shown in Fig. 10.38. Each link is composed of an OA. Total power targets are $P_1^0 = 1.5$ mW, $P_2^0 = 2.5$ mW and $P_3^0 = 1.5$ mW, respectively. Channels 1 and 2 are added on link 1 and channel 3 is added on link 2. We use following wavelengths for each channel: 1535.04 nm on Light source 1 (LS 1), 1537.40 nm on Light source 2 (LS 2) and 1533.47 nm on Light source 3 (LS 3). Channel 3 is dropped after link 2. This is realized by using a filter which is used to perform wavelength selection. Channels 1 and 2 are transmitted through two filters, respectively.

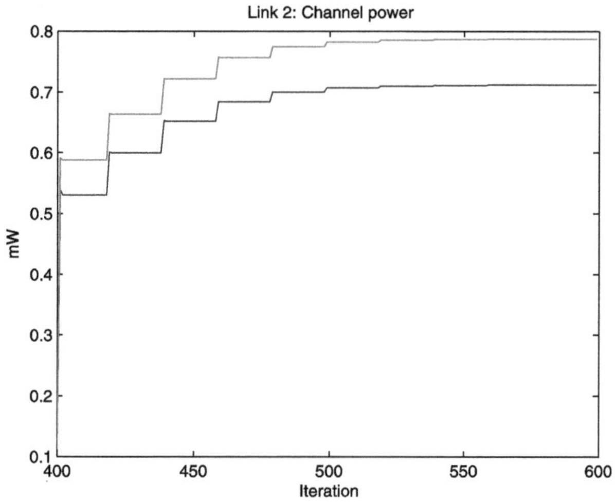

Fig. 10.34 Mesh case: channel input power on link 2

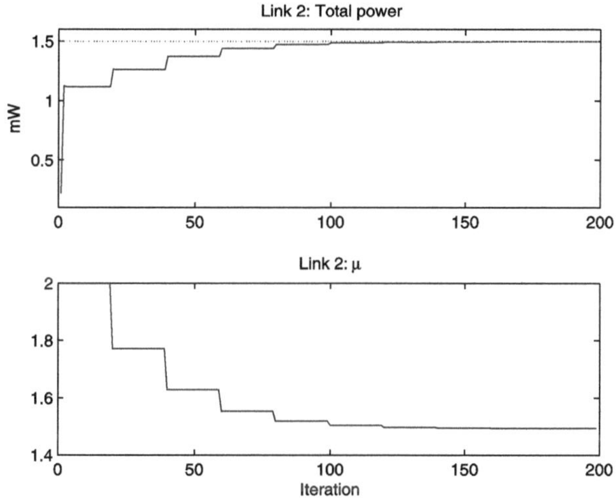

Fig. 10.35 Mesh case: total power and link price on link 2

We partition this three-link game into two stage games ($K = 2$): Stage 1 is composed of links 1 and 2 and stage 2 is link 3. Diagonal elements of each stage system matrix $\widehat{\Gamma}_k$ are derived from stage OSNR mode. Particularly in this configuration,

$$\widehat{\Gamma}_{1_{i,i}} = \frac{ASE_{1,i}}{P_1^0} + \frac{ASE_{2,i}}{P_2^0}, \quad i = 1, 2,$$

$$\widehat{\Gamma}_{1_{3,3}} = \frac{ASE_{2,3}}{P_2^0},$$

$$\widehat{\Gamma}_{2_{i,i}} = \frac{ASE_{3,i}}{P_3^0}, \quad i = 1, 2,$$

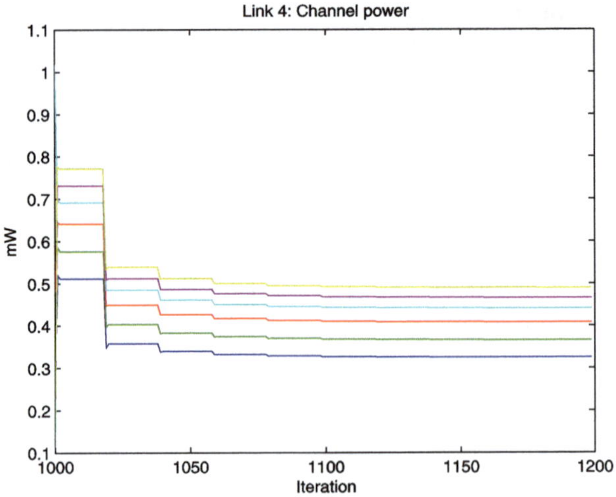

Fig. 10.36 Mesh case: channel input power on link 4

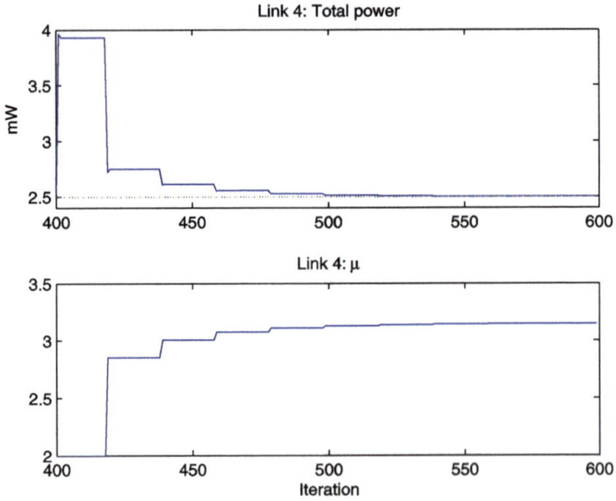

Fig. 10.37 Mesh case: total power and link price on link 4

where

$$ASE_{l,i} = 2n_{sp}(G_{l,i} - 1)h\nu_i B,$$

with $G_{l,i}$ the gain of OA at the wavelength of channel i on link l.

The value of the dynamic adjustment parameter $\gamma_{k,i}$ is in fact the attenuation value of channel i on stage k, which is physically bounded.

The iterative hierarchical algorithm developed in Chap. 8 is implemented, which is composed of a channel algorithm and a link algorithm. At each stage k, the following channel algorithm is used:

$$u_{k,i}(n+1) = \frac{\beta_{k,i}}{\alpha_{k,i} + \boldsymbol{\mu}_{k,r(i)}(t)} - \left(\frac{1}{OSNR_{k,i}(n)} - \frac{1}{OSNR_{k',i}} - \widehat{\Gamma}_{k_{i,i}}\right)\frac{u_{k,i}(n)}{a_{k,i}},$$

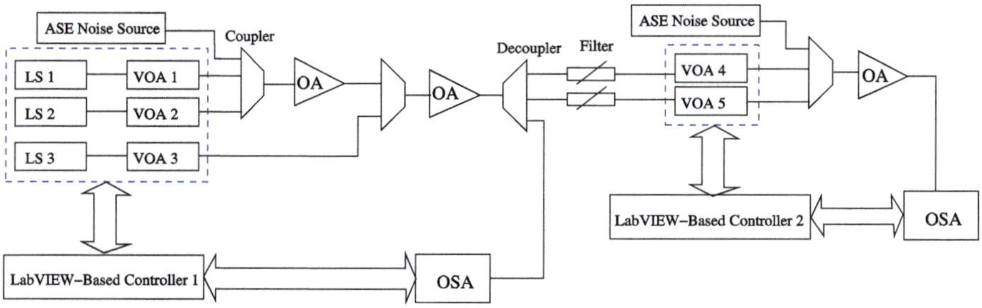

Fig. 10.38 ONTS configuration setup: multiple links

Fig. 10.39 Evolution of input power on stage 1

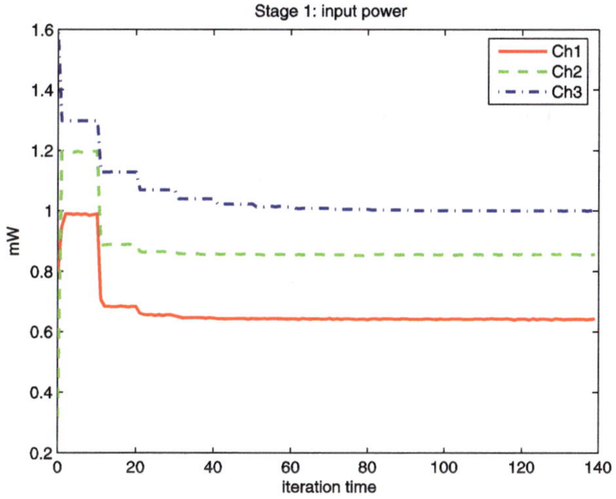

where stage k' is the precedent of stage k and $OSNR_{k',i}$ and $\boldsymbol{\mu}_{k,r(i)}(t)$ are invariable during the channel iteration in stage k. Then after every N_k iterations of the channel algorithm, the new link price is generated according to the following link algorithm:

$$\boldsymbol{\mu}_k(t+1) = \left[\boldsymbol{\mu}_k(t) - \eta_k\left(\triangle P_k^0 - E_k\,\mathbf{u}_k(\bar{\boldsymbol{\mu}}_k(t))\right)\right]^+,$$

where $\triangle P_k^0$ is a modified coupled power constraint on stage k and

$$\bar{\boldsymbol{\mu}}_k(t) = \left[\boldsymbol{\mu}_k(t) - \eta_k\left(\triangle P_k^0 - E_k\,\mathbf{u}_k(\boldsymbol{\mu}_k(t))\right)\right]^+$$

In the experiment, the game on Stage 1 is played first. Figure 10.39 shows the evolution in iteration time of channel input power on Stage 1.

For every $N_1 = 10$ iteration, link 1 and link 2 adjust their prices simultaneously (link algorithm) and then channels readjust their powers. Total power and prices evolutions are shown in Fig. 10.40. After Stage 1 settles down, the game on Stage 2 starts to play. For every $N_2 = 7$ iteration, link 3 adjusts its price. The evolutions are shown in Figs. 10.41 and 10.42. Snapshots of the values of OSNR and attenuation

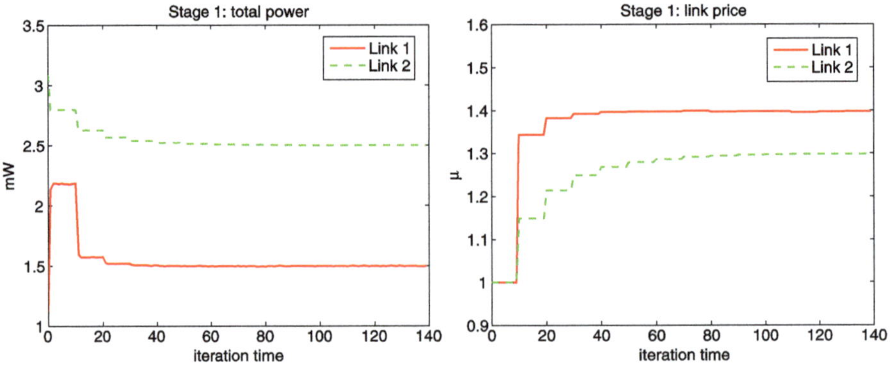

(a) Total power (b) Price

Fig. 10.40 Evolution of total power and price on stage 1

Fig. 10.41 Evolution of input power on stage 2

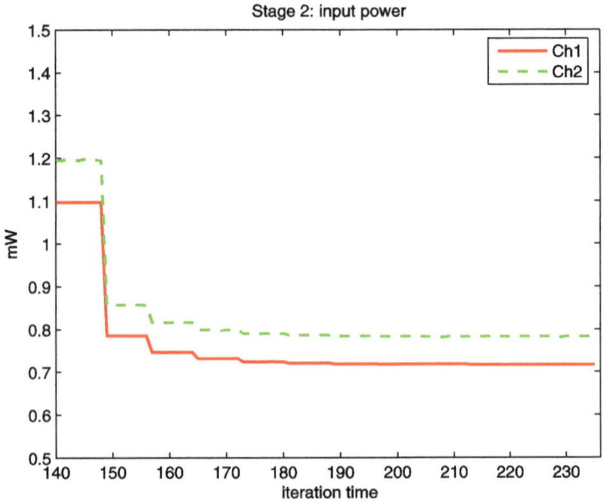

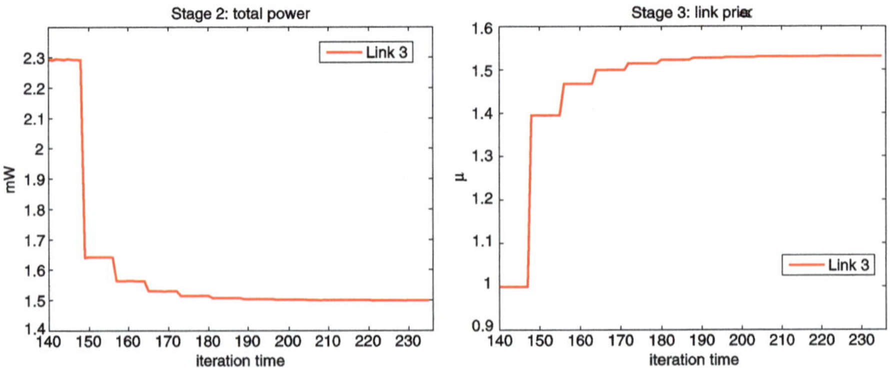

(a) Total power (b) Price

Fig. 10.42 Evolution of total power and price on stage 2

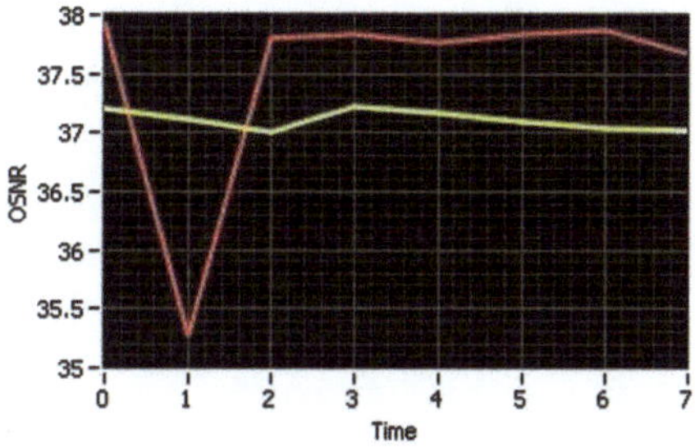

(a) OSNR

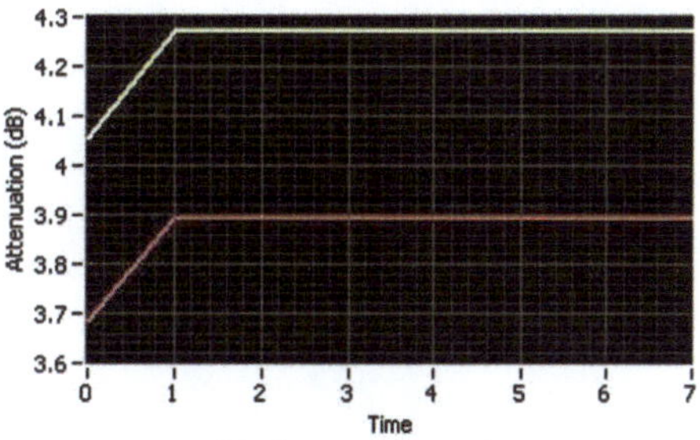

(b) Attenuation value

Fig. 10.43 Snapshot: partial evolution of OSNR and attenuation value on stage 2

during the iteration on Stage 2 are shown in Fig. 10.43. The final values of adjustable parameters, i.e., the attenuation values are

$$\gamma^* = \begin{bmatrix} 0 & 0 & 0 \\ 4.454 & 4.073 & 0 \end{bmatrix}$$

Part III
Robustness, Delay Effects, and Other Problems

Chapter 11
Robustness and Delay Effects on Network Games

Abstract This chapter presents results on robustness of network control algorithms derived from game-theoretic formulations in the presence of time delay. Mesh optical networks are distributed over large surface areas. Any realistic OSNR model must account for these delays. Sufficiently large time delays may destabilize the closed-loop systems that implement these game-theoretic-inspired control algorithms. These algorithms need to have been appropriately adjusted so as to ensure closed-loop stability in the presence of time delays. We consider delay effects in network games without constraints, followed by games with constraints. Then we study a delayed primal–dual algorithm and perform a two time-scale stability analysis. We conclude by considering robustness and delay effects combined.

11.1 Introduction

In the previous chapters we presented game-theoretic formulations and, based on these, network control algorithms towards channel OSNR optimization. We discussed pricing strategies and selection of parameters that ensure optimal convergence to the NE point without time delays. Due to large distances and propagation delays, any realistic OSNR model must account for these delays. The figure below shows a typical mesh topology with the control plane being the layer where control algorithms will be implemented to adjust network parameters (Fig. 11.1). These parameters are given as reference values to the transport layer (physical layer). Links between nodes span usually hundreds or thousands of kilometers hence propagation time delay is non-negligible. Control algorithms need to be appropriately modified to ensure closed-loop stability in the presence of time delays. In this chapter, we address this problem.

We first present a multi-link OSNR model with time delay and the corresponding closed-loop system based on game-theoretic algorithms. Next we consider delay effects in network games without constraints, followed by a treatment for games with coupled constraints. We present an analysis for the delayed primal–dual algorithm and a time-scale stability analysis for point-to-point topologies. We also indicate how this can be extended to multi-link topologies. The last section is focused on robustness to a combination of delay effects and uncertainties. Stability conditions

L. Pavel, *Game Theory for Control of Optical Networks*,
Static & Dynamic Game Theory: Foundations & Applications,
DOI 10.1007/978-0-8176-8322-1_11,

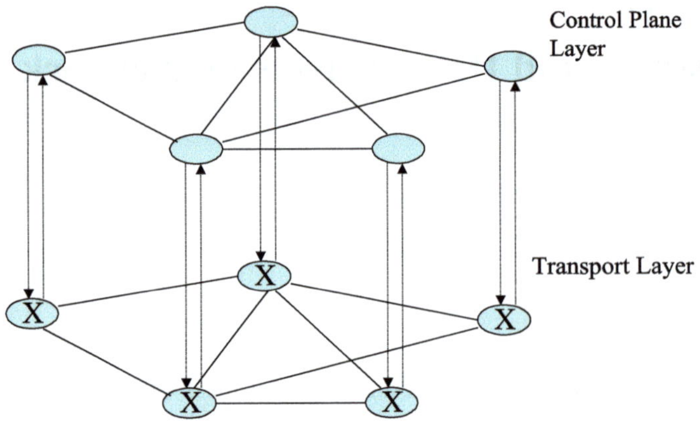

Fig. 11.1 Example general optical network

are presented based on a frequency domain analysis and a Lyapunov–Razumikhin function analysis. The material in this chapter is based on [145, 149, 150].

11.2 System Model

In this section we consider a multi-link OSNR model with time delay and the corresponding closed-loop system based on game-theoretic algorithms. The OSNR models developed in Chap. 6 in Lemmas 6.2 and 6.9 (see (6.16), (6.36)) do not account for time delays. Let us define the notation to express time delays for arbitrary multi-link topology and for single links. Consider a general mesh network topology with time delays. Each channel $OSNR_i$, y_i value is measured in real-time, and fed back to its respective source to be processed by the control algorithms at Tx and adjust the channel power u_i.

Let us denote by $\tau_{i,j}^f$ the forward propagation delay of a signal from its source, $j \in \mathcal{M}$, to an OSNR output i. Similarly let τ_i^b denote the backward propagation delay of a signal from the OSNR output, i, to its associated source, i. Note that a signal from any channel, j, may interfere at the OSNR output of any other channel i, so two indices are needed for the forward propagation delay, $\tau_{i,j}^f$. However, even in the mesh configuration, the OSNR output of a channel i is fed back only to the source of channel i, and not to any other channel. Thus, for the backward propagation delay, τ_i^b a single index is needed. Let us denote the combined round-trip delay by $\tau_{i,j} = \tau_{i,j}^f + \tau_i^b$. Round-trip propagation delay from a source i to its OSNR output i, y_i, and back to the source i is denoted by $\tau_{i,i}$.

With these notations, from Lemmas 6.2 and 6.9, the following time-delayed OSNR model is obtained.

Lemma 11.1 *Consider the ith channel in an arbitrary multi-link optical network, $i \in \mathcal{M}$. Its OSNR value, fed back from Rx to Tx, evaluated at time t is denoted by*

$OSNR_i$ or y_i and is given as

$$y_i(t) = \frac{u_i(t - \tau_{i,i})}{n_{0,i} + \sum_{j \in \mathcal{M}} \Gamma_{i,j} u_j(t - \tau_{i,j})} \tag{11.1}$$

where $\Gamma_{i,j}$, elements of the full $(n \times n)$ system matrix Γ, are defined as (6.36).

Note that (11.1) is the OSNR signal that is measured at Rx and then fed back to Tx. Thus both the forward and the backward time delays are present, since the OSNR value is fed back to the transmitter.

For a single-link case, as in Lemma 6.2, time-delay notation can be simplified such that $\tau = \tau^f + \tau^b$, where τ is the round-trip time delay of the link, τ^f is the forward propagation delay, and τ^b is the backward propagation delay. Thus, from Lemma 6.2, we obtain the following lemma.

Lemma 11.2 *Consider the ith channel in a point-to-point optical link, $i \in \mathcal{M}$. Its OSNR value, fed back from Rx to Tx, evaluated at time t is denoted by $OSNR_i$ or y_i and is given as*

$$y_i(t) = \frac{u_i(t - \tau)}{n_{0,i} + \sum_{j \in \mathcal{M}} \Gamma_{i,j} u_j(t - \tau)}, \quad i \in \mathcal{M} \tag{11.2}$$

where Γ is the $(n \times n)$ link matrix with elements $\Gamma_{i,j}$, given by (6.15).

Next let us revisit the iterative algorithm for channel power. From Chap. 7, recall that the algorithm (7.62) ensures that the closed-loop system converges to the NE equilibrium when no time delays are considered. In order to find out under what conditions this holds in the presence of time delay let us introduce tunable control parameters, ρ_i, for each channel i, into the control algorithm (7.62) and convert into it to continuous-time. Thus, for each channel i we subtract $u_i(n)$ from each side of (7.62). The n time index is normalized with respect to the time period T_0, i.e. a time period of T_0 elapses between n and $n+1$ iterations, i.e., $t_n = nT_0$, where t_n is the total elapsed time at the nth algorithm update. We also introduce parameter ρ_i, $0 < \rho_i \leq 1$, multiplying the right-hand side of (7.62), and we approximate the left-hand side by $\frac{du_i(t)}{dt} = u_i(n+1) - u_i(n)$, to obtain

$$\frac{du_i(t)}{dt} = \rho_i \left\{ \frac{\beta_i}{\alpha_i} - \frac{1}{a_i} \left(\frac{1}{y_i(t)} - \Gamma_{i,i} + a_i \right) u_i(t) \right\} \tag{11.3}$$

where t is the independent continuous-time variable. The above approximation assumes that the algorithm is updated quickly enough that we may work with t rather than discrete time periods t_n. Note that this continuous-time approximation in (11.3) is relatively simplistic, and may be replaced by other approaches. For example, bilinear transformation could be used to create a mapping between continuous-time and discrete-time domain. Alternatively, one could study the stability of (7.62) and

(11.1) within the discrete-time domain. However, as the majority of time-delay stability results in literature in continuous-time, this is the approach we take also.

Consider further that (11.3) is modified by substituting $u_i(t)$ for $u_i(t-\tau_{i,i})$, which yields

$$\frac{du_i(t)}{dt} = \rho_i \left\{ \frac{\beta_i}{\alpha_i} - \frac{1}{a_i} \left(\frac{1}{y_i(t)} \right) u_i(t-\tau_{i,i}) + \left(\frac{\Gamma_{i,i}}{a_i} - 1 \right) u_i(t) \right\} \tag{11.4}$$

This is used so as to simplify the closed-loop system's structure. We explicitly write this as the following assumption.

Assumption 11.1 *Each channel, i, has knowledge of its own round-trip time delay, $\tau_{i,i}$.*

Assumption 11.1 is realistic in practice since round-trip time delays, which are due to signal propagation times, are measurable. Furthermore, the decentralized nature of the control algorithm (11.3) is also preserved since this added information is local. Because the control algorithm is implemented in software, this substitution as in (11.4) is allowed. Furthermore, the channel algorithm running at the channel sources may store the signal powers, $u(t)$, over time. Thus, with $\tau_{i,i}$ known, it is used to reference the appropriate past channel power $u(t-\tau_{i,i})$. Hence, (11.4) is a modified channel algorithm of (7.62) to handle time delays more effectively. Note that while the control algorithm (11.4) can be modified, the OSNR model (11.1), which is the physical model of the optical network, cannot be modified.

11.3 Delay Effects in Network Games Without Constraints

In this section we consider games without coupled constraints so that we apply only the primal algorithm (channel) without the dual algorithm (link price). Let us first obtain a closed-loop system that incorporates time delays by substituting the time-delayed OSNR model (11.1) into (11.4). This yields

$$\frac{du_i(t)}{dt} = \rho_i \left\{ \frac{\beta_i}{\alpha_i} - \frac{1}{a_i} \left(n_{0,i} + \sum_{j \in \mathcal{M}} \Gamma_{i,j} u_j(t-\tau_{i,j}) \right) + \left(\frac{\Gamma_{i,i}}{a_i} - 1 \right) u_i(t) \right\} \tag{11.5}$$

or, after shifting it around the equilibrium point u^*, (7.11) (see Chap. 7), the equivalent form is

$$\frac{du_i}{dt} = -\frac{\rho_i}{a_i} \left((a_i - \Gamma_{i,i}) u_i(t) + \sum_{j \in \mathcal{M}} \Gamma_{i,j} u_j(t-\tau_{i,j}) \right) \tag{11.6}$$

This (11.6) is the closed loop that is analyzed next. Specifically, we give sufficient conditions for stability of the time-delay closed-loop system (11.6). These conditions are given in terms of tunable parameters ρ_i and a_i. The approach is based on an analysis as in [100], but unlike [100], the system does not have any symmetry in the system matrix Γ that can be exploited.

First, let us map (11.6) into the form (A.33) (see the appendix) in order to study its asymptotic stability as in Definition A.32. For convenience, we restate (A.33) using the state variables and system parameters from (11.6) as

$$\dot{u}(t) = A_0 u(t) + \sum_{k=1}^{m^2} A_k u(t - \tau_k), \qquad \tau_k \geq 0 \tag{11.7}$$

for m channels in the system. Starting from (11.6), assume $a_i > \Gamma_{i,i}$ for all i, let $\varepsilon_i = \frac{\rho_i}{a_i}(a_i - \Gamma_{i,i})$, and define

$$A_0 = \text{diag}(-\varepsilon_i)$$

In order to define A_k one can proceed as follows. Notice in (11.6), for channel i, there may be up to m delay terms with the form $-\frac{\rho_i}{a_i}\Gamma_{i,j}u(t - \tau_{i,j})$ for $j = 1, \ldots, m$. Each time delay, $\tau_{i,j}$, may be unique for a total of m^2 time delays for all $i, j = 1, \ldots, m$. Thus, to rewrite (11.6) into the form (11.7), each coefficient $-\frac{\rho_i}{a_i}\Gamma_{i,j}$, associated with the time-delayed state $u(t - \tau_{i,j})$, occupies one element of the $m \times m$ matrix A_k for the associated time-delayed state $u(t - \tau_k)$. Let $k = (i - 1) \times m + j$ for $i, j = 1, \ldots, m$ and let A_k be an $(m \times m)$ matrix of zeros except for the (i, j)th element corresponding to the k index which is equal to $-\frac{\rho_i}{a_i}\Gamma_{i,j}$. Furthermore, let $\tau_k = \tau_{i,j}$. With these A_0, A_k, and τ_k (11.6) is mapped directly to (11.7).

Then by applying Laplace transform to (11.7) we obtain

$$\left(sI - A_0 - \sum_{k=1}^{m^2} A_k e^{-s\tau_k}\right) U(s) = u(0) + \sum_{k=1}^{m^2} \left[\int_{-\tau_k}^{0} A_k e^{-st}\, dt\right] e^{-s\tau_k} \tag{11.8}$$

where $U(s)$ is the Laplace transform of $u(t)$. Similarly to (A.34), let

$$\Delta(s) = \left(sI - A_0 - \sum_{k=1}^{m^2} A_k e^{-s\tau_k}\right) \tag{11.9}$$

The roots of the characteristic equation $\det(\Delta(s)) = 0$ determine the stability of (11.7) or (11.8) (see Definition A.32). Consider $\Delta(s)$ in an equivalent form

$$\Delta(s) = (sI - A_0)\left(I + (sI - A_0)^{-1} \sum_{k=1}^{m^2} -A_k e^{-s\tau_k}\right)$$

or

$$\Delta(s) = (sI - A_0)\left(I + \text{diag}\left(\frac{1}{s + \varepsilon_i}\right) F(s)\right)$$

where $F(s) = \sum_{k=1}^{m^2} -A_k e^{-s\tau_k}$. Finally, we write

$$\Delta(s) = (sI - A_0)\big(I + L(s)\big) \tag{11.10}$$

where $L(s) = \text{diag}(\frac{1}{s+\varepsilon_i})F(s)$. Thus $\Delta(s)$ in (11.10) is expressed in a unity feedback form with loop transfer function $L(s)$.

This loop $(m \times m)$ transfer function, $L(s)$, has the (i, j)th element

$$L_{i,j}(s) = \rho_i \frac{\Gamma_{i,j}}{a_i} \frac{e^{-\tau_{i,j}s}}{s + \rho_i(1 - \frac{\Gamma_{i,i}}{a_i})} \tag{11.11}$$

Stability of (11.8) is equivalent to $\Delta^{-1} \in H_\infty$ (see Definition A.32), where H_∞ denotes the Hardy space of all complex-valued functions, analytic and bounded in the open RHP, $\text{Re}[s] > 0$. Since $\varepsilon_i > 0$ by assumption, A_0 is stable, so by (11.10), it follows that $\Delta^{-1} \in H_\infty$ if $(I + L)^{-1} \in H_\infty$, where L is as above. Based on this the following result, modified from [100] for $\varepsilon_i \neq 0$, gives sufficient conditions for stability of (11.8).

Proposition 11.3 *Suppose*

(i) $F(0) - A_0$ *has all its eigenvalues in the open RHP, and*
(ii) *for every nonzero* ω *and every* $\bar{\mu} \in (0, 1]$, $I + \bar{\mu}L(j\omega)$ *is invertible.*

Then $(I + L)^{-1} \in H_\infty$.

We now state and prove a stability result (Theorem 11.6) for the closed-loop system resulting from games with no coupled constraints; part (i) gives a time-delay dependent stability condition based on the tuning parameters ρ_i; part (ii) provides a conservative stability condition independent of time delay. Before that we give two useful lemmas.

Lemma 11.4 (See [145]) *Let* $L_{i,i}(j\omega)$ *be defined as*

$$L_{i,i}(j\omega) = \rho_i \frac{\Gamma_{i,i}}{a_i} \frac{e^{-\tau_{i,i}j\omega}}{j\omega + \rho_i(1 - \frac{\Gamma_{i,i}}{a_i})} \tag{11.12}$$

Furthermore, define $\omega = \omega^*$ *as the frequency value that minimizes* $|\bar{\mu}L_{i,i}(j\omega) - (-1)|$ *over all frequencies for all* $\bar{\mu} \in (0, 1]$. *Then,*

$$1 - \bar{\mu}\big|L_{i,i}(j\omega)\big| \le \big|\bar{\mu}L_{i,i}\big(j\omega^*\big) - (-1)\big| \quad \forall\omega \in \big[0, \omega^*\big],\ \forall\bar{\mu} \in (0, 1] \tag{11.13}$$

or equivalently, $\forall\omega \in [0, \omega^*]$, $\forall\bar{\mu} \in (0, 1]$

$$1 - \frac{\Gamma_{i,i}}{a_i} \frac{\bar{\mu}\rho_i}{\sqrt{\omega^2 + \rho_i^2(1 - \Gamma_{i,i}/a_i)^2}} \le \big|\bar{\mu}L_{i,i}\big(j\omega^*\big) - (-1)\big| \tag{11.14}$$

Proof of Lemma 11.4 is immediate, based on the triangle inequality and the fact that the magnitude of $L_{i,i}(j\omega)$ is monotone with respect to ω.

Lemma 11.5 (See [145]) *Let $L_{i,i}(j\omega)$ be defined as in* (11.12) *and $\tilde{\omega} = \frac{\pi}{2\tau_{i,i}}$. Then $\tilde{\omega}$ is guaranteed to satisfy* (11.13), *i.e. $\tilde{\omega} \leq \omega^*$, if*

$$\frac{0.68244}{(1 - \frac{\Gamma_{i,i}}{a_i})\tau_{i,i}} \leq \rho_i$$

Theorem 11.6 (i) *If the system design parameters a_i are selected such that*

$$\sum_{j \neq i} \Gamma_{i,j} < a_i < \Gamma_{i,i} + \sum_j \Gamma_{i,j}, \quad \forall i \tag{11.15}$$

$$\Gamma_{i,i} < a_i, \quad \forall i \tag{11.16}$$

then the closed-loop time-delay system, (11.4), (11.1) *is stable if the tuning parameters ρ_i are selected such that*

$$\frac{0.68244}{(1 - \frac{\Gamma_{i,i}}{a_i})\tau_{i,i}} \leq \rho_i, \quad \forall i \tag{11.17}$$

$$0 < \rho_i < \frac{\pi}{2\tau_{i,i}(1 - \frac{\Gamma_{i,i}}{a_i})\sqrt{\Gamma_{i,i}^2/(a_i - \sum_j \Gamma_{i,j})^2 - 1}}, \quad \forall i \tag{11.18}$$

(ii) *The closed-loop time-delayed system,* (11.4), (11.1) *is stable for any time delays if the parameters a_i are selected such that*

$$\Gamma_{i,i} + \sum_j \Gamma_{i,j} < a_i, \quad \forall i \tag{11.19}$$

Proof We prove that $(I + L(s))^{-1} \in H_\infty$ by applying Proposition 11.3 to $L(s)$. To show condition (i), we apply Gershgorin's theorem to $F(0) + \text{diag}(\varepsilon_i)$. Note, based on the left side of (11.15), that we have

$$\rho_i \frac{\Gamma_{i,i}}{a_i} + \rho_i - \rho_i \frac{\Gamma_{i,i}}{a_i} > \sum_{j \neq i} \rho_i \frac{\Gamma_{i,j}}{a_i}, \quad \forall i$$

or equivalently, $\sum_{j \neq i} F_{i,j}(0) < |F_{i,i}(0) + \varepsilon_i|$, and condition (i) follows. Next, we show that condition (ii) is satisfied for $L(j\omega)$, (11.11) for $s = j\omega$. We apply Gershgorin's theorem to show that no Gershgorin disc of $\bar{\mu} L(j\omega)$ contains the value -1 for any $\bar{\mu} \in (0, 1]$. This is ensured if for all $\bar{\mu} \in (0, 1]$

$$\frac{\bar{\mu}\rho_i}{\sqrt{\omega^2 + \rho_i^2(1 - \Gamma_{i,i}/a_i)^2}} \sum_{j \neq i} \frac{\Gamma_{i,j}}{a_i} < \left|\bar{\mu} L_{i,i}(j\omega) + 1\right| \tag{11.20}$$

$\forall i$, $\forall \omega \geq 0$ where $L_{i,i}(j\omega)$ is defined in (11.12). The RHS of (11.20) is the distance from the critical point, -1, to the Nyquist curve of $\bar{\mu} L_{i,i}(j\omega)$. It can be shown that if

$$\frac{1}{(1-\frac{\Gamma_{i,i}}{a_i})} \sum_{j \neq i} \frac{\Gamma_{i,j}}{a_i} < \left| \bar{\mu} L_{i,i}(j\omega) + 1 \right|, \quad \forall i, \ \forall \omega \geq 0 \tag{11.21}$$

holds, see (11.21), then (11.20) is satisfied.

(i) Let ω^* be the value of ω that minimizes the RHS of (11.21). Then if the following holds:

$$\frac{1}{(1-\frac{\Gamma_{i,i}}{a_i})} \sum_{j \neq i} \frac{\Gamma_{i,j}}{a_i} < \left| \bar{\mu} L_{i,i}(j\omega^*) + 1 \right|, \quad \forall i \tag{11.22}$$

it follows that (11.21), hence (11.20) holds, so $-1 \notin \text{eig}(\bar{\mu} L(j\omega))$. We rewrite the RHS of (11.22) into a more useful form based on Lemmas 11.4 and 11.5. From Lemmas 11.4 and 11.5, $\tilde{\omega} = \frac{\pi}{2\tau_i}$ satisfies (11.14), as long as (11.17) is satisfied. Next, (11.15), (11.17), and (11.18) guarantee that

$$\frac{1}{(1-\frac{\Gamma_{i,i}}{a_i})} \sum_{j \neq i} \frac{\Gamma_{i,j}}{a_i} < 1 - \frac{\bar{\mu} \rho_i \Gamma_{i,i}}{a_i \sqrt{\pi^2/(4\tau_{i,i}^2) + \rho_i^2 (1 - \Gamma_{i,i}/a_i)^2}} \tag{11.23}$$

Thus by (11.23) and (11.14) it follows that (11.22) and (11.21) hold, and hence $-1 \notin \text{eig}(\bar{\mu} L(j\omega))$. To show (11.23), note that $\bar{\mu} = 1$ gives the smallest value on the RHS of (11.23). After some manipulation, this can be rewritten as

$$\rho_i^2 \left(1 - \frac{\Gamma_{i,i}}{a_i}\right)^2 \left[\frac{\Gamma_{i,i}^2}{(a_i - \sum_j \Gamma_{i,j})^2} - 1\right] < \frac{\pi^2}{4\tau_{i,i}^2} \tag{11.24}$$

By (11.15) the third term on the LHS of (11.24) is positive and (11.24) can be rearranged as in (11.18). Then (11.24) holds along with (11.23) and therefore $-1 \notin \text{eig}(\bar{\mu} L_{ii}(j\omega))$. By Proposition 11.3, the τ-dependent stability is proved.

(ii) We return to (11.21). On the RHS, we have $|\bar{\mu} L_{i,i}(j\omega) + 1| \geq 1 - |L_{i,i}(0)|$ based on $0 < \bar{\mu} \leq 1$ and the monotonicity of $|L_{i,i}(j\omega)|$ with respect to ω. Thus, $|\bar{\mu} L_{i,i}(j\omega) + 1| \geq 1 - \Gamma_{i,i}/(a_i - \Gamma_{i,i})$ and from condition (11.19) it follows that (11.21) holds, which implies (11.20) holds, and $-1 \notin \text{eig}(\bar{\mu} L(j\omega))$.

□

Corollary 11.7 *Both* (11.17) *and* (11.18) *can only be satisfied simultaneously in the range*

$$\frac{\Gamma_{i,i}}{\delta'} + \sum_j \Gamma_{i,j} \leq a_i \tag{11.25}$$

where $\delta' = \sqrt{(\frac{\pi}{2(0.68244)})^2 + 1} = 2.5096$.

11.4 Delay Effects in Network Games with Constraints

In this section we consider the effect of time delay in network games with coupled constraints. We consider a single link for simplicity and both channel and link algorithm as in Chap. 7. For multi-link topologies extension of the results in Chap. 8 can be similarly to incorporate time delay and the reader is referred to [148]. We present stability conditions for single links with time delays via primal–dual control algorithms based on [149]. We consider primal–dual algorithms (or channel-link algorithms) as developed for game with coupled constraints in Chap. 7. Compared to the analysis in the previous section, the new component of pricing algorithm (link algorithm) ensures that channel powers are below the threshold for nonlinear effects. The added dynamics of the pricing or link algorithm produces a nonlinear closed-loop system that requires a more advanced stability analysis than that presented in the previous section. We present a singular perturbation approach [69] (see the appendix for a review). This approach provides the means to analyze stability of systems that operate on two separate time-scales.

First we formulate the continuous-time closed-loop system that incorporates time delay. We follow with a Lyapunov–Razumikhin analysis combined with a singular perturbation approach applied to this closed-loop system. The main stability result is stated as Theorem 11.10. Material is based on [149] and [146], where detailed proofs can be found. An approach based on Lyapunov–Krasovskii stability analysis can be found in [150].

11.4.1 Delayed Primal–Dual Algorithm

In this section, we derive the closed-loop expression for the channel-price or primal–dual algorithms (7.62) and (7.61) applied to the time-delayed OSNR model (11.2). We then apply a coordinate shift that simplifies the foregoing stability analysis in the following section.

Let us first briefly review the primal–dual algorithms developed in Chaps. 7 and 8. Recall that the channel algorithm (7.62) is located at the transmitters (Tx) and the price algorithm or link algorithm (7.61) is located at the receivers (Rx) at the output of the link. The control algorithm of the ith channel adjusts the channel powers, u_i, towards OSNR optimization using its OSNR value and channel price μ as feedback. The link algorithm measures the total sum of the channel powers and computes a channel price, μ, based on the threshold for nonlinear effects. The channel price is fed back to the channel algorithms at Tx.

Let us consider the closed-loop system obtained by using the channel and price algorithms and the OSNR model (11.2). Recall, from Chaps. 7 and 8 that the channel algorithm operates on a faster time-scale reflected by the fact that its index, n, is related to the index of the link algorithm, $\bar{k}$, via $n = K\bar{k}$, where K is large, e.g., $K = 100$. In continuous-time, we denote the "fast" time variable by t and the "slow" time variable by $\hat{t}$. The two time-scales are related by $\hat{t} = \varepsilon t$, where $\varepsilon = \frac{1}{K}$ is small, and we will see later that the stability of the closed-loop system is ensured if ε is sufficiently small. The time delays on the two time-scales are related similarly by $\hat{\tau} = \varepsilon\tau$. A smaller ε value means a more effective time-scale decoupling between the channel and the price (link) algorithm.

Let us begin with the continuous-time channel algorithm (11.3) derived in Chap. 7, where $\alpha_i = \mu(\hat{t}) = \mu(\varepsilon t)$ is the channel price, rewritten below as (11.26)

$$\frac{du_i(t)}{dt} = \rho_i \left\{ \frac{\beta_i}{\mu(\varepsilon t)} - \frac{1}{a_i}\left(\frac{1}{y_i(t)} - \Gamma_{i,i} + a_i \right) u_i(t) \right\} \tag{11.26}$$

Now, substituting the time-delayed model (11.2) into (11.26) and including the backward propagation delay, $\hat{\tau}^b = \varepsilon\tau^b$, into the channel price, $\mu(\hat{t}) = \mu(\varepsilon t)$, yields

$$\frac{du_i(t)}{dt} = \rho_i \left\{ \frac{\beta_i}{\mu(\varepsilon(t-\tau^b))} \right.$$
$$\left. - \frac{1}{a_i}\left(\frac{n_{0,i} + \sum_{j\in\mathcal{M}} \Gamma_{i,j} u_j(t-\tau)}{u_i(t-\tau)} - \Gamma_{i,i} + a_i \right) u_i(t) \right\} \tag{11.27}$$

Note that (11.27) and (11.26) are on the "fast" time-scale, t. For the single-link case here, $\tau = \tau_{i,i}$, since there is only one round-trip time delay. The price is adjusted on the "slow" time-scale; let us rewrite (11.27) in this slow time-scale, $\hat{t}$, so that both the channel and link/price dynamics are on the same time-scale. Using $\hat{t} = \varepsilon t$, from (11.27) we obtain

$$\varepsilon \frac{d\hat{u}_i(\hat{t})}{d\hat{t}} = \rho_i \left\{ \frac{\beta_i}{\mu(\hat{t}-\hat{\tau}^b)} \right.$$
$$\left. - \frac{1}{a_i}\left(\frac{n_{0,i} + \sum_j \Gamma_{i,j} \hat{u}_j(\hat{t}-\hat{\tau})}{\hat{u}_i(\hat{t}-\hat{\tau})} - \Gamma_{i,i} + a_i \right) \hat{u}_i(\hat{t}) \right\} \tag{11.28}$$

where we used $\frac{d\hat{t}}{dt} = \varepsilon$ and $\hat{u}$ denotes u on the slow time-scale $\hat{t}$. This form is used to apply standard singular perturbation theory (see Sect. A.9, (A.19)). If we divide both sides of (11.28) by ε, then the $\frac{1}{\varepsilon}$ term on the RHS is a large gain. Thus, the ε term embodies the fast dynamics of the channel algorithm in the slow time-scale.

Let us move now to the price or link algorithm (7.61) and convert it to a continuous-time version. Using the same approximation $\frac{d\mu(\hat{t})}{d\hat{t}} = \mu(\bar{k}+1) - \mu(\bar{k})$, we obtain in the slow time-scale, $\hat{t}$

$$\dot{\mu}(\hat{t}) = \eta\left(\sum_{j=1} \hat{u}_j(\hat{t}) - P_0 \right) \tag{11.29}$$

Finally let us incorporate the forward time delay that acts on the input powers to get

$$\dot{\mu}(\hat{t}) = \eta\left(\sum_{j=1} \hat{u}_j(\hat{t}-\hat{\tau}^f) - P_0 \right) \tag{11.30}$$

As before one can modify (11.28) to eliminate $\hat{u}_i(\hat{t}-\hat{\tau})$ in the denominator by design (see also (11.4) for a similar technique). If we keep a record of past power input, $\hat{u}_i(\hat{t}-\hat{\tau})$, we can then appropriately replace $\hat{u}_i(\hat{t})$ with $\hat{u}_i(\hat{t}-\hat{\tau})$ to obtain

$$\varepsilon\dot{\hat{u}}_i(\hat{t}) = \rho_i\left\{\frac{\beta_i}{\mu(\hat{t}-\hat{\tau}^b)} - \frac{1}{a_i}\left(n_{0,i} + \sum_j \tilde{\Gamma}_{i,j}\hat{u}_j(\hat{t}-\hat{\tau})\right)\right\} \tag{11.31}$$

where $\tilde{\Gamma}_{i,j}$ is $\Gamma_{i,j}$ modified on the diagonal (scaling by a_i).

Thus the closed-loop time-delay system is represented by (11.31) and (11.30). Let us proceed by shifting (11.30) and (11.31) about the equilibrium point (u^*, μ^*). In vector form this gives

$$\dot{x}(\hat{t}) = \eta 1_{\text{row}} z(\hat{t}-\hat{\tau}^f) \tag{11.32}$$

$$\varepsilon\dot{z}(\hat{t}) = \rho\left\{-\beta\frac{x(\hat{t}-\hat{\tau}^b)}{\mu^*(x(\hat{t}-\hat{\tau}^b)+\mu^*)} - \tilde{\Gamma}_a z(\hat{t}-\hat{\tau})\right\} \tag{11.33}$$

where z and x denote in vector form the shifted variables, $z_i = \hat{u}_i - u_i^*$ and $x = \mu - \mu^*$. In the foregoing $\tilde{\Gamma}_a$ is a matrix with elements $\frac{\tilde{\Gamma}_{i,j}}{a_i}$, β is a column matrix with elements β_i, 1_{row} is a row vector with all elements equal to 1, $\rho = \text{diag}(\rho_i)$. Note that $\tilde{\Gamma}_a$ is invertible because of the diagonal dominance condition (see Chap. 7).

Now we can see that (11.32) and (11.33) is written in a standard singular perturbation form. Let

$$h(x) = \tilde{\Gamma}_a^{-1}\beta\left(\frac{-x}{\mu^*(x+\mu^*)}\right) \tag{11.34}$$

be the isolated root of (11.33) and use the coordinate shift

$$\hat{\mathbf{z}}(\hat{t}) = z(\hat{t}-\hat{\tau}^f) - h(x(\hat{t})) \tag{11.35}$$

into (11.32). This yields

$$\dot{x}(\hat{t}) = \eta 1_{\text{row}}\left(\hat{\mathbf{z}}(\hat{t}) + h(x(\hat{t}))\right) \tag{11.36}$$

Next, taking the derivative of (11.35) with respect to time, $\hat{t}$, yields

$$\dot{\hat{\mathbf{z}}}(\hat{t}) = \dot{z}(\hat{t}-\hat{\tau}^f) - \frac{dh(x)}{dx}\dot{x}(\hat{t})$$

and using (11.33) it follows that

$$\varepsilon\dot{\hat{\mathbf{z}}}(\hat{t}) = \rho\left\{-\beta\frac{x(\hat{t}-\hat{\tau})}{\mu^*(x(\hat{t}-\hat{\tau})+\mu^*)} - \tilde{\Gamma}_a z(\hat{t}-\hat{\tau}-\hat{\tau}^f)\right\} - \varepsilon\frac{dh(x)}{dx}\dot{x}(\hat{t})$$

On the right-hand side, using (11.34) and then again (11.35) yields

$$\varepsilon\dot{\hat{\mathbf{z}}} = -\rho\tilde{\Gamma}_a\hat{\mathbf{z}}(\hat{t}-\hat{\tau}) - \varepsilon\frac{dh(x)}{dx}\dot{x}(\hat{t}) \tag{11.37}$$

Finally we rewrite (11.36) and (11.37) as

$$\dot{x}(\hat{t}) = f\big(\hat{\mathbf{z}}(\hat{t}) + h\big(x(\hat{t})\big)\big) \tag{11.38}$$

$$\varepsilon\dot{\hat{\mathbf{z}}}(\hat{t}) = g\big(\hat{\mathbf{z}}(\hat{t} - \hat{\tau})\big) - \varepsilon\frac{dh(x)}{dx} f\big(\hat{\mathbf{z}}(\hat{t}) + h\big(x(\hat{t})\big)\big) \tag{11.39}$$

where

$$f\big(\hat{\mathbf{z}}(\hat{t}) + h\big(x(\hat{t})\big)\big) := \eta 1_{\text{row}}\big(\hat{\mathbf{z}}(\hat{t}) + h\big(x(\hat{t})\big)\big) \tag{11.40}$$

$$g\big(\hat{\mathbf{z}}(\hat{t} - \hat{\tau})\big) := -\rho\tilde{\Gamma}_a\hat{\mathbf{z}}(\hat{t} - \hat{\tau}) \tag{11.41}$$

Thus, the coordinate shifted system to be analyzed is given by (11.38) and (11.39), where x is on the fast and $\hat{\mathbf{z}}$ is on the slow time-scale. Let us introduce the reduced system and the boundary-layer system (see the appendix). These are simpler to analyze and can be used to study the full system (11.38) and (11.39).

To obtain the reduced system, we set $\hat{\mathbf{z}} = 0$ in (11.38), that is, we assume that the fast dynamics converge instantly to steady-state value. This yields

$$\dot{\hat{x}}(\hat{t}) = f\big(h\big(\hat{x}(\hat{t})\big)\big) \tag{11.42}$$

where we denoted its solution by $\hat{x}(\hat{t})$, which is distinct from $x(\hat{t})$, solution of (11.38). Let us use the same notation

$$\hat{\mathbf{z}}\big(\hat{t}(t)\big) = (\hat{\mathbf{z}} \circ \hat{t})(t) := \tilde{\mathbf{z}}(t) \tag{11.43}$$

where $\tilde{\mathbf{z}} = \hat{\mathbf{z}} \circ \hat{t}$, and recalling $\hat{\tau} = \varepsilon\tau$ we get $\hat{\mathbf{z}}(\hat{t} - \hat{\tau}) = \tilde{\mathbf{z}}(t - \tau)$. Towards obtaining the boundary-layer system, we rewrite (11.39) back on the fast time-scale, t, to explicitly account for its fast dynamics. Thus, we substitute (11.43), $\hat{t} = \varepsilon t$ and $\hat{\tau} = \varepsilon\tau$ into (11.39) to get

$$\frac{d\tilde{\mathbf{z}}(t)}{dt} = g\big(\tilde{\mathbf{z}}(t - \tau)\big) - \varepsilon\frac{dh(x(\varepsilon t))}{dx} f\big(\tilde{\mathbf{z}}(t) + h\big(x(\varepsilon t)\big)\big) \tag{11.44}$$

Finally, we set $\varepsilon = 0$, that is we assume that $\hat{t}$ and $x(\hat{t})$ are "frozen" (since they evolve on the slow time-scale). Denoting the solution by $\check{\mathbf{z}}(t)$, we get

$$\frac{d\check{\mathbf{z}}(t)}{dt} = g\big(\check{\mathbf{z}}(t - \tau)\big) \tag{11.45}$$

We can write (11.42) and (11.45) in their explicit forms by using (11.40), (11.41)

$$\dot{\hat{x}}(\hat{t}) = \eta 1_{\text{row}}\tilde{\Gamma}_a^{-1}\beta\left(\frac{-\hat{x}}{\mu^*(\hat{x} + \mu^*)}\right) \tag{11.46}$$

$$\frac{d\check{\mathbf{z}}}{dt} = -\rho\tilde{\Gamma}_a\check{\mathbf{z}}(t - \tau) \tag{11.47}$$

Thus we can see that the reduced system (11.46) is a nonlinear, scalar system with no time delay. On the other hand, the boundary-layer system (11.47) is a linear time-delay system. Since the two systems are decoupled, they are much simpler to study than the full system (11.38) and (11.39). The full system (11.38), (11.39) or (11.32), (11.33), is studied based on these simplified subsystems and a composite Lyapunov function.

11.4.2 Time-Scale Stability Analysis: Point-to-Point Topologies

In this section, we present conditions for stability of the closed-loop system (11.32), (11.33) in the presence of time delays, based on the reduced and boundary-layer systems (11.46) and (11.47). We apply a Lyapunov–Razumikhin stability analysis (see Sect. A.11.2). This Lyapunov–Razumikhin stability analysis does not rely on functionals and the stability conditions are simple in form, but relatively conservative.

The following lemmas are used to prove the main result, Theorem 11.10.

Lemma 11.8 *The boundary-layer system* (11.47) *is exponentially stable if*

$$\tau < \frac{\underline{\sigma}(\rho\tilde{\Gamma}_a + \tilde{\Gamma}_a^T\rho)}{2\sqrt{\bar{\sigma}((\rho\tilde{\Gamma}_a)^2((\rho\tilde{\Gamma}_a)^2)^T)}} \tag{11.48}$$

where $\tilde{\Gamma}_a$ *is defined as in* (11.33), $\rho = \operatorname{diag}(\rho_i)$.

Proof Since the boundary-layer system (11.47) is a linear, time-delay system, we can use Razumikhin theory. From Theorem A.35, given a system of the form

$$\frac{d\check{\mathbf{z}}}{dt} = A_1\check{\mathbf{z}}(t-\tau) \tag{11.49}$$

we ensure asymptotic stability if we satisfy the LMI (A.40), in fact ensuring exponential stability.

For the boundary-layer system, (11.47), $A_1 = -\rho\tilde{\Gamma}_a$. The LMI (A.40) then becomes

$$\begin{pmatrix} -\frac{1}{\tau}[P\rho\tilde{\Gamma}_a + \tilde{\Gamma}_a^T\rho P] + (\hat{\alpha}_0+\hat{\alpha}_1)P & 0 & -P\rho\tilde{\Gamma}_a\rho\tilde{\Gamma}_a \\ 0 & -\hat{\alpha}_0 P & 0 \\ -\tilde{\Gamma}_a^T\rho\tilde{\Gamma}_a^T\rho P & 0 & -\hat{\alpha}_1 P \end{pmatrix} < 0 \tag{11.50}$$

We rewrite (11.50) using the Schur complement, we set $Q = P^{-1}$, and after some manipulation one can obtain

$$-\frac{1}{\tau}[\rho\tilde{\Gamma}_a Q + Q\tilde{\Gamma}_a^T\rho] + (\hat{\alpha}_0+\hat{\alpha}_1)Q + \frac{1}{\hat{\alpha}_1}(\rho\tilde{\Gamma}_a)^2 Q\big((\rho\tilde{\Gamma}_a)^2\big)^T < 0 \tag{11.51}$$

Notice that the second and third terms of (11.51) are positive definite. In addition, notice that there always exists a Q to satisfy (11.51) given a sufficiently small τ. Thus, we know that a solution exists, the only question remaining is the bound on τ that gives such a result.

We can simplify (11.51) by a priori choosing $Q = vI$, where v is a constant multiplier, and I is the identity matrix. Rearranging (11.51), and setting $\hat{\alpha}_0 = 0$ yields

$$\tau\left[\hat{\alpha}_1 I + \frac{1}{\hat{\alpha}_1}(\rho\tilde{\Gamma}_a)^2\left((\rho\tilde{\Gamma}_a)^2\right)^T\right] < \rho\tilde{\Gamma}_a + \tilde{\Gamma}_a^T\rho \tag{11.52}$$

Following rather lengthy manipulations leads to (11.48) as a sufficient condition for (11.52). □

Lemma 11.9 *The reduced system* (11.46) *is exponentially stable if*

$$a_i > \frac{1}{2}\sum_{j\neq i}(\Gamma_{i,j} + \Gamma_{j,i}) \tag{11.53}$$

The following result gives a stability result based on Lyapunov–Razumikhin theorem (see the appendix, Sect. A.11.2).

Theorem 11.10 *For the singularly perturbed system* (11.32) *and* (11.33), *there exists* $\varepsilon^* > 0$ *such that, for* $0 < \varepsilon < \varepsilon^*$, *the origin is exponentially stable if* (11.48) *and* (11.53) *are satisfied.*

Proof We use the singular perturbation approach reviewed in the appendix, Sect. A.9, but modified to handle time delays (see Sect. A.11.2). We prove that the system (11.38) and (11.39) is exponentially stable via a Lyapunov function based on the reduced and boundary-layer systems, (11.46) and (11.47), respectively.

Consider the reduced system (11.46). If (11.53) holds, it can be shown that the Lyapunov function $V(\hat{x}) = \frac{1}{2}\hat{x}^2$ applied to (11.46) satisfies

$$\frac{\partial V}{\partial \hat{x}} f\big(h(\hat{x})\big) \leq -\frac{\eta 1_{\text{row}}\tilde{\Gamma}_a^{-1}\beta}{\mu^*(\mu^* + r_1)}\|\hat{x}\|_2^2 \leq -k_*\|\hat{x}\|_2^2 \tag{11.54}$$

for $\hat{x} \leq r_1$, where $r_1 > -\mu^*$, and $k_* > 0$ is a constant. By (11.53), $1_{\text{row}}\tilde{\Gamma}_a^{-1}\beta > 0$. Note that we can make r_1 arbitrarily large, but (11.54) is satisfied locally for $\hat{x}$.

Consider the boundary-layer system (11.47) and select the Lyapunov function $W(\check{\mathbf{z}}) = \check{\mathbf{z}}^T\check{\mathbf{z}}$. By (11.48) in Lemma 11.8, this is exponential stable and it can be shown that

$$\begin{aligned}\frac{\partial W}{\partial \check{\mathbf{z}}} g\big(\check{\mathbf{z}}(t-\tau)\big) = 2\check{\mathbf{z}}^T\big(-\rho\tilde{\Gamma}_a\check{\mathbf{z}}(t-\tau)\big) \leq -k_{*2}\big\|\check{\mathbf{z}}(t)\big\|_2^2 \\ \left\|\frac{\partial W}{\partial \check{\mathbf{z}}}\right\|_2 = 2\|\check{\mathbf{z}}\|_2\end{aligned} \tag{11.55}$$

for k_{*2} sufficiently small.

Moreover, the following relations hold:

$$\left|f\big(\hat{\mathbf{z}}+h(x)\big)-f\big(h(x)\big)\right| = \|\eta 1_{\text{row}}\hat{\mathbf{z}}\|_2 \le \eta\sqrt{m}\|\hat{\mathbf{z}}\|_2 \tag{11.56}$$

$$\left|f\big(h(x)\big)\right| \le \frac{\eta 1_{\text{row}}\tilde{\Gamma}_a^{-1}\beta}{\mu^*(\mu^*+r_2)}\|x\|_2 \tag{11.57}$$

$$\left\|\frac{\partial h(x)}{\partial x}\right\|_2 \le \left\|\tilde{\Gamma}_a^{-1}\beta\right\|_2 \frac{1}{(\mu^*+r_2)^2} \le k_2 \tag{11.58}$$

$$\left\|h(x)\right\|_2 \le \left\|\tilde{\Gamma}_a^{-1}\beta\right\|_2 \frac{\|x\|_2}{\mu^*(\mu^*+r_2)} \le k_3\|x\|_2 \tag{11.59}$$

for $x \ge r_2$, where $r_2 > -\mu^*$, and $k_2 > 0$ and $k_3 > 0$ are constants. By (11.53), $1_{\text{row}}\tilde{\Gamma}_a^{-1}\beta > 0$. Thus, we have the overall range, $r_2 \le x \le r_1$. Note that r_2 can be selected arbitrarily close to $-\mu^*$. Thus, the inequalities (11.56)–(11.59) are valid locally over x.

We analyze the stability of the full closed-loop system (11.38), (11.39) via a composite Lyapunov function obtained from these two above. This composite Lyapunov function is defined as $\chi(x,\hat{\mathbf{z}}) = V(x) + W(\hat{\mathbf{z}}) = \frac{1}{2}x^2 + \hat{\mathbf{z}}^T\hat{\mathbf{z}}$, hence is quadratic. Thus, we immediately satisfy the first condition in Theorem A.33.

Next we use the composite Lyapunov function $\chi(x,\hat{\mathbf{z}})$ for (11.38) and (11.39), and exploit the Lyapunov inequalities (11.54), (11.55) and Lipschitz properties (11.56)–(11.58). The rest of the proof follows as in Theorem 11.4 [69]. We take the time-derivative of $\chi(x,\hat{\mathbf{z}})$ along the trajectory of the system (11.38) and (11.39),

$$\dot{\chi} = \frac{\partial V}{\partial x} f\big(\hat{\mathbf{z}}+h(x)\big) - \frac{\partial W}{\partial \hat{\mathbf{z}}}\frac{\partial h}{\partial x} f\big(\hat{\mathbf{z}}+h(x)\big) + \frac{1}{\varepsilon}\frac{\partial W}{\partial \hat{\mathbf{z}}} g\big(\hat{\mathbf{z}}(\hat{t}-\hat{\tau})\big) \tag{11.60}$$

where the general functions f, g and h, are defined in (11.40), (11.41), and (11.34), respectively. Next, we use the norm inequalities (11.54), (11.55), and (11.56)–(11.58), applied to each term on the RHS of (11.60). For the first term of (11.60), we apply (11.56), (11.57), and (11.54) to get

$$\frac{\partial V}{\partial x} f\big(\hat{\mathbf{z}}+h(x)\big) \le L_3\|x\|_2\|\hat{\mathbf{z}}\|_2 - c_3\|x\|_2^2 \tag{11.61}$$

where L_3 and c_3 are constants. Using (11.55) the third term of (11.60) can be rewritten as

$$\frac{1}{\varepsilon}\frac{\partial W}{\partial \hat{\mathbf{z}}} g\big(\hat{\mathbf{z}}(\hat{t}-\hat{\tau})\big) \le \frac{-1}{\varepsilon} k_{*2}\|\hat{\mathbf{z}}\|_2^2 \tag{11.62}$$

Finally, we apply (11.55)–(11.58) to the second term of (11.60) to get

$$\frac{\partial W}{\partial \hat{\mathbf{z}}}\frac{\partial h}{\partial x} f\big(\hat{\mathbf{z}}+h(x)\big) \le b_4\|\hat{\mathbf{z}}\|_2^2 + L_4\|x\|_2\|\hat{\mathbf{z}}\|_2 \tag{11.63}$$

where b_4 and L_4 are constants. By substituting (11.61)–(11.63) back into (11.60), we get

$$\dot{\chi} \le - \begin{pmatrix} \|x\|_2 \\ \|\hat{\mathbf{z}}\|_2 \end{pmatrix}^T \begin{pmatrix} \tilde{d}_1 & -\frac{\tilde{d}_5}{2} \\ -\frac{\tilde{d}_5}{2} & (\tilde{d}_3/\varepsilon) - \tilde{d}_4 \end{pmatrix} \begin{pmatrix} \|x\|_2 \\ \|\hat{\mathbf{z}}\|_2 \end{pmatrix} \tag{11.64}$$

where $\tilde{d}_1 = c_3$, $\tilde{d}_3 = k$, $\tilde{d}_4 = b_4$, and $\tilde{d}_5 = L_3 + L_4$. The d_i matrix on the RHS of (11.64) is positive definite if

$$\varepsilon < \frac{4\tilde{d}_1\tilde{d}_3}{4\tilde{d}_1\tilde{d}_4 + \tilde{d}_5^2} = \varepsilon^*$$

Thus, (11.64) satisfies the second condition in Theorem A.33. The system (11.38) and (11.39) is exponentially stable. □

The time-delay bound (11.48) is very simple in form and offers a clear intuition for selecting the design parameters a_i and ρ_i. This time-delay bound (11.48) is inversely proportional to the control gain, ρ_i and is proportional to the design parameters a_i. The trade-off for stability is a slower convergence time for the control algorithms.

11.5 Robustness and Delay Effects Combined

This section extends the results to include uncertainties and multiple time delays, based on [147] and [148]. We first present a perturbed OSNR model with time delays, where we apply a norm-bounded uncertainty framework to the system gains, the input powers and the transmitter noise. This model is presented as (11.68). We then augment the single-link analysis from the previous section to include both additive uncertainty and input multiplicative uncertainties.

11.5.1 OSNR Model with Time Delays and Uncertainties

Until now the OSNR models have assumed that the system gains, $\Gamma_{i,j}$, are constant (see Lemmas 11.1 and 11.2). In fact due to slow parameter drift over long periods of time and changes in amplified spontaneous emission (ASE), they are time-varying. The slow parameter drift occurs due to the aging of network hardware. Uncertainties exist in time delays $\tau_{i,j}^f$ and τ_i^b for $i, j \in \mathcal{M}$ due to external environmental effects on the fibers and system reconfigurations. Fiber lengths may expand or contract due to environmental effects, or the network may be reconfigured without remodeling the network and redesigning all of the control algorithms. Finally, uncertainty can exist in the transmitter noise at the sources due to slow parameter drift. These uncertainties are captured in the following OSNR model.

Slow parameter drift in optical networks over long periods of time may be modeled as additive uncertainty in the Γ matrix. Additive uncertainty accurately models the independent changes that occur to each $\Gamma_{i,j}$ element over time. Let the uncertainty set for the perturbed system $\Gamma_\Delta \in \Omega_\Gamma$ be defined as

$$\Omega_\Gamma = \{\Gamma + \Delta\Gamma \mid \|\Delta\Gamma_{i,\text{row}}\|_2 \le \hat{q}_i,\ \forall i\} \tag{11.65}$$

where Γ is a matrix with elements $\Gamma_{i,j}$, $\Delta\Gamma$ is a matrix with elements $\Delta\Gamma_{i,j}$, $\Delta\Gamma_{i,\text{row}}$ is an $(m \times m)$ matrix with the ith row equal to the ith row of $\Delta\Gamma$ and the remaining terms equal to zero, and $\hat{q}_i$ is an independent uncertainty bound.

Due to system reconfigurations, or external environmental effects on the link lengths, time-delay values in the input signals may change over time. These changes result in the channel powers varying from their predicted values. The actual channel powers and the predicted channel powers are proportional to each other. Thus, we may model the signal uncertainty as an input multiplicative uncertainty. We define the uncertain input, $u_{\Delta j} \in \Omega_j$, where

$$\Omega_j = \{u_j + \delta_j u_j \mid \|\delta_j\|_2 \le \bar{q}_j\}, \quad \forall j \tag{11.66}$$

where $\bar{q}_j$ is the independent uncertainty for input channel j. Note $\bar{q}_j < 1$.

The transmitter noise may drift over time as the hardware ages. Assume that the signal powers change proportionally, modeled via a proportional additive uncertainty and let us define the uncertain transmitter noise, $m_{\Delta,0,i} \in \Omega_{ni}$, where

$$\Omega_{ni} = \{n_{0,i} + \delta_{ni} n_{0,i} \mid \|\delta_{ni}\|_2 \le q_{ni}\}, \quad \forall i \tag{11.67}$$

with q_{ni} being the independent uncertainty bound on the transmitter noise for channel i. Note that $q_{ni} < 1$ since the noise power $m_{0,i}$ is positive.

Then the OSNR model that captures multiple time delays and uncertainties (11.65), (11.66), and (11.67) follows immediately from Lemma 11.1. This perturbed OSNR model for the ith channel in a multi-link network is given as

$$y_{\Delta i}(t) = \frac{u_{\Delta i}(t - \tau_{i,i})}{n_{\Delta,0,i} + \sum_{j \in \mathcal{M}_{\ell_i}} \Gamma_{\Delta i,j} u_{\Delta j}(t - \tau_{i,j})} \tag{11.68}$$

where $\Gamma_\Delta \in \Omega_\Gamma$, $u_{\Delta j} \in \Omega_j$, and $m_{\Delta,0,i} \in \Omega_{ni}$.

The special case of a single link with only additive system uncertainty, i.e. $\bar{q}_j = q_{ni} = 0$ will also be considered. For this situation, define the perturbed system matrix, $\Gamma_\Delta \in \Omega_S$, where

$$\Omega_S = \{\Gamma + \Delta\Gamma \mid \|\Delta\Gamma\|_2 \le \hat{q}\} \tag{11.69}$$

where Γ is a matrix with elements $\Gamma_{i,j}$ defined in (6.15), $\Delta\Gamma$ is a matrix with elements $\Delta\Gamma_{i,j}$, and $\hat{q}$ is an independent uncertainty bound, distinct from $\hat{q}_i$ defined in (11.65). The uncertainty set (11.69) is similar to the uncertainty set (11.65), except the uncertainty bounds in (11.65) apply to each channel output separately, while (11.65) has one independent bound for all output.

11.5.2 Single-Link Analysis

Consider the continuous-time control algorithms developed based on the link algorithm (7.61) and the channel algorithm (7.62). The following analysis applies to a single link with time delays and uncertainties as modeled in (11.68). Conditions that ensure the stability of the closed-loop system are presented.

Using a similar procedure to rewrite (7.61) and (7.62) in their continuous-time forms yields

$$\dot{\mu}(\hat{t}) = \eta\left(\sum_{j=1} u_{\text{in},j} - P_0\right) \tag{11.70}$$

$$\varepsilon\frac{d\hat{u}_i(\hat{t})}{d\hat{t}} = \rho_i\left\{\frac{\beta_i}{\mu_{\text{in},i}} - \frac{1}{a_i}\left(\frac{1}{y_{\text{in},i}(\hat{t})} - \Gamma_{i,i} + a_i\right)\hat{u}_i(\hat{t} - \hat{\tau})\right\} \tag{11.71}$$

where $u_{\text{in},j}$ is the signal power transmitted to the link algorithm, $\mu_{\text{in},i}$ is the channel price transmitted to the source i, and $y_{\text{in},i}$ is the OSNR signal transmitted back to the source i. The "*in*" subscript is used to abstract various signals as input to the control algorithms without the details of the time delays or uncertainties. This allows us to substitute for $u_{\text{in},j}$, $\mu_{\text{in},i}$ and $y_{\text{in},i}$ appropriately later in the analysis. We use this notation throughout the remainder of this chapter.

Consider (11.68) for the single-link case, i.e.,

$$y_{\Delta i}(\hat{t}) = \frac{\hat{u}_{\Delta i}(\hat{t} - \hat{\tau})}{n_{\Delta,0,i} + \sum_{j\in\mathcal{M}_{\ell_i}} \Gamma_{\Delta i,j}\hat{u}_{\Delta j}(\hat{t} - \hat{\tau})} \tag{11.72}$$

where $\hat{u}_{\Delta j} \in \Omega_j$, $m_{\Delta,0,i} \in \Omega_{ni}$, and $\Gamma_\Delta \in \Omega_S$, where Ω_j, Ω_{ni} and Ω_S are defined in (11.66), (11.67), and (11.69), respectively. Note that we use $\Gamma_\Delta \in \Omega_S$ for the single-link case rather than $\Gamma_\Delta \in \Omega_\Gamma$, where Ω_Γ is defined in (11.65).

Substitute (11.72) into (11.71), where $y_{\text{in},i} = y_{\Delta i}(\hat{t})$ and $\mu_{\text{in},i} = \mu(\hat{t} - \hat{\tau}^b)$ to obtain the closed-loop uncertain system with time delays,

$$\begin{aligned}\varepsilon\dot{\hat{u}}_i(\hat{t}) = \rho_i\Bigg\{&\frac{\beta_i}{\mu(\hat{t} - \hat{\tau}^b)} + \left(\frac{\Gamma_{i,i}}{a_i} - 1\right)\hat{u}_i(\hat{t} - \hat{\tau}) \\ &- \frac{1}{a_i}\left(n_{\Delta,0,i} + \sum_{j\in\mathcal{M}_{\ell_i}} \Gamma_{\Delta i,j}\hat{u}_{\Delta j}(\hat{t} - \hat{\tau})\right)\frac{\hat{u}_i(\hat{t} - \hat{\tau})}{\hat{u}_{\Delta i}(\hat{t} - \hat{\tau})}\Bigg\}\end{aligned} \tag{11.73}$$

In (11.70), let $u_{\text{in},j} = (1 + \delta_j)\hat{u}_j(\hat{t} - \hat{\tau}^f)$ to get

$$\dot{\mu}(\hat{t}) = \eta\left(\sum_{j\in\mathcal{M}} (1 + \delta_j)\hat{u}_j(\hat{t} - \hat{\tau}^f) - P_0\right) \tag{11.74}$$

Thus, (11.73) and (11.74) represent the uncertain closed-loop system with time delay for a single link. Next, we use the fact that $\hat{u}_{\Delta j} = \hat{u}_j + \delta_j\hat{u}_j = (1 + \delta_j)\hat{u}_j$ since

$\hat{u}_j$ is scalar and make the following substitution, which we state as the following assumption.

Assumption 11.2 *In the denominator of the last term of* (11.73), *we replace the uncertainty* $(1+\delta_i)$ *in* $\hat{u}_{\Delta i}$ *with the worst case situation* $(1-\bar{q}_i)$.

This Assumption 11.2 can be justified as follows: as a_i increases in (11.73), the magnitude of the last term approaches zero. A very large a_i essentially eliminates the *OSNR* feedback, which results in a stable, but suboptimal open-loop system response. The smaller the a_i value, the larger the *OSNR* feedback, and the less stable the close-loop system. However, we are still interested in minimizing the a_i values, since the transient response becomes faster. Thus, replacing $(1+\delta_i)$ by $(1-\bar{q}_i)$ in the denominator of the last term in (11.73), where $(1+\delta_i)$ is proportional to a_i, produces a worst case stability condition. Note that this substitution replaces a system with nonlinear uncertainties with a system with linear uncertainties. Thus, (11.73) becomes

$$\varepsilon\dot{\hat{u}}_i(\hat{t}) = \rho_i\left\{\frac{\beta_i}{\mu(\hat{t}-\hat{\tau}^b)} - \frac{n_{\Delta,0,i}}{(1-\bar{q}_i)a_i} - \sum_{j\in\mathcal{M}_{\ell_i}} \bar{\Gamma}_{\Delta i,j}(1+\delta_j)\hat{u}_j(\hat{t}-\hat{\tau})\right\} \tag{11.75}$$

where $\bar{\Gamma}_{\Delta i,j} = \bar{\Gamma}_{i,j} + \Delta\Gamma_{i,j}/(1-\bar{q}_i)a_i$, such that

$$\bar{\Gamma}_{i,j} = \begin{cases} \frac{1}{1-\bar{q}_i}, & i=j \\ \frac{\Gamma_{i,j}}{(1-\bar{q}_i)a_i}, & i\neq j \end{cases} \tag{11.76}$$

Rewrite (11.75) and (11.74) using $x=\mu-\mu^*$ and $z_i=\hat{u}_i-u_i^*$, where μ^* and u_i^* are the equilibrium points of (11.75) and (11.74)

$$\dot{x} = \eta 1_{\text{row}}\big(I+\text{diag}(\delta_j)\big)z\big(\hat{t}-\hat{\tau}^f\big) \tag{11.77}$$

$$\varepsilon\dot{z} = \rho\left\{\beta\frac{-x(\hat{t}-\hat{\tau}^b)}{\mu^*(x(\hat{t}-\hat{\tau}^b)+\mu^*)} - \bar{\Gamma}_\Delta\big(I+\text{diag}(\delta_j)\big)z(\hat{t}-\hat{\tau})\right\} \tag{11.78}$$

where z is a vector with elements z_i, $\rho=\text{diag}(\rho_i)$, β is a column vector with elements β_i, and 1_{row} is a row vector of 1's. Equations (11.77) and (11.78) represent the perturbed, closed-loop system with time delay shifted about the equilibrium points (μ, u^*). Note that $m_{\Delta,0,i}$ is no longer present.

Next, we rewrite (11.77) and (11.78), using the coordinate shift,

$$\hat{\mathbf{z}}_1(\hat{t}) = z\big(\hat{t}-\hat{\tau}^f\big) - h_1\big(x(\hat{t})\big) \tag{11.79}$$

where the function $h_1(x(\hat{t}))$ is the isolated root of the RHS of (11.78),

$$h_1\big(x(\hat{t})\big) = \big(I+\text{diag}(\delta_i)\big)^{-1}\bar{\Gamma}_\Delta^{-1}\beta\frac{-x(\hat{t})}{\mu^*(x(\hat{t})+\mu^*)} \tag{11.80}$$

and $I + \text{diag}(\delta_i)$ is invertible, $\bar{\Gamma}_\Delta = \bar{\Gamma} + \text{diag}(\frac{1}{(1-\bar{q}_i)a_i})\Delta\Gamma$, $\bar{\Gamma}$ is a matrix with elements defined in (11.76). For now, we assume that $\bar{\Gamma}_\Delta$ is invertible. In fact, we impose this condition later in Lemma 11.9. Proceeding as in the previous section we can obtain the reduced and the boundary-layers systems as

$$\dot{\hat{x}}_1(\hat{t}) = \eta 1_{\text{row}} \bar{\Gamma}_\Delta^{-1} \beta \frac{-\hat{x}_1(\hat{t})}{\mu^*(\hat{x}_1(\hat{t}) + \mu^*)} \tag{11.81}$$

$$\frac{d\check{\mathbf{z}}_1}{dt} = -\rho \bar{\Gamma}_\Delta \big(I + \text{diag}(\delta_i)\big) \check{\mathbf{z}}_1(t-\tau) \tag{11.82}$$

The reduced system (11.81) is scalar and nonlinear with no time delay. Notice that the input multiplicative uncertainty terms have canceled out. The boundary-layer system (11.82) is linear with multiple time delays. The two systems are decoupled from each other.

The following lemma presents the LMI that ensures exponential stability for the boundary-layer system (11.82). A detailed proof can be found in [147].

Lemma 11.11 *The boundary-layer system* (11.82) *is exponentially stable if there exist symmetric, positive definite matrices P and S_1 such that*

$$\begin{pmatrix} \bar{H}_1 & \bar{H}_2 \\ \bar{H}_2^T & \bar{H}_3 \end{pmatrix} < 0 \tag{11.83}$$

where $\bar{\lambda}_0 > 0$, $k_i > 0$, $k_1 > 0$ with $\bar{k}_i < k_1$ for all i,

$$\bar{H}_1 = \begin{pmatrix} M_{n2} & -PA_0^2 \\ -(A_0^T)^2 P & -S_1 + \bar{\lambda}_0 (G^*)^T G^* \end{pmatrix}$$

$$\bar{H}_2 = \begin{pmatrix} PE^* & \frac{1}{\tau}(G^*)^T & -PA_0E^* \\ 0 & -A_0^T(G^*)^T & \bar{\lambda}_0 (G^*)^T D^* \end{pmatrix}$$

$$\bar{H}_3 = \begin{pmatrix} -I & (D^*)^T & 0 \\ D^* & -I & -G^*E^* \\ 0 & -(E^*)^T(G^*)^T & -\bar{\lambda}_0(I - (D^*)^T D^*) \end{pmatrix}$$

$A_0 = -\rho\bar{\Gamma}$, $E_2 = -\rho \, \text{diag}(\frac{1}{(1-\bar{q}_i)a_i})$ *and*

$$M_{n2} = \frac{1}{\tau}(PA_0 + A_0^T P) + S_1, \qquad E^* = \big(\sqrt{\hat{q}} E_2 k_1 A_0 \, \text{diag}(\sqrt{\bar{q}_i}\bar{k}_i)\big)$$

$$G^* = \left(\frac{\sqrt{\hat{q}}}{k_1} I \, \text{diag}\left(\frac{\sqrt{\bar{q}_i}}{\bar{k}_i}\right)\right), \qquad D^* = \begin{pmatrix} 0 & \frac{\sqrt{\hat{q}}}{k_1} \text{diag}(\sqrt{\bar{q}_i}) \\ 0 & 0 \end{pmatrix}$$

One can remark that if we set $\hat{q} = 0$ and $\bar{q}_i = 0$ (no uncertainty), and we select $S_1 = \hat{\alpha} P$, where $\hat{\alpha}$ is a scalar variable, we can recover the LMI used to derive the time-delay bound (11.48).

The following lemma ensures the stability of the reduced system (11.81) via the adjustment of the channel parameter a_i.

Lemma 11.12 *The uncertain reduced system* (11.81) *is exponentially stable if*

$$a_i > \frac{1}{2}\sum_{j\neq i}(\Gamma_{i,j} + \Gamma_{j,i}) + \hat{q}, \quad \forall i \tag{11.84}$$

One can note that the bound on the channel uncertainty term $(1 - \bar{q}_i)$ is not present. Only the bound on the system matrix uncertainty $\hat{q}$ is present. If we eliminate the uncertainty, i.e. $\hat{q} = 0$, we exactly recover (11.53).

We now state the main stability theorem for the single-link case based on Lemmas 11.11 and 11.12.

Theorem 11.13 *Consider the uncertain, singularly perturbed system* (11.77) *and* (11.78), *with uncertainty sets defined in* (11.65), (11.66), *and* (11.67). *There exists an* $\varepsilon^* > 0$ *such that for* $0 < \varepsilon < \varepsilon^*$ *the origin is asymptotically stable if* (11.83) *and* (11.84) *hold.*

The proof of asymptotic stability can be done using a composite Lyapunov functional based on the reduced and boundary-layer system functionals. For details the reader is referred to [147] and [144].

11.6 Notes

In this chapter we presented stability conditions for the closed-loop system in the presence of time delays based on frequency domain and on Lyapunov–Razumikhin analysis. The time-delay bound (11.48) is very simple in form. A Lyapunov–Krasovskii analysis which utilizes Lyapunov functionals could also be applied. Lyapunov functionals significantly complicate the system analysis, and they produce more complicated stability conditions. However, the Lypaunov–Krasovskii analysis produces less conservative stability conditions than the Lyapunov–Razumikhin analysis. The reader is referred to [150] for such an approach.

Chapter 12
Games for Routing and Path Coloring

Abstract This chapter provides and overview of routing and path coloring problems in all-optical networks as noncooperative games. We focus on oblivious payment functions, that is, functions that charge a player according to its own strategy only. We review results on the relation between such games and online routing and path coloring. In particular, these results show that the Price of Anarchy of such games is lower-bounded by, and in several cases precisely equal to, the competitive ratio of appropriate modifications of the First Fit algorithm.

12.1 Introduction

In this chapter we present other types of game formulations in optical networks. One of the problems that can be considered in a game-theory framework has to do with the problem of routing a number of communication requests in WDM all-optical networks. Specifically, we focus on routing and wavelength assignment as a class of problems. Recall that communication requests are carried out by assigning a path in the network (routing) as well as a transmission wavelength (or color). Wavelength division multiplexing (WDM) allows several requests to be routed through the same link(s) of the network, and carry them out simultaneously by assigning a different wavelength to each request [165, 167].

Given an optical network topology and a set of communication requests another set of questions arise in addition to optimizing channel performance (e.g. OSNR) as treated until now. In case when the specific routing of the requests is not given, the routing and path coloring Routing and Path Coloring (RPC) problem asks for both a routing and a wavelength (color) assignment minimizing the number of colors. See Fig. 12.1 for a request routed over two different paths. This RPC problem is also called the RWA assignment problem [125, 126]. On the other hand, if the routing of the requests is given, the Path Coloring (PC) or wavelength assignment (WA) problem asks for the minimum number of colors (wavelengths) required such that requests sharing a common link are assigned different colors. Other optimization questions can be stated by introducing additional parameters and constraints [165]. A large number of results have been concentrated on the complexity and approximability questions for these optimization problems [39, 51, 99, 166]. A survey of early results can be found in [52] and references therein.

L. Pavel, *Game Theory for Control of Optical Networks*,
Static & Dynamic Game Theory: Foundations & Applications,
DOI 10.1007/978-0-8176-8322-1_12,

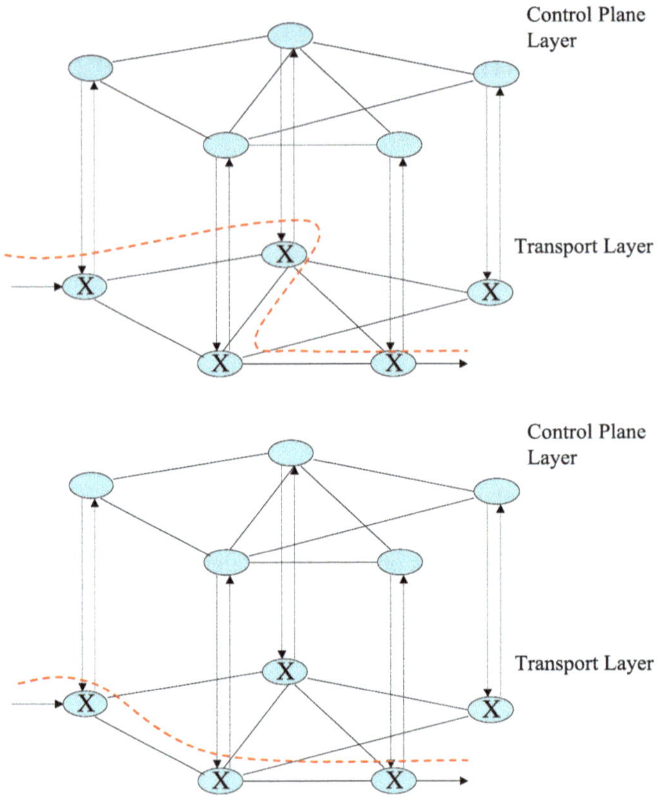

Fig. 12.1 Network layers diagram

A recent research direction in this context considers game-theoretic criteria [73, 130, 160, 162]. As illustrated at various points in this monograph, one can formulate a noncooperative game between independent players, herein as a model of an optimization problem that arises in the context of routing. These players have their own objectives and they do not necessarily have to follow a centralized protocol, or they can manipulate this protocol (e.g. by providing false information) for their own goals.

In the rest of this chapter we discuss approaches to the PC (WA) and RPC (RWA) problems in all-optical networks modeled as noncooperative games. In the previous chapters each channel was a player in a game and the games considered were in the class of continuous action games with coupled utilities and coupled constraints. Herein we formulate games where each routing request (pair of source-destination nodes) is a player, and a strategy of a player consists of a path from the source to the destination and a wavelength/frequency (color), hence in the class of finite strategy games (see Chap. 4). Such an approach is presented in [53]. Other approaches are those in which the ISPs act as brokers or players between the TSP and the end user [162].

Since each communication request is considered a player, an individual cost or payment function is associated as follows: each player is charged a cost depending on the (routing and color) choices of all players (including her own choices).

As in a standard game setup, given a set of choices for all players we say that the game is in equilibrium if no player can decrease her own cost by changing her choices, hence we talk of a Nash equilibrium.

Although Nash has shown that each noncooperative game has a mixed Nash equilibrium, the existence of a pure one is an open question for many games, as mentioned in Chap. 4. Moreover, due to the selfish behavior of the players, such a pure equilibrium does not necessarily optimize a global objective goal, i.e., it is not socially optimal. Such a social goal is also known as the social welfare cost that we mentioned in Chap. 9. For the problems studied in this chapter, given a set of requests, the social cost can be defined as the number of colors needed for (routing and) coloring these requests.

The efficiency of Nash equilibria is measured by the Price of Anarchy which is defined as the ratio of the social cost of the worst Nash equilibrium over the optimal centralized solution [73], and reflects the loss in the global performance due to lack of coordination between players.

The chapter is organized as follows: In the next section we describe the formal model and the notation used, following [28, 86]. Then we examine the relation between the solutions obtained by online and offline algorithms for PC and RPC and the Nash equilibria for the corresponding noncooperative game noncooperative games. We review some results for selfish PC and RPC, respectively, based on payment functions that yield Nash equilibria, as well as upper and lower bounds for the Price of Anarchy in both cases.

12.2 Game-Theoretic Model

Consider an optical network modeled as an undirected graph $G = (N_g, E_g)$, where nodes in N_g represent sites (OXC/OADM), $|N_g| = N$ and undirected edges in E_g are bidirectional optical fiber links between the sites (see also Fig. 12.2). Recall from Chap. 6 that communication is realized by using different wavelengths (colors) along the optical fiber lines (edges).

Assume that a set of communication requests R is given where each request r is a source-destination pair of nodes of G, i.e., $r = (x, y)$, $x, y \in N_g$.

A point-to-point communication requires to establish a uniquely colored path between the two nodes, whose color (wavelength) is different from the colors (wavelengths) of all other paths sharing one of it sedges (no color collisions). This is a version of the graph coloring problem [99].

We consider that a set of colors (wavelengths) $\{\lambda_1, \ldots, \lambda_k\}$ is represented as integers in the interval $C = [1, \ldots, k]$, where k is the maximum number of available colors. If we were to relate this to the number of channels m used in previous chapters we would have $m \le k$. Without loss of generality assume that the number of colors to be no more than $|R|$, i.e., $k \le |R|$.

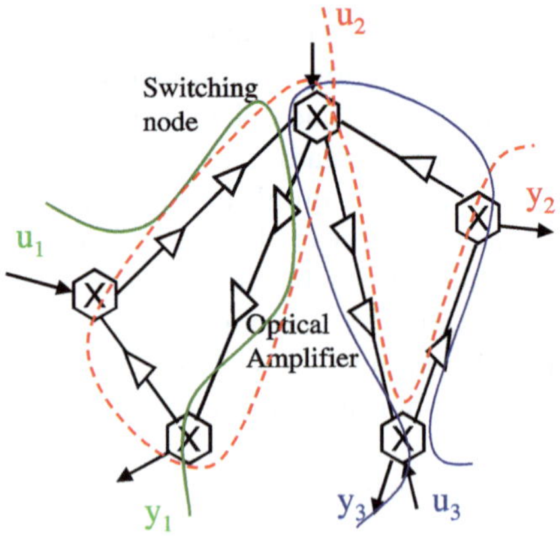

Fig. 12.2 Network diagram

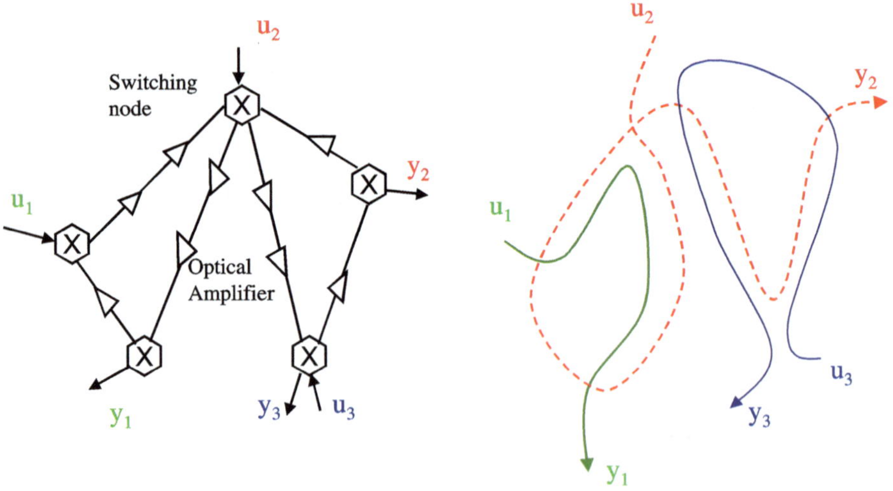

Fig. 12.3 Network graph diagram (1)

A solution to the request is a set of $|R|$ paths of G, denoted by $\mathcal{P}_i$, $i = 1, \dots, |R|$, and a set of wavelengths (colors), denoted by λ_i, $i = 1, \dots, |R|$, each wavelength associated with a path (see Figs. 12.2, 12.3 and 12.4). By a path between two nodes x, y we understand a set of consecutive edges that begin in node x and end in node y. If $\mathcal{P}_i \cap \mathcal{P}_j \neq$ for $i \neq j$ then $\lambda_i \neq \lambda_j$. This represents the assumption that if any two requests are routed through edge-intersecting paths, i.e., they use the same optical link at some point along their paths, then they must use different wavelengths (colors) (no color collisions).

In the case when the actual routing of requests in R is given in advance (predetermined), this means we can simply consider that a set of paths P is given instead of R and only the wavelengths (colors) have to be decided. Therefore, an instance of the RPC problem is denoted by (G, R), while an instance of the PC problem is

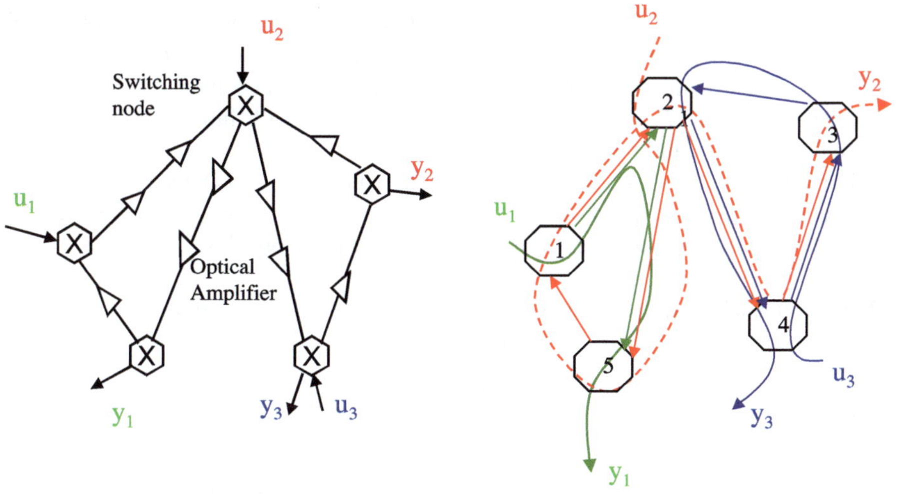

Graph G(Ng, Eg) |Ng| =5, |Eg| =6

Nodes/Vertices =OXC, Edges = links (dyn. MIMO)

Fig. 12.4 Network graph diagram (2)

denoted by (G, P), and in this last case the players only have to choose a color for their paths.

Consider a selfish or noncooperative RPC (PC) game on G where each player i issues a request r_i (a path, respectively), hence we identify a player with a request. A strategy σ_i for player i is a pair $\sigma_i = (\mathcal{P}_i, \lambda_i)$, where $\mathcal{P}_i$ is a simple path connecting the endpoints of r_i and λ_i is a color assigned to $\mathcal{P}_i$. In the case of PC game this strategy is just λ_i. Let S_i denote all possible strategies of player i. Note that for each player these possible strategies depend on the topology of graph G and the number of colors allowed.

Under the assumption that the number of colors is restricted to be no more than $|R|$, this means that there is a finite number of strategies for each player. Thus a game with $|R|$ players is defined by a finite set of strategies S_i, $i = 1, \ldots, |R|$.

Towards formulating the game, for each player i we define a payoff function (utility function) $\mathcal{U}_i$, $U_i : S_1 \times \cdots \times S_{|R|} \to N$, function that herein maps $S_1 \times \cdots \times S_{|R|}$ to the integers. Such a payoff (utility) is typically associated with a cost or payment function J_i, $J_i : S_1 \times \cdots \times S_{|R|} \to N$, and here we consider that this done via the relationship $\mathcal{U}_i = -J_i$, $i = 1, \ldots, |R|$.

The elements of $S_1 \times \cdots \times S_{|R|}$ are called action combinations or states [28]. Note that unlike the setup in the previous chapter, e.g., Chap. 7, herein the strategies are finite as in Chap. 3 and mapping is to the integers (unlike a continuous action set Ω_i for games in Chap. 7).

From now on, for simplicity we consider games where all players have the same cost function or payment function, J. Also, herein the graph plays an important role so instead of the set of players we use the graph G in the game notation (see below).

Let S-RPC and S-PC denote the class of Selfish-RPC and class of Selfish-PC games (pre-determined routing), respectively. Then by $\mathcal{G}(G, R, J)$ we denote a *game in S-RPC* with input graph G, set of requests R, and payment function J. Similarly, by $\mathcal{G}(G, P, J)$ we denote a game in S-PC with input graph G, set of routed requests P, and payment function J.

For a game $\mathcal{G}(G, R, J)$ (and similarly for a game $\mathcal{G}(G, P, J)$), let a *pure-strategy* profile, or simply strategy profile, be defined as $\sigma = \{\sigma_1, \dots, \sigma_{|R|}\}$, one strategy for each player. Let $\mathcal{S}$ denote the set of all possible strategy profiles.

Definition 12.1 A (pure) strategy profile $\sigma^* \in \mathcal{S}$ is a *pure-strategy Nash Equilibrium* (NE) if for each player i we have

$$J(\sigma_1^*, \dots, \sigma_i^*, \dots, \sigma_{|R|}^*) \leq J(\sigma_1^*, \dots, \sigma_i', \dots, \sigma_{|R|}^*)$$

for any strategy $\sigma_i' \in S_i$.

A game may not have pure-strategy Nash equilibria but if we extend the game to include as mixed-strategies all possible distributions on S_i, then a mixed-strategy Nash equilibrium (NE) is always guaranteed to exist [95] (see Theorem 3.15 in Chap. 3). In order to look at the efficiency (performance) of an NE, let $NE(\mathcal{G})$ denote the set of all NEs, $NE(\mathcal{G}) \subset \mathcal{S}$.

Definition 12.2 The *social cost* $J_s(\sigma)$ of a strategy profile $\sigma \in \mathcal{S}$ is defined as the number of colors used for (routing and) coloring, if no color collisions appear; otherwise $J_s(\sigma) = \infty$.

Let J_s^* denote the socially optimal cost, that is,

$$J_s^* = \min_{\sigma \in \mathcal{S}} J_s(\sigma)$$

where $\mathcal{S}$ is the set of possible strategy profiles. Note that J_s^* coincides with the cost of an optimal solution of the corresponding RPC (PC) problem.

Now consider an NE point, $\sigma^* \in NE(\mathcal{G})$, where $NE(\mathcal{G})$ denotes the set of all NEs.

Definition 12.3 The *Price of Anarchy* (PoA) of a game $\mathcal{G}(G, R, J)$ (or $\mathcal{G}(G, P, J)$) is defined as the worst-case number of colors used in a NE (social cost) divided by J_s^*, that is,

$$\textit{Price of Anarchy (PoA)} = \frac{\max_{\sigma^* \in NE(\mathcal{G})} J_s(\sigma^*)}{J_s^*}$$

The *Price of Stability* (PoS) of a game $\mathcal{G}(G, R, J)$ (or $\mathcal{G}(G, P, J)$) is defined as the best-case number of colors used in a NE (social cost) over J_s^*, that is,

$$\textit{Price of Stability (PoS)} = \frac{\min_{\sigma^* \in NE(\mathcal{G})} J_s(\sigma^*)}{J_s^*}$$

The assumption of being oblivious collision-free is typically used as defined next with respect to the cost or payment function.

Definition 12.4 Consider a (routing and) path coloring game $\mathcal{G}(G, R, J)$ (or $\mathcal{G}(G, P, J)$). A cost or payment function J is called *oblivious collision-free* if:

(a) it guarantees that in a Nash Equilibrium (NE) no color collisions occur, and
(b) it charges a player (who does not collide with other players) according to the player's own strategy only.

Note that condition (a) can be satisfied if one charges a very large amount to players that use the same color and share links of the network. For simplicity the condition that guarantees collision-free Nash Equilibria is omitted from the descriptions of the payment functions, since functions we consider here are oblivious collision-free.

Let us observe that for any instance of S-RPC with oblivious collision-free payment function, if σ^* is a NE, then the following holds:

$$J_s(\sigma^*) \leq |R|$$

Similarly for any instance of S-PC with oblivious collision-free payment function, if σ^* is a NE, then the following hold:

$$J_s(\sigma^*) \leq |P|$$

Thus

$$PoA \leq \frac{|R|}{J_s^*}$$

and, respectively,

$$PoA \leq \frac{|P|}{J_s^*}$$

in the S-PC case.

12.3 Cost Functions and PoA

As in a typical game setup, the question we are interested in is whether Nash equilibria exist for such selfish RPC (PC) games, and if so, what is their performance (efficiency). In this section we present an approach based on [28, 42] in which selfish or noncooperative players can configure their communications so as to minimize the cost paid for the service.

As in the previous section we model an optical network as an undirected graph $G = (N_g, E_g)$ where nodes in N_g represent sites, and undirected edges in E_g represent bidirectional optical fiber links between the sites. During a series of time steps, at each of which only one player can move to a better configuration, a Nash equilibrium is eventually reached. In such a setting, we are interested in suitable payment functions covering the network costs that actually induce Nash equilibria (NE) with the best possible efficiency (lowest Price of Anarchy PoA). We focus on the classical scenario of minimizing the total number of used wavelengths (see social cost J_s defined above), that is, the approach of optimizing the optical spectrum.

Definition 12.5 A game $\mathcal{G}(G, R, J)$ is said to be *convergent* if, starting from a generic configuration, and at each stage letting a player to move to a better configuration, a Nash equilibrium (NE) is always reached.

Notice that, conversely, a non-convergent game might admit a Nash equilibrium.

As before recall that for a $r = (x, y)$ request (player) the utility function is $\mathcal{U} = -J(x, y)$, where J is the individual cost function (payment function), which is assumed the same for all players/requests. In what follows this cost or payment function J will be associated with a pricing function f that is introduced below. Let λ_r or $\lambda(x, y)$ denote the color (wavelength) used by the request $r = (x, y)$.

The following assumption is made to model the scenario of minimizing J_s, i.e., the total number of colors used: for each color (wavelength) λ_r there exists an associated (positive) cost (pricing), i.e., there exists a non-decreasing pricing function $f : C \to R_+$ associating a (positive) cost to every color λ_r or. This setup of using the pricing function f models in fact the increasing cost incurred by a network provider when implementing a routing that uses up to a given wavelength.

Specifically, we consider four possible (cost) payment functions [28]:

1. $J_1(x, y) = f(\lambda(x, y))$ or $J_1(r) = f(\lambda_r)$
   This means that for a request $r = (x, y)$ the amount charged (payment function) is the price of the color that request uses to communicate, price that goes according to f.
2. $J_2(x, y) = \max_{e \in R(x,y)} \max_{(a,b) \in R | e \in R(a,b)} f(\lambda(a, b))$
   This means that a request $r = (x, y)$ pays for the maximum price of all colors used by any other request sharing an edge with the path used by (x, y).
3. $J_3(x, y) = \sum_{e \in R(x,y)} \max_{(a,b) \in R | e \in R(a,b)} f(\lambda(a, b))$
   This means that a request $r = (x, y)$ pays for the sum over all the edges belonging to the used path of the maximum price of the colors through the edge.
4. $J_4(x, y) = \sum_{e \in R(x,y)} \max_{(a,b) \in R | e \in R(a,b)} \frac{f(\lambda(a,b))}{|(a,b) \in R | e \in R(a,b)|}$
   This means that a request $r = (x, y)$ pays for the sum over all the edges belonging to the used path of the maximum price of the colors through the edge, divided by the number of request sharing the edge.

These natural cost functions are used as building blocks for the definition of the players' cost or payment functions. The combination of this introduced pricing functions f with the strategies of maximization or of summation leads to a number of possible cost or payment functions J.

In all these cases, since the pricing function f is non-decreasing, players have an incentive to choose small colors so as to possibly minimize the overall number of used colors. Unfortunately, the results of [26–28] show that these cost or payment functions either are not convergent or yield the worst possible Price of Anarchy (PoA), i.e., they converge to an NE equilibrium in which each player uses a different wavelength (color).

In fact only two of these cost (payment) functions guarantee convergence to a pure NE (see results below). These are the cases where each player pays for its own color hence J_1, and where it pays for the maximum color used by any other overlapping player J_2, respectively. As for performance, the PoA is as high as $|R|$ even for rings. More precisely, the following theorems hold (for proofs see [28]).

Theorem 12.6 (See [28]) *The game $\mathcal{G}(G, R, J_1)$ converges to a Nash equilibrium (NE) in at most $|R|^2$ moves.*

This result shows that a NE is reached in polynomial time. However, the payment function J_1 is unable in general to guarantee good performance. As shown by the following theorem, there exist graphs G for which the Price of Anarchy can be very high.

Theorem 12.7 (See [28]) *For any pricing function f, there exist G and R such that the Price of Anarchy of game $\mathcal{G}(G, R, J_1)$ is PoA $= |R|$.*

While negative, this result can be justified as follows. Given a routing, finding a minimum assignment of wavelengths to satisfy the paths in the routing is equivalent to finding a minimum coloring on the graph induced by the routing. It has been shown that the problem of coloring a graph $G = (N_g, E_g)$ cannot be approximated within $|N_g|^{1/7}$ [23], and since an NE achieved in polynomial time is an approximated solution to this graph coloring, this implies that the Price of Anarchy cannot be better than $|R|^{1/7}$.

For the second type of cost function J_2 the following two results hold.

Theorem 12.8 (See [28]) *The game $\mathcal{G}(G, R, J_2)$ converges to a Nash equilibrium (NE) in a finite number of moves.*

Theorem 12.9 (See [28]) *For any pricing function f, there exist G and R such that the Price of Anarchy of game $\mathcal{G}(G, R, J_2)$ is PoA $= |R|$.*

For the last two types of cost function, J_3 and J_4, it can be shown that the associated games do not even converge in general [28].

Thus the next logical step is to restrict the graph topology, by considering chains (point-to-point topologies), trees, or rings, or alternatively restricting the communication patterns. Indeed the following improved results can be shown.

Theorem 12.10 (See [42]) *There exist cost or payment functions inducing convergent games with a Price of Anarchy (PoA) of* 25.72 *in chains,* 51.44 *in rings and* $O(\log(|R|))$ *in trees, all converging in $|R|^2$ steps.*

Finally, it is possible to improve the above result for rings and chains by forcing the agents to simulate the behavior of the online algorithm proposed by Slusarek [140]. In particular, the following theorem holds.

Theorem 12.11 (See [42]) *There exist cost or payment functions inducing convergent games with a Price of Anarchy (PoA) of* 6 *in rings and* 3 *in chains.*

Another aspect that can be considered is such a RPC game formulation is related to the different information levels of local knowledge that players may have for computing their payments [28]. In the *complete level of information* each player knows all other players' routing and coloring strategies. In the *intermediate level of information* each player only knows which colors are used on any edge of the network, while in the *minimal level of information* each player knows which colors are used only on edges along paths that the player can choose. For the complete level it can be shown that the PoA is 1 in chains and 2 in rings, under cost or payment functions specifically constructed according to the corresponding algorithms [28].

The existence of Nash equilibria and the complexity of recognizing and computing a Nash equilibrium (NE) for selfish RPC under several payment functions are considered also in [53]. These results indicate that recognizing a Nash equilibrium can be done in polynomial time, when each player pays for its own color, when it pays for the maximum color used by any other overlapping player and when it pays for the most loaded edge that it uses. On the other hand, when the player pays for all the different colors appearing along its path, recognizing a Nash equilibrium is NP-complete.

12.4 Solution to Online S-RPC as Nash Equilibria

Consider now an *online* version of RPC problem whereby requests arrive as an ordered sequence $\mathbf{R} = \{r_1, r_2, \ldots, r_{|R|}\}$. Such an online instance of RPC is denoted by $(G, \mathbf{R})$. Similarly, an instance of online PC is denoted by $(G, \mathbf{P})$, where $\mathbf{P}$ is a sequence of paths ordered by arrival time. In this section we review results that explore the relation of Nash equilibria (NE) for S-PC and S-RPC games and various oblivious collision-free payment functions, to the solutions obtained by *online* and *offline* algorithms for $(G, \mathbf{R})$.

Upon arrival of a request r_i, $r_i \in \mathbf{R}$ an online algorithm decides a path $\mathcal{P}_i$ and a color assignment λ_i to r_i so that no color collisions appear on any edge of paths that are already colored (that is, corresponding to requests r_j with $j < i$).

We assume that the algorithm has no knowledge of requests that are going to appear later (that is, requests r_j with $j > i$). The objective is to minimize the number of colors used.

The simplest online algorithm for PC is the *First-Fit algorithm* or *greedy algorithm* [126]. This algorithm colors each request r_i with the smallest available color, provided that no color collisions occur. For online RPC an appropriate version is the following version of First-Fit: a path and color for request r_i is chose in such a way that no color collisions occur and the color assigned to r_i is the minimum possible.

In the context of a setup where a cost or payment function is specified also, the following useful generalization of First-Fit for RPC can be defined. Consider a cost function J which specifies a cost for each path and wavelength (color) assignment $(\mathcal{P}, \lambda)$ to a request r_i, taking into account the path and color assignment to previous requests r_j, $j < i$. Then, First-Fit with criterion J (or $FF(J)$ for short) assigns to each request r_i the path $\mathcal{P}_i$ and color λ_i that minimize $J(r_i, \mathcal{P}, \lambda)$, assuming that ties are broken arbitrarily. Notice that here we have added the two arguments $\mathcal{P}, \lambda$ in addition to r in the notation of the payment function to indicate this dependency.

For example, the standard First-Fit for RPC described above can be seen as $FF(J)$, where $J(r_i, \mathcal{P}, \lambda) = \lambda$ if $\mathcal{P}$ does not overlap with any path of color λ, otherwise $J(r, \mathcal{P}, \lambda) = \infty$. A similar generalization of First-Fit can be defined for PC [86]. Note that in the above descriptions, formally for each payment function J, the path-color (or just wavelength (color)) assignment for the previous requests r_j, $j < i$, should also appear as argument of function J.

The relation between selfish routing and coloring (S-RPC) games and the corresponding *online* (centralized) RPC problems is shown in the following two results.

Lemma 12.12 (See [86]) *Consider a game $\mathcal{G}(G, R, J)$ in S-RPC (S-PC) where J is an oblivious collision-free payment function. For any ordering* $\mathbf{R}$ *of R, an execution of $FF(J)$ algorithm on (G, R) gives a strategy profile which is a Nash Equilibrium (NE) for $\mathcal{G}(G, R, J)$.*

Proof Consider the path-color assignment obtained by an execution of $FF(J)$ on (G, R). By definition, a request r_i cannot be assigned a path-color combination of lower cost unilaterally, otherwise $FF(J)$ would have chosen that path-color combination for r_i (see above description of First-Fit algorithm). This can be also justified as follows: if such a different assignment were possible then it does not cause color collisions with respect to the path-color assignment of all other requests. Therefore, it certainly does not cause any color collision with respect to previous requests r_j, $j < i$; hence, upon arrival of r_i, $FF(J)$ would have chosen this lower cost assignment. □

Lemma 12.13 (See [28]) *Consider a game $\mathcal{G}(G, R, J)$ in S-RPC (S-PC) where J is collision-free and non-decreasing on the players' color (hence also oblivious). For every strategy profile σ^* that is a Nash Equilibrium for $\mathcal{G}(G, R, J)$, there is an ordering* $\mathbf{R}$ *of R such that there is an execution of $FF(J)$ algorithm on (G, R) yielding the same path-color assignment to R as σ^*.*

This result gives the explicit relationship of an NE to an online solution via FF. Moreover, any (routing and) coloring solution of RPC (PC) can be converted to a Nash Equilibrium (NE) for the corresponding game in S-RPC (S-PC, respectively), with at most the same number of colors, via the so called Nash conversion algorithm [86]. This leads to the following result.

Theorem 12.14 (See [86])

1. *The Price of Anarchy (PoA) for any game $\mathcal{G}(G, R, J)$ ($\mathcal{G}(G, P, J)$, respectively) in S-RPC (S-PC), where J is oblivious collision-free, is at least as large as the competitive ratio of $FF(J)$ for RPC (PC, respectively).*
2. *The Price of Anarchy (PoA) for any game $\mathcal{G}(G, R, J)$ ($\mathcal{G}(G, P, J)$, respectively) in S-RPC (S-PC), where J is oblivious collision-free and is a non-decreasing*

function of the players' color, is equal to the competitive ratio of First-Fit for RPC (PC, respectively).

3. *The Price of Stability (PoS) for any game $\mathcal{G}(G, R, J)$ ($\mathcal{G}(G, P, J)$, respectively) in S-RPC (S-PC), where J is oblivious collision-free and is a non-decreasing function of the players' color, is equal to* 1.

For the full proof, see [86]. Here we give a short summary.

Proof For 1, by Lemma 12.12, each execution of $FF(J)$ leads to a path-color (wavelength) assignment which is a Nash equilibrium (NE) for a game $\mathcal{G}(G, R, J)$. Hence the social cost J_s of that NE is equal to the number of colors used by $FF(J)$. Dividing by J_s^* we get the claim. For 2, let σ^* be an NE of the highest social cost. By Lemma 12.13, it follows that there is an execution of $FF(J)$ on the corresponding RPC (PC) instance that requires the same number of colors as σ^*. Dividing by J_s^* we find that the competitive ratio of $FF(J)$ is at least as large as the Price of Anarchy. Claim 2 follows by combining this with 1. Part 3 can be shown by considering the optimal coloring and converting it to an NE by using the Nash conversion algorithm. □

Some specific cost functions can be defined based on which some refinements of the PoA results be obtained [86].

Specifically consider the cost or payment function $J(r, \mathcal{P}, \lambda) = \lambda \cdot N + \mathit{length}(\mathcal{P})$, where N is the number of nodes in the ring. Under this function a player r always selects the smallest possible color even if it requires to follow the longest one of its two possible alternative paths.

Theorem 12.15 (See [86]) *The cost function $J(r, \mathcal{P}, \lambda) = \lambda \cdot N + \mathit{length}(\mathcal{P})$ used for S-RPC games in rings leads to a Price of Anarchy (PoA) equal to $\frac{|R|}{2J_s^*} + 1$, where R is the given set of requests.*

Even if this function does not achieve a low PoA, its PoA is half of that of any payment function that charges according to the value of the color only (see results above from [26, 28]).

Next, consider the cost or payment function $J(r, \mathcal{P}, \lambda) = \mathit{length}(\mathcal{P}) \cdot |R| + \lambda$, where R is the given set of requests. Under this function a player r always selects the shortest one of its two possible alternative paths even if it requires to take a larger color. For this case the following result holds.

Theorem 12.16 (See [86]) *The cost function $J(r, \mathcal{P}, \lambda) = \mathit{length}(\mathcal{P}) \cdot |R| + \lambda$ used for S-RPC games in rings leads to a Price of Anarchy (PoA) such that $FF_{chain} + 1 \leq PoA \leq 5.06 \log N + 10$, where N is the number of nodes in the ring.*

As seen above, this second payment function (which favors shortest path) achieves a PoA which does not depend on the number of requests but only on the number of nodes of the ring (logarithmically). It is still an open question whether this upper bound can be further improved. Also worth mentioning is the fact that all these functions require only local color information, namely to know which colors are used along edges that can be used by a player (minimal level of information according to the classification in [28] and [27]).

12.5 Notes

We presented an overview of a class of games for routing and path coloring problems in all-optical networks. We focused on games where each routing request (pair of source-destination nodes) is a player, and a strategy of a player consists of a path from the source to the destination and a wavelength/frequency (color), hence in the class of finite strategy games. We reviewed results on the relation between such games and online routing and path coloring. Other approaches are those in which the ISPs act as brokers or players between the TSP and the end user [162].

Chapter 13
Summary and Conclusions

The goal of this monograph has been to study optical networks as examples of communication networks and to understand how control algorithms can be designed for optical networks from a game theoretic perspective. From a game-theory and control perspective, there are a multitude of problems to be tackled in optical networks and the field is still in its infancy. An important problem is how to control channel transmission performance in optical networks while incorporating physical layer impairments. Of particular interest have been the OSNR optimization problems with link capacity constraints and/or channel OSNR targets. Game-theoretic approaches have been introduced to study such problems.

An optical network can be considered as a dynamic multiuser environment, in which signal of a channel is regarded as interfering for all others on the same fiber link. Power control is a key issue in designing an interference-limited multiuser communication network system. A major portion of this monograph has been focused on this first type of problems by using game-theoretic approaches. Another part of the book has discussed game-theoretic approaches for other challenging problems such as routing and wavelength assignment (path coloring) (RWA) or (RPC) in optical networks [99, 125, 165].

Throughout the book our focus has been on analysis of dynamic systems arising from game formulations with non-separable player utilities and with coupled as well as propagated (modified) constraints, in the class of continuous strategy games. It turns out that this is the case for a large class of games in optical networks, where the utility of each player is affected by decisions of other players, and action sets are no longer orthogonal (Chaps. 7 and 8).

The monograph has been organized as follows. Chapter 2 has provided an overview of basic concepts in game theory, mostly focused on noncooperative (Nash) game theory. Chapter 3 focused on matrix games, while Chap. 4 on games with continuous action sets and cost functions. Chapter 5 presented some relatively new theoretical results for continuous Nash games with coupled constraints, i.e., coupled action sets. Chapter 6 presented an overview of basic background on transmission in optical networks and on general topologies. The concepts of OSNR and link power capacity constraint in optical networks have been introduced. In Chap. 7

L. Pavel, *Game Theory for Control of Optical Networks*,
Static & Dynamic Game Theory: Foundations & Applications,
DOI 10.1007/978-0-8176-8322-1_13,

a basic formulation of a game framework towards solving the OSNR optimization problem in optical networks has been presented, restricted to single point-to-point optical links, as the simplest network topology. A Nash game played among channels has been employed towards maximizing OSNR firstly without coupled link capacity constraint. Then for incorporating the coupled power link capacity constraint, two approaches have been considered—an indirect and a direct one, based on the Lagrangian extension and decomposition theoretical results in Chap. 5.

Approaches on how to deal with games in more complicated network topologies have been studied in Chap. 8. In network configurations, coupled constraints are propagated along fiber links and constraint functions become complicated from the end-to-end point of view. The physical system constraint in optical networks provides an elimination approach to reduce constraints to a simple structure (less-coupled constraints). Specifically a partitioned game approach was introduced. The multi-link structure can be partitioned into stages where each stage has a single-sink structure (with either a single link or multiple links) and channel powers on each stage are adjustable. A partitioned Nash game is composed of ladder-nested stage Nash games. Instead of maximization of channel OSNR from Tx to Rx (end to end), we considered minimization of channel OSNR degradation between stages. We showed that the partition directly led to the convexity of link capacity constraints on each stage. For the case when channel powers are adjustable at each optical switch each link is a stage after partition. The convexity of constraints propagated along links is automatically satisfied and Lagrangian extension and decomposition results can be applied directly. Chapter 9 has addressed the efficiency of Nash game and methods to improve the efficiency by tuning the parameters in cost functions. Extensive simulations in MATLAB and practical experimental implementations in an optical network testbed system (ONTS) have been presented in Chap. 10.

Stability properties of game and pricing dynamics have been investigated under the assumption of time-scale separation using Lyapunov theory and singular perturbation approach, and considering time-delay effects (Chap. 11). As communication networks grow in geographical size and complexity, the effects of time delay become more pronounced and cannot be ignored. Chapter 11 has addressed the impact of time delay in optical network control algorithms derived from game-theoretic formulations. In Chap. 12 other game formulations have been addressed, specifically for routing and the wavelength-assignment (path coloring) RWA or RPC problem, these ones being in the class of finite strategy games.

The material presented in this monograph generates several possible directions for future work. Some open questions are related to *asynchronism*. Algorithms in Chaps. 7 and 8 are two-timescale algorithms. The channel and link algorithms converge to optimal solutions for the synchronous updating of channels' power and links' price, respectively. An asynchronous setting of algorithms can better resemble the reality of large-scale networks, in the sense that updates at channels and links in complex network topologies do not occur synchronously.

Another issue is related to *dynamics*. Optical networks are operated in a dynamic environment, where network reconfiguration (channels add/drop) is being performed while other existing channels are still in service. The inherent dynamics of a fiber link composed of cascaded dynamical optical elements makes the optical network dynamics an important aspect both at the data layer and the physical layer. Such a problem has been addressed in [114]. In mesh network topologies, channel routes can be changed by optical switches and closed loop structures can be formed, for example, quasi-ring structures in Chap. 8. Strictly speaking, time delay and asynchronism are aspects of dynamics. The material presented in this monograph considers the network after reconfiguration. In other words, system parameters are stationary between any updates. From the dynamical point of view, study of the scalability of algorithms is an open direction.

Finally I would like to end on the same note mentioned in the introduction: I believe there is a lot of scope for fundamental research in optical networks from a control- and game-theoretic perspective. Hopefully this book will provide the motivation for many other developments and much research in this area.

Appendix A
Supplementary Material

A.1 Notations

We present below some notations used throughout the monograph.

Let $\mathbf{a} = [a_i]$ and $\mathbf{b} = [b_i]$ be n-dimensional vectors. We write $\mathbf{a} \geq \mathbf{b}$ if all $a_i \geq b_i$ and $\mathbf{a} > \mathbf{b}$ if all $a_i > b_i$. We use diag($\mathbf{a}$) to denote a diagonal matrix whose diagonal entries are elements of vector $\mathbf{a} = [a_i]$. A superscript T denotes the transpose operation.

For a twice continuously differentiable function $f : \mathbb{R}^m \to \mathbb{R}$, its *gradient* at $\mathbf{u}$, denoted by $\nabla f(\mathbf{u})$, is defined as

$$\nabla f(\mathbf{u}) = \begin{bmatrix} \frac{\partial f(\mathbf{u})}{\partial u_1} \\ \cdots \\ \frac{\partial f(\mathbf{u})}{\partial u_m} \end{bmatrix}, \tag{A.1}$$

while the Hessian of function f, denoted by $\nabla^2 f$, is given as

$$\nabla^2 f(\mathbf{u}) = \begin{bmatrix} \frac{\partial^2 f(\mathbf{u})}{\partial u_1^2} & \frac{\partial^2 f(\mathbf{u})}{\partial u_1 \partial u_2} & \cdots & \frac{\partial^2 f(\mathbf{u})}{\partial u_1 \partial u_m} \\ \frac{\partial^2 f(\mathbf{u})}{\partial u_2 \partial u_1} & \frac{\partial^2 f(\mathbf{u})}{\partial u_2^2} & \cdots & \frac{\partial^2 f(\mathbf{u})}{\partial u_2 \partial u_m} \\ \vdots & \vdots & \ddots & \vdots \\ \frac{\partial^2 f(\mathbf{u})}{\partial u_m \partial u_1} & \frac{\partial^2 f(\mathbf{u})}{\partial u_m \partial u_2} & \cdots & \frac{\partial^2 f(\mathbf{u})}{\partial u_m^2} \end{bmatrix} \tag{A.2}$$

For a vector-valued function $\mathbf{f} : \mathbb{R}^m \to \mathbb{R}^m$, with components $f_1, \ldots, f_m$, which are twice continuously differentiable, its *pseudo-gradient* at $\mathbf{u}$, denoted by $\overline{\nabla}\mathbf{f}(\mathbf{u})$ is defined as a column vector of the first-order partial derivatives of $f_i(\mathbf{u})$ with respect to u_i, $\frac{\partial f_i(\mathbf{u})}{\partial u_i}$,

$$\overline{\nabla}\mathbf{f}(\mathbf{u}) := \begin{bmatrix} \frac{\partial f_1(\mathbf{u})}{\partial u_1} \\ \cdots \\ \frac{\partial f_m(\mathbf{u})}{\partial u_m} \end{bmatrix}, \tag{A.3}$$

L. Pavel, *Game Theory for Control of Optical Networks*,
Static & Dynamic Game Theory: Foundations & Applications,
DOI 10.1007/978-0-8176-8322-1, © Springer Science+Business Media New York 2012

with elements from the diagonal of the Jacobian matrix of **f**. The Jacobian of $\overline{\nabla}\mathbf{f}(\mathbf{u})$ with respect to **u** is denoted by $\overline{\nabla}^2\mathbf{f}(\mathbf{u})$,

$$\overline{\nabla}^2\mathbf{f}(\mathbf{u}) := \begin{bmatrix} \frac{\partial^2 f_1(\mathbf{u})}{\partial u_1^2} & \frac{\partial^2 f_1(\mathbf{u})}{\partial u_1 \partial u_2} & \cdots & \frac{\partial^2 f_1(\mathbf{u})}{\partial u_1 \partial u_m} \\ \frac{\partial^2 f_2(\mathbf{u})}{\partial u_2 \partial u_1} & \frac{\partial^2 f_2(\mathbf{u})}{\partial u_2^2} & \cdots & \frac{\partial^2 f_2(\mathbf{u})}{\partial u_2 \partial u_m} \\ \vdots & \vdots & \ddots & \vdots \\ \frac{\partial^2 f_m(\mathbf{u})}{\partial u_m \partial u_1} & \frac{\partial^2 f_m(\mathbf{u})}{\partial u_m \partial u_2} & \cdots & \frac{\partial^2 f_m(\mathbf{u})}{\partial u_m^2} \end{bmatrix} \tag{A.4}$$

For a two-argument continuously differentiable function $f : \mathbb{R}^m \times \mathbb{R}^m \to \mathbb{R}$, with arguments $(\mathbf{u}; \mathbf{x})$, its gradient is defined as

$$\nabla f(\mathbf{u}; \mathbf{x}) = \left[\nabla_u^T f(\mathbf{u}; \mathbf{x}) \quad \nabla_x^T f(\mathbf{u}; \mathbf{x})\right]^T \tag{A.5}$$

where $\nabla_u f(\mathbf{u}; \mathbf{x})$ and $\nabla_x f(\mathbf{u}; \mathbf{x})$ are the gradients with respect to the first argument $\mathbf{u} = [u_1, \dots, u_m]$ and the second argument $\mathbf{x} = [x_1, \dots, x_m]$, respectively, as defined in (A.1).

A.2 Standard Optimization Review

We review basic concepts in optimization, mainly drawn from [24]. Covered topics include unconstrained and constrained optimization problems, Lagrange multiplier and the duality approach.

A mathematical optimization problem has the form

$$\begin{array}{ll} \min & J_0(\mathbf{u}) \\ \text{subject to} & \mathbf{u} \in \Omega, \end{array} \tag{A.6}$$

where $\mathbf{u} = [u_1, \dots, u_m]^T$ is the variable, $\Omega \subseteq \mathbb{R}^m$ is the general *constraint set*. The problem (A.6) is called an *unconstrained* optimization problem if $\Omega \mathbb{R}^m$. The objective function $J_0 : \Omega \to \mathbb{R}$ is the *cost function*.

A vector satisfying the constraints in (A.6) is called a *feasible* vector for this problem. A feasible vector $\mathbf{u}^{\text{opt}}$ is a *local minimum* of $J_0(\mathbf{u})$ if there exists an $\varepsilon > 0$ such that

$$J_0\left(\mathbf{u}^{\text{opt}}\right) \le J_0(\mathbf{u}), \quad \forall \mathbf{u} \in \Omega \text{ with } \left\|u - \mathbf{u}^{\text{opt}}\right\| < \varepsilon$$

where $\|\cdot\|$ denotes the Euclidean norm. A feasible vector $\mathbf{u}^{\text{opt}}$ is called a *global minimum* of $J_0(\mathbf{u})$, or a solution of the problem (A.6) if

$$J_0\left(\mathbf{u}^{\text{opt}}\right) \le J_0(\mathbf{u}), \quad \forall \mathbf{u} \in \Omega$$

and $\mathbf{u}^{\text{opt}}$ is called *strict* if the inequality above is strict for $\mathbf{u} \neq \mathbf{u}^{\text{opt}}$.

The following is a standard result (combining Proposition 1.1.1 and Proposition 1.1.2 in [24]) regarding the solution of (A.6).

Proposition A.1 (See [24]) *Let* $\mathbf{u}^{\mathrm{opt}}$ *be a local minimum of* $J_0 : \Omega \to \mathbb{R}$. *Assume that* J_0 *is continuously differentiable in an open set* $\mathcal{S} \subseteq \Omega$ *containing* $\mathbf{u}^{\mathrm{opt}}$. *Then*

$$\nabla J_0\big(\mathbf{u}^{\mathrm{opt}}\big) = 0 \quad \textit{(first-order necessary condition)} \tag{A.7}$$

If in addition $J_0(\mathbf{u})$ *is twice continuously differentiable within* $\mathcal{S}$, *then*

$$\nabla^2 J_0\big(\mathbf{u}^{\mathrm{opt}}\big) \textit{ is positive semidefinite} \quad \textit{(second-order necessary condition)} \tag{A.8}$$

If in addition $\mathcal{S}$ *is convex and* $J_0(\mathbf{u})$ *is a convex function over* $\mathcal{S}$, *then* (A.7) *is a necessary and sufficient condition for* $\mathbf{u}^{\mathrm{opt}} \in \mathcal{S}$ *to be a global minimum of* $J_0(\mathbf{u})$ *over* $\mathcal{S}$.

The problem (A.6) is called a *constrained* optimization problem if Ω is a strict subset of $\mathbb{R}^m$, $\Omega \subset \mathbb{R}^m$. Throughout this section, it is assumed that Ω is convex. The following result follows directly from Proposition 2.1.1 in [24].

Proposition A.2 *If* $\Omega \subset \mathbb{R}^m$ *is a convex and compact set and* $J_0 : \Omega \to \mathbb{R}$ *is a strictly convex function, then the problem* (A.6) *admits a unique global minimum.*

For *constrained* optimization problems, a specific structure of $\Omega \subset \mathbb{R}$ constructed by inequalities is taken into account in this monograph. That is,

$$\begin{array}{lll} \min & J_0(\mathbf{u}) & \\ \text{subject to} & g_r(\mathbf{u}) \le 0, & r = 1, \ldots, R, \end{array} \tag{A.9}$$

where constraints $g_1(\mathbf{u}), \ldots, g_R(\mathbf{u})$ are real-valued continuously differentiable functions defined from $\mathbb{R}^m$ to $\mathbb{R}$. In a compact form the general constraint can be written as

$$\Omega\big\{\mathbf{u} \in \mathbb{R}^m \;\big|\; g_r(\mathbf{u}) \le 0, r = 1, \ldots, R\big\}$$

We denote this optimization problem (A.9) by $OPT(\Omega, J_0)$.

For any feasible vector $\mathbf{u}$, the *set of active inequality constraints* is denoted by

$$A(\mathbf{u}) = \big\{r \;\big|\; g_r(\mathbf{u}) = 0, r = 1, \ldots, R\big\} \tag{A.10}$$

If $r \notin A(\mathbf{u})$, the constraint $g_r(\mathbf{u})$ is *inactive* at $\mathbf{u}$. A feasible vector $\mathbf{u}$ is said to be *regular* if the active inequality constraint gradients $\nabla g_r(\mathbf{u})$, $r \in A(\mathbf{u})$, are linear independent.

The *Lagrangian function* $L : \mathbb{R}^{m+R} \to \mathbb{R}$ for problem (A.9) is defined as

$$L(\mathbf{u}, \boldsymbol{\mu}) = J_0(\mathbf{u}) + \sum_{r=1}^{R} \mu_r g_r(\mathbf{u}), \tag{A.11}$$

where μ_r, $r = 1, \ldots, R$, are scalars. The following proposition (Proposition 3.3.1 in [24]) states necessary conditions for solving $OPT(\Omega, J_0)$ in terms of the Lagrangian function defined in (A.11).

Proposition A.3 (Karush–Kuhn–Tucker (KKT) Necessary Condition) *Let* $\mathbf{u}^{\mathrm{opt}}$ *be a local minimum of the problem* (A.9). *Assume that* $\mathbf{u}^{\mathrm{opt}}$ *is regular. Then there exists a unique vector* $\boldsymbol{\mu}^* = (\mu_1^*, \dots, \mu_R^*)$, *called a* Lagrange multiplier *vector, such that*

$$\nabla_u L(\mathbf{u}^{\mathrm{opt}}, \boldsymbol{\mu}^*) = 0 \tag{A.12}$$

$$\mu_r^* \geq 0, \quad \forall r = 1, \dots, R \tag{A.13}$$

$$\mu_r^* = 0, \quad \forall r \notin A(\mathbf{u}^{\mathrm{opt}}) \tag{A.14}$$

Condition (A.14) in Proposition A.3 is called the complementary slackness condition.

The next result (Proposition 3.3.4 in [24]) reviews general sufficient conditions, in terms of the Lagrangian function (A.11), for the following constrained optimization problem:

$$\begin{array}{ll} \min & J_0(\mathbf{u}) \\ \text{subject to} & g_r(\mathbf{u}) \leq 0, \quad r = 1, \dots, R, \\ & \mathbf{u} \in \mathcal{U}, \end{array} \tag{A.15}$$

where $\mathcal{U} \subseteq \mathbb{R}^m$. The conditions are general since differentiability and convexity of J_0 and g_r, $r = 1, \dots, R$, are not required. Meanwhile, $\mathcal{U}$ may be a strict subset of $\mathbb{R}^m$, $\mathcal{U} \subset \mathbb{R}^m$.

Proposition A.4 (General Sufficiency Condition) *Consider the problem* (A.15). *Let* $\mathbf{u}^{\mathrm{opt}}$ *be a feasible vector, together with a vector* $\boldsymbol{\mu}^* = [\mu_1^*, \dots, \mu_R^*]^T$ *that satisfies*

$$\mu_r^* \geq 0, \quad \forall r = 1, \dots, R$$

$$\mu_r^* = 0, \quad \forall r \notin A(\mathbf{u}^{\mathrm{opt}})$$

and assume $\mathbf{u}^{\mathrm{opt}}$ *minimizes the Lagrangian function* $L(\mathbf{u}, \boldsymbol{\mu}^*)$ (A.11) *over* $\mathbf{u} \in \mathcal{U}$, *denoted as*

$$\mathbf{u}^{\mathrm{opt}} \in \arg\min_{\mathbf{u} \in \mathcal{U}} L(\mathbf{u}, \boldsymbol{\mu}^*) \tag{A.16}$$

Then $\mathbf{u}^{\mathrm{opt}}$ *is a global minimum of the problem* (A.15).

Remark A.5 If in addition, J_0 and g_r, $r = 1, \dots, R$, are also convex and $\mathcal{U} = \mathbb{R}^m$, then the Lagrangian function $L(\mathbf{u}, \boldsymbol{\mu})$ is convex with respect to $\mathbf{u}$. Therefore, by Proposition A.1, (A.16) is equivalent to the first-order necessary condition (A.12) in Proposition A.3. Thus conditions (A.12)–(A.14) are also sufficient.

A.3 Diagonally Dominant and M-matrices

Definition A.6 (See [94]) Let $A := [a_{ij}]$ be an $m \times m$ matrix. The matrix A is said to be diagonally dominant if

$$|a_{ii}| \geq \sum_{j=1, j\neq i}^{m} |a_{ij}|, \quad \forall i = 1, \ldots, m$$

It is said to be strictly diagonally dominant if

$$|a_{ii}| > \sum_{j=1, j\neq i}^{m} |a_{ij}|, \quad \forall i = 1, \ldots, m$$

Some useful results are shown in the following theorem, adapted from [59].

Theorem A.7 (See [59], p. 349) *Let the $m \times m$ matrix $A = [a_{ij}]$ be strictly diagonally dominant. Then A is invertible and*

(a) *If all main diagonal entries of A are positive, then all the eigenvalues of A have positive real part.*
(b) *If A is Hermitian and all main diagonal entries of A are positive, then all the eigenvalues of A are real and positive.*

Lemma A.8 *Let A be an $m \times m$ real matrix with all main diagonal entries positive. Then A is positive definite if A and A^T are both strictly diagonally dominant.*

Proof If A and A^T are both strictly diagonally dominant, then it follows that

$$a_{ii} > \sum_{j=1, j\neq i}^{m} |a_{ij}| \quad \text{and} \quad a_{ii} > \sum_{j=1, j\neq i}^{m} |a_{ji}|$$

Thus for matrix $A_s := A + A^T$,

$$2a_{ii} > \sum_{j=1, j\neq i}^{m} \left(|a_{ij}| + |a_{ji}|\right) \geq \sum_{j=1, j\neq i}^{m} |a_{ij} + a_{ji}|$$

Thus A_s is also strictly diagonally dominant. From Theorem A.7, it follows that A_s is positive definite. Therefore A is positive definite in the sense that the symmetric part, $\frac{1}{2}A_s$, is positive definite. □

Definition A.9 A square matrix is called a Z-matrix if all off-diagonal entries are less than or equal to zero.

Definition A.10 (See [43], Theorem 5.1) Let the $m \times m$ matrix $A = [a_{ij}]$ be a Z-matrix. A is called an M-matrix if it satisfies any one of the following conditions.

1. There exists an $m \times 1$ vector v with non-negative entries such that $Av > 0$.
2. Every real eigenvalue of A is positive.
3. There exist a non-negative $m \times m$ matrix B and a number $\lambda > \rho(B)$, where $\rho(B)$ is the spectral radius of B, such that $A = \lambda I - B$.
4. The real part of any eigenvalue of A is positive.
5. A is non-singular and the inverse of A is non-negative.
6. $Av \geq 0$ implies $v \geq 0$, where v is an $m \times 1$ vector.

Theorem A.11 *Let the $m \times m$ matrix $A = [a_{ij}]$ be an M-matrix. Then*

1. *there exist a non-negative $m \times m$ matrix B and a number $\lambda > \rho(B)$, where $\rho(B)$ is the spectral radius of B, such that $A = \lambda I - B$;*
2. *specifically when $\lambda = 1$, $A = I - B$, $\rho(B) < 1$, and $(I - B)^{-1} = \sum_{k=0}^{\infty} B^k$ exists and is positive component-wise.*

Proof The proof for the first part can be found in [43] where Theorem 5.1 is proved. In the case when $\lambda = 1$, $\rho(B) < 1$ leads to the convergence of $\sum_{k=0}^{\infty} B^k$. Furthermore,

$$\sum_{k=0}^{\infty} B^k(I - B) = \left(I + B + B^2 + \cdots\right)(I - B) = I$$

Thus it follows that $(I - B)^{-1} = \sum_{k=0}^{\infty} B^k$. □

A.4 Maximum Theorem

Definition A.12 Let $\mathcal{S}$ be a (nonempty) subset of $\mathbb{R}^n$. The set of all nonempty subsets of $\mathcal{S}$ is called the power set of $\mathcal{S}$, denoted by $\mathcal{P}(\mathcal{S})$.

Definition A.13 Let Θ and $\mathcal{S}$ be subsets of $\mathbb{R}^l$ and $\mathbb{R}^n$, respectively. A correspondence Φ from Θ to $\mathcal{S}$ is a map that associates with each element $\theta \in \Theta$ a (nonempty) subset $\Phi(\theta) \subset \mathcal{S}$.

Definition A.14 Let Φ be a correspondence from Θ to $\mathcal{S}$ by $\Phi : \Theta \to \mathcal{P}(\mathcal{S})$. Φ is said to be upper-semi-continuous at a point $\theta \in \Theta$ if for all open sets $\mathcal{V}$ such that $\Phi(\theta) \subset \mathcal{V}$, there exists an open set $\mathcal{U}$ containing θ, such that $\theta' \in \mathcal{U} \cap \Theta$ implies $\Phi(\theta') \subset \mathcal{V}$. Φ is upper-semi-continuous if it is at any point of Θ.

Example A.15 The correspondence Φ is defined by

$$\Phi(x) := \begin{cases} \{0\}, & \text{if } x \neq 0 \\ [-1, +1], & \text{if } x = 0 \end{cases}$$

and is upper-semi-continuous at 0 (Fig. A.1).

Fig. A.1 Example: upper-semi-continuous

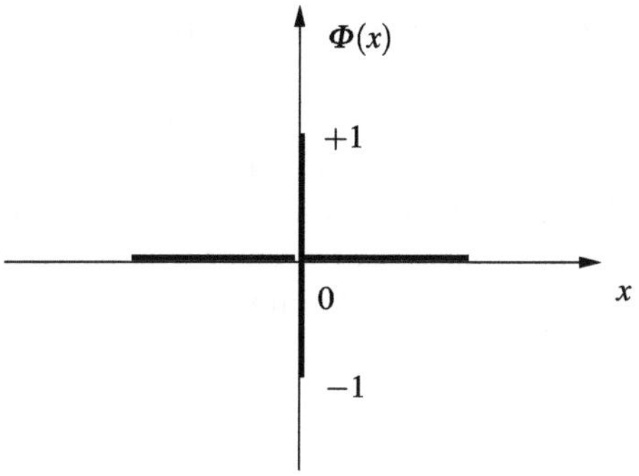

Remark A.16 Every upper-semi-continuous single valued correspondence is a continuous function.

Theorem A.17 (Berge's Maximum Theorem) *Let $f : \mathcal{X} \times \mathcal{Y} \to \mathbb{R}$ be a continuous function and $\Phi : \mathcal{X} \to \mathcal{Y}$ be a nonempty, compact-valued, continuous correspondence. Then*

1. $f^* : \mathcal{X} \to \mathbb{R}$ *with*
$$f^*(x) := \max\{f(x, y) : y \in \Phi(x)\}$$
is a continuous function;
2. $\Phi^* : \mathcal{X} \to \mathcal{Y}$ *with*
$$\Phi^* := \arg\max\{f(x, y) : y \in \Phi(x)\} = \{y \in \Phi(x) : f(x, y) = f^*(x)\}$$
is a compact-valued, upper-semi-continuous correspondence.

A.5 Fixed-Point Theorems

We introduce the two fundamental fixed-point theorems used in existence proofs.

Theorem A.18 (See [20], Brouwer's Fixed-point Theorem) *If $\mathcal{S}$ is a compact and convex subset of $\mathbb{R}^n$ and f is a continuous function mapping $\mathcal{S}$ into itself, then there exists at least one $x \in \mathcal{S}$ such that $f(x) = x$.*

Theorem A.19 (See [20], Kakutani Fixed-point Theorem) *Let $\mathcal{S}$ be a compact and convex subset of $\mathbb{R}^n$, and let f be an upper-semi-continuous correspondence which assigns to each $x \in \mathcal{S}$ a closed and convex subset of $\mathcal{S}$. Then there exists some $x \in \mathcal{S}$ such that $x \in f(x)$.*

A.6 Projection Theorem

Projection method is widely used in developing iterative algorithms solving constrained optimization problems. Each time when an update jumps outside the feasible set $\mathcal{X}$, the algorithm can project it back to the set $\mathcal{X}$.

The projection of a vector x onto a nonempty, closed, and convex set $\mathcal{X}$ is defined with respect to the Euclidean norm, denoted by $[\cdot]^+$, i.e.,

$$[x]^+ := \arg\min_{z \in \mathcal{X}} \|z - x\|$$

The properties of the projection are presented in the following theorem.

Theorem A.20 (See [25], Projection Theorem) *Let $\mathcal{X}$ be a nonempty, closed, and convex subset of $\mathbb{R}^n$.*

1. *For every $x \in \mathbb{R}^n$, there exists a unique $z \in \mathcal{X}$ that minimizes $\|z - x\|$ over all $z \in \mathcal{X}$ and is denoted by $[x]^+$.*
2. *Given some $x \in \mathbb{R}^n$, a vector $x^* \in \mathcal{X}$ is equal to $[x]^+$ if and only if*

$$(z - x^*)^T (x - x^*) \le 0, \quad \forall z \in \mathcal{X}$$

3. *The mapping $f : \mathbb{R}^n \to \mathcal{X}$ defined by $f(x) = [x]^+$ is continuous and nonexpansive, that is,*

$$\left\| [x]^+ - [y]^+ \right\| \le \|x - y\|, \quad \forall x, y \in \mathbb{R}^n$$

A.7 Lipschitz Continuity

Lipschitz continuity is a smoothness condition for functions which is stronger than regular continuity. For a function $f(x) : D \to \mathbb{R}^n$, where D is a compact subset of $\mathbb{R}^m$, the *Lipschitz condition* is defined as

$$\left\| f(x) - f(y) \right\| \le L \|x - y\|, \tag{A.17}$$

where $\|\cdot\|$ denotes Euclidean norm. The positive constant L is called a *Lipschitz constant.*

Definition A.21 A function $f(x)$ is said to be Lipschitz continuous if there exists a constant $L \ge 0$ such that for all $x, y \in D$, the Lipschitz condition (A.17) is satisfied.

The function $f(x)$ is called locally Lipschitz continuous if each point of D has a neighborhood D_0 such that $f(x)$ satisfies the Lipschitz condition for all points in D_0 with some Lipschitz constant L_0.

A locally Lipschitz continuous function $f(x)$ on D is Lipschitz continuous on every compact (closed and bounded) set $W \subseteq D$.

The Lipschitz property is weaker than continuous differentiability, as stated in the next proposition.

Proposition A.22 (Lemma 3.2, [69]) *If a function $f(x)$ and $\frac{\partial f}{\partial x}(x)$ are continuous on D, then $f(x)$ is locally Lipschitz continuous on D.*

A.8 Variational Inequalities

Variational inequalities can be used to study equilibrium problems. Given a subset $X \subset \mathbb{R}^n$ and a function $f : \mathbb{R}^n \to \mathbb{R}^n$, the variational inequality problem associated with X is to find a vector $x^* \in X$ such that

$$\left(x - x^*\right)^T f\left(x^*\right) \geq 0, \quad \forall x \in X$$

A short notation of this problem is by VI(X, f).

A necessary and sufficient condition for a vector x^* to be a solution of VI(X, f) is given below.

Proposition A.23 (Proposition 5.1, [25], p. 267) *Let γ be a positive scalar and let G be a symmetric positive definite matrix. A vector x^* is a solution of VI(X, f) if and only if*

$$\left[x^* - \gamma G^{-1} f\left(x^*\right)\right]_G^+ = x^*,$$

where $[\cdot]_G^+$ is the projection on X with respect to norm $\|x\|_G = (x^T G x)^{1/2}$.

The existence of a solution is ensured by the following proposition.

Proposition A.24 (Proposition 5.2, [25], p. 268) *Suppose that X is compact and that $f : \mathbb{R}^n \to \mathbb{R}^n$ is continuous. Then there exists a solution to VI(X, f).*

A.9 Singular Perturbation Theory

Singular perturbation theory [69] is used to study the stability of systems that have dynamics on two different time-scales. Consider a general nonlinear system without time-delays of the form

$$\begin{aligned} \dot{x}(\hat{t}) &= f\left(x(\hat{t}), z(\hat{t})\right) \\ \varepsilon \dot{z}(\hat{t}) &= g\left(x(\hat{t}), z(\hat{t})\right) \end{aligned} \tag{A.18}$$

where $\hat{t}$ is the independent time variable, $x \in \Re^n$, $z \in \Re^p$, $\varepsilon > 0$ is a real scalar, and f and g are locally Lipschitz around the origin with $f(0,0) = 0$ and $g(0,0) = 0$. The ε term is considered to be "small", i.e. take the limit as $\varepsilon \to 0$. We define by

$\dot{v} = \frac{dv(\hat{t})}{d\hat{t}}$. The system (A.18) is defined on two different time-scales, one fast, and one slow.

We describe (A.18) in further detail as follows. We denote by $\hat{t}$ the "slow" time-scale, and we denote by t the "fast" time-scale. The two time-scales are related to each other by $\hat{t} = \varepsilon t$. A nonlinear system that is in singular perturbation form, as presented in (A.18), may be time-decoupled into two subsystems that are easier to analyze than the full nonlinear system. One subsystem is defined on the slow time-scale, and the other subsystem is defined on the fast time-scale. The system on the slow time-scale is called the reduced system, while the system on the fast time-scale is called the boundary-layer system. A Lyapunov stability analysis may be applied to the reduced and boundary-layer systems separately. The Lyapunov functions of the reduced and boundary-layer systems may be combined into a composite Lyapunov function that applies to the full, singularly perturbed system. This composite Lyapunov function can be used to prove the stability of the full system.

Next we show how to obtain the reduced system and the boundary-layer system of (A.18) as in [69]. Assume there is at least one isolated root of $0 = g(x, z)$ and let $z = h(x)$ be one such root that vanishes at $x = 0$. For convenience, we define the coordinate shift $\hat{z} = z - h(x)$. Thus, the coordinate shifted system becomes

$$\dot{x} = f\big(x, \hat{z} + h(x)\big) \tag{A.19}$$

$$\varepsilon\dot{\hat{z}} = g\big(x, \hat{z} + h(x)\big) - \varepsilon\frac{\partial h}{\partial x} f\big(x, \hat{z} + h(x)\big) \tag{A.20}$$

where x and $\hat{z}$ are functions of the slow time variable $\hat{t}$. Note that the ε term appears in the second equation because it is written with respect to the slow time variable, $\hat{t}$. If we rewrite this equation as a function of the fast time variable t, the ε term disappears.

The reduced model is derived from (A.19) by setting $\hat{z} = 0$, which is equivalent to assuming the fast dynamics, $\varepsilon\dot{\hat{z}}$, converge instantly. This produces the following reduced system:

$$\dot{x} = f\big(x, h(x)\big)$$

Notice that this system is on the slow time-scale, $\hat{t}$.

The boundary-layer system is derived from (A.20). We rewrite (A.20) with respect to the fast time-scale, t, as follows. Let $t = \frac{\hat{t}}{\varepsilon}$. Note that setting $\varepsilon = 0$ freezes the $\hat{t}$ and $x(\hat{t})$ variables with respect to the fast time-scale t so that $\hat{t}$ and x appear as constants. Alternatively, we can think of the variables on the t time-scale converging instantaneously, since $t = \frac{\hat{t}}{\varepsilon}$. We introduce the notation $\hat{t}(t) := \varepsilon t$ for the scaling by ε. Thus, we define $\tilde{z}(t) = \hat{z}(\hat{t}(t)) = (\hat{z} \circ \hat{t})(t)$, where $\circ$ denotes composition and $y = \hat{z} \circ \hat{t}$. Thus, when we set $\varepsilon = 0$ in (A.20), and rewrite it with respect to the t time-scale, we obtain

$$\frac{d\tilde{z}(t)}{dt} = g\big(x(\hat{t}), \tilde{z}(t) + h\big(x(\hat{t})\big)\big)$$

where $\frac{d\tilde{z}}{dt} = \varepsilon \frac{d\hat{z}}{d\hat{t}}$. Thus, we define the reduced and boundary-layer systems as follows.

Definition A.25 Consider the system (A.18). Let the reduced system be defined as

$$\dot{x} = f\big(x(\hat{t}), h\big(x(\hat{t})\big)\big)$$

and the boundary system be defined as

$$\frac{d\tilde{z}(t)}{dt} = g\big(x(\hat{t}), \tilde{z}(t) + h\big(x(\hat{t})\big)\big)$$

where $h(x)$ is the isolated root of the RHS of (A.18) that vanishes at $x = 0$, t and $\hat{t}$ are the fast and slow time-scales in the system, respectively, $\tilde{z} = \hat{z} \circ \hat{t}$ and $\hat{z} = z - h(x)$ is the coordinate shift.

We define exponential stability as follows [69].

Definition A.26 Consider the general autonomous system

$$\dot{v} = f(t, v) \tag{A.21}$$

where $f : [0, \infty) \times D \to \Re^n$ is piecewise continuous in time t and locally Lipschitz in v on $[0, \infty) \times D$ and $D \subset \Re^n$ is a domain that contains the origin $v = 0$. The equilibrium point $v = 0$ of (A.21) is exponentially stable if there exist positive constants c_1, c_2, and c_3 such that

$$\|v(t)\| \le c_2 \|v(t_0)\| e^{-c_3(t-t_0)}, \quad \forall \|v(t_0)\| < c_1,\ \forall t \ge t_0 \tag{A.22}$$

and globally exponentially stable if (A.22) is satisfied for any initial state $v(t_0)$.

We now introduce the formal singular perturbation theorem for analyzing the stability of (A.18). The following result is Theorem 11.4 simplified from [69] and restated as Theorem A.27. Assume that (A.18) satisfies all necessary assumptions outlined below.

Theorem A.27 *Consider the singularly perturbed system*

$$\dot{x} = f\big(x(\hat{t}), z(\hat{t})\big) \tag{A.23}$$

$$\varepsilon \dot{z} = g\big(x(\hat{t}), z(\hat{t})\big) \tag{A.24}$$

where $x \in \Re^n$, $z \in \Re^p$, $\hat{t} \in \Re$ *and* ε *is a real scalar.*

Assume the following are satisfied for all

$$(\hat{t}, x, \varepsilon) \in [0, \infty) \times B_r \times [0, \varepsilon_0]$$

where $B_r = \{x \in \Re^n \mid \|x\| \le r\}$:

- $f(0,0)=0$ *and* $g(0,0)=0$.
- *The equation* $0=g(x,z)$ *has an isolated root* $z=h(x)$ *such that* $h(0)=0$.
- *The functions* f,g,h *and their partial derivatives up to the second order are bounded for* $z-h(x)\in B_\rho$.
- *The origin of the reduced system* $\dot{x}=f(x(\hat{t}),h(x(\hat{t})))$ *is exponentially stable.*
- *The origin of the boundary-layer system* $\frac{d\tilde{z}(t)}{dt}=g(x(\hat{t}),\tilde{z}(t)+h(x(\hat{t})))$ *is exponentially stable, uniformly in* $(\hat{t},x)$, *i.e., there exist constants* c_1, c_2, *and* c_3 *such that* $\|\tilde{z}(t)\|\le c_1\|\tilde{z}(0)\|e^{-c_2 t}, \forall\|\tilde{z}(0)\|<c_3, \forall(\hat{t},x), \forall t\ge 0$.

Then there exists $\varepsilon^*>0$ *such that for all* $\varepsilon<\varepsilon^*$, *the origin of* (A.23)–(A.24) *is exponentially stable.*

The high level proof of Theorem A.27 proceeds as follows. First, a Lyapunov function for each of the reduced and boundary-layer systems is found. Next, the two functions are summed together to produce a composite Lyapunov function that is applied to the full system, (A.23)–(A.24). Using the properties of the Lyapunov functions, as well as bounds on f, g, and h around the origin, exponential stability is proved. The details of the proof of Theorem A.27 are found in [69].

A.10 Time-Delay Stability Results

A.11 Time-Delay Stability Theory

In this section, we review results on stability of time-delay systems. We first define the continuous norm, the retarded functional differential equation, and the definitions of stability from [55]. We then present the Lyapunov–Razumikhin theorems [55]. These theorems do not rely on functionals, but on typical Lyapunov functions, and as such they produce simpler, more conservative stability criteria. Lyapunov–Krasovskii theory relies on functional equations which give less conservative stability criteria at the expense of both analytical and computational complexity and it is not covered here.

A.11.1 Definitions

We define $\mathcal{C}([-\tau,0],\Re^n)$ as the set of all continuous functions mapping $[-\tau,0]$ to $\Re^n$. Let $\mathcal{C}=\mathcal{C}([-\tau,0],\Re^n)$. The general form of a retarded functional differential equation is defined as

$$\frac{dx}{dt}=f(t,x_t) \tag{A.25}$$

where $x_t\in\mathcal{C}$ is the function mapping $[t-\tau,t]$ to $\Re^n$, and $f:\Re\times\mathcal{C}\to\Re^n$.

The following norm is used in the definition of stability for time-delay systems.

Definition A.28 The continuous norm is defined as

$$\|x_t\|_c = \max_{-\tau\le\theta\le 0} \big\|x(t+\theta)\big\|_2$$

where $x_t \in \mathcal{C}$ is the function mapping $[t-\tau, t]$ to $\Re^n$.

Note that $\|x(t)\|_2 \le \|x_t\|_c$. We now define stability for (A.25).

Definition A.29 For the system (A.25), the origin is:

- *stable* if for any $t_0 \in \Re$ and any $\hat{\varepsilon} > 0$, there exists a $\hat{\delta} = \hat{\delta}(t_0, \hat{\varepsilon}) > 0$ such that $\|x_{t_0}\|_c < \hat{\delta}$ implies $\|x(t)\| < \hat{\varepsilon}$ for $t \ge t_0$.
- *asymptotically stable* if it is stable and for any $t_0 \in \Re$ and any $\hat{\varepsilon} > 0$, there exists a $\hat{\delta}_a = \hat{\delta}_a(t_0, \hat{\varepsilon}) > 0$ such that $\|x_{t_0}\|_c < \hat{\delta}_a$ implies $\lim_{t\to 0} x(t) = 0$.
- *uniformly stable* if it is stable and $\hat{\delta}(t_0, \hat{\varepsilon})$ can be chosen independently of t_0.
- *uniformly asymptotically stable* if it is uniformly stable and there exists a $\hat{\delta}_a > 0$ such that for any $\eta_a > 0$, there exists a $T = T(\hat{\delta}_a, \eta_a)$ such that $\|x_{t_0}\|_c < \hat{\delta}$ implies $\|x(t)\| < \eta_a$ for $t \ge t_0 + T$ and $t_0 \in \Re$.

We define $\mathcal{C}_h$ as the space of continuous functions, ϕ, mapping $[-\tau, 0]$ to $\Re^n$ such that $\|\phi\|_c < h$, where h is a positive real number. Thus, $\mathcal{C}_h$ is the open ball of $\mathcal{C}$ of radius h. We use the following definition of exponential stability.

Definition A.30 The origin of (A.25) is exponentially stable if there exist positive real numbers h, a, and b such that for every initial condition $\phi \in \mathcal{C}_h$ the solution $x_t(\phi)$ of (A.25) exists for all $t \ge 0$ and furthermore satisfies

$$\big\|x_t(\phi)\big\|_c \le a e^{-bt} \|\phi\|_c$$

The above results apply to general retarded functional differential equations of the form (A.25). We next specialize stability criteria for linear time-delay systems. Consider the system of the form

$$\dot{x}(t) = a_0 x(t) + a_1 x(t-\tau) \tag{A.26}$$

where $x(t), \tau, a_0, a_1 \in \Re$. The initial condition for (A.26) is defined as the function $\phi : [-\tau, 0] \to \Re$ such that

$$x(t) = \phi(t) \quad \forall t \in [-\tau, 0] \tag{A.27}$$

The Laplace transform of (A.26) with initial conditions ϕ exists, and may be written as follows

$$X(s) = \frac{1}{\Delta(s)} \left\{ \phi(0) + a_1 \int_{-\tau}^{0} e^{-s(v+\tau)} \phi(v)\, dv \right\} \tag{A.28}$$

where

$$\Delta(s) = s - a_0 - a_1 e^{-\tau s} \tag{A.29}$$

is the characteristic quasipolynomial of (A.26). Let $\Phi(t)$ denote the inverse Laplace transform of $1/\Delta(s)$. We call $\Phi(t)$ the fundamental solution of (A.26). Notice that with the initial condition $\phi(0)=1$, and $\phi(v)=0$ for all $v\in[-\tau,0)$, then $\Phi(t)$ is the solution of (A.26). The growth of the general solution of (A.26) is related to the exponential growth of the fundamental solution $\Phi(t)$, which is determined by the poles of the system (A.28), i.e., the solutions of the characteristic equation

$$\Delta(s)=0 \tag{A.30}$$

In general, (A.30) has an infinite number of solutions. Denote by $\mathrm{Re}[s]$ the real component of the complex value s. The following proposition from [55] gives conditions for stability of (A.26).

Proposition A.31 *For any $\alpha\in\Re$, there are a finite number of poles with real parts greater than α. Let s_i for $i=1,2,\dots$ be the poles of the system* (A.26), *i.e., the solutions of* (A.30), *and let*

$$\alpha_0=\max_i \mathrm{Re}[s_i] \tag{A.31}$$

Then for any $\alpha>\alpha_0$, there exists a $L>0$ such that the solution of (A.26) *with the initial condition* (A.27) *satisfies the inequality*

$$\left|x(t)\right|\le Le^{\alpha t}\|\phi\|_c \tag{A.32}$$

By Proposition A.31, for the solution of (A.26) to approach zero as $t\to\infty$ for any initial condition, it is sufficient that the poles of the system have negative real parts. This stability result is similar to the stability result for systems without time-delay. In fact, this condition is both necessary and sufficient [55].

The above discussion for (A.26) may be generalized to linear systems of the following form:

$$\dot{x}(t)=A_0x(t)+\sum_{k=1}^{K}A_kx(t-\tau_k),\qquad \tau_k\ge 0 \tag{A.33}$$

where $A_0, A_k\in\Re^{n\times n}$, and $\tau_k\in\Re$ for $k=1,\dots,K$. The stability of (A.33) is determined by the characteristic quasipolynomial

$$p\big(s;e^{-\tau_1 s},\dots,e^{-\tau_K s}\big)=\det\big(\Delta(s)\big) \tag{A.34}$$

where

$$\Delta(s)=sI-A_0-\sum_{k=1}^{K}A_ke^{-\tau_k s}$$

We say (A.33) is asymptotically stable if and only if $p(s;e^{-\tau_1 s},\dots,e^{-\tau_K s})$ has no root in the closed RHP, or $\bar{C}_+$. This definition of stability is stated formally as follows [55].

Definition A.32 The system (A.33) is said to be asymptotically stable if and only if its characteristic quasipolynomial (A.34) satisfies

$$p\big(s; e^{-\tau_1 s}, \dots, e^{-\tau_K s}\big) \neq 0, \quad \forall s \in \bar{C}_+ \tag{A.35}$$

The following section presents the Lyapunov–Razumikhin stability theorems which provide alternative stability criteria for linear time-delay systems.

A.11.2 Razumikhin Stability Theory

We review the Lyapunov–Razumikhin theorem and a set of stability theorems for linear systems with time-delays [55]. This theorem does not depend on Lyapunov functionals, but on the traditional Lyapunov functions used in systems without delays.

Consider the linear time-invariant system of the form

$$\frac{dx(t)}{dt} = A_0 x(t) + A_1 x(t-\tau) \tag{A.36}$$

where A_0 and A_1 are $(n \times n)$ real matrices.

The following theorem, known as the restricted Lyapunov–Razumikhin theorem [55] (Proposition 5.1, p. 149) gives conditions for asymptotic stability.

Theorem A.33 *The time-delay system* (A.36) *with maximum time-delay* τ *is asymptotically stable if there exists a quadratic Lyapunov function* $V : \Re^n \to \Re$ *such that for some* $\psi_1 > 0$, $\psi_2 > 0$ *it satisfies*

$$\psi_1 \|x\|_2^2 \leq V(x) \leq \psi_2 \|x\|_2^2$$

where the time derivative along the system trajectory $\frac{dV(x)}{dt}$ *satisfies*

$$\frac{dV(x)}{dt} \leq -\psi_1 \big\|x(t)\big\|_2^2 \tag{A.37}$$

if $\forall p > 1, \exists t, \forall \xi \in [-\tau, 0]$ *such that*

$$V\big(x(t+\xi)\big) \leq pV\big(x(t)\big)$$

Notice that Theorem A.33 closely resembles the Lyapunov stability theorem [69]. The first condition is the same as in the Lyapunov stability theorem. The second condition requires that the time derivative of the Lyapunov function be non-positive which is also similar to the Lyapunov stability theorem. The third requirement is an extra condition that only appears for time-delayed systems. We explain this extra condition as follows. The negation of the third condition of Theorem A.33 is

$$\exists p > 1,\ \forall t,\ \exists \xi \in [-\tau, 0] \quad V\big(x(t+\xi)\big) > pV\big(x(t)\big) \tag{A.38}$$

Thus, for some $p > 1$, at every instant of time, t, the value $V(x(t))$ is less than a past instant. Intuitively, this means the state trajectory is approaching its equilibrium point, and hence, the system is stable. Thus, we only need to ensure the first two conditions of Theorem A.33 are satisfied if the third condition holds. If the third condition does not hold, then the system is implicitly stable. An interesting note is that the Lyapunov function and its time derivative are compared to the square of the norm of the state, which implies exponential stability.

Theorem A.33 presents a general set of stability conditions that do not exploit the structure of (A.36), i.e., A_0 and A_1 do not appear in the stability criteria. In addition, it is not obvious how to pick a Lyapunov function to satisfy the stability criteria in Theorem A.33. We show that in the foregoing work, the Lyapunov function $V = x^T Px$ may be used.

The remainder of this section is dedicated to delay-independent and delay-dependent stability criteria specific to the structure of the linear system (A.36). We first outline time-delay-independent stability conditions, which provide the simplest and most conservative stability conditions (Proposition 5.3 in [55]). These are provided below as Proposition A.34.

Proposition A.34 *The system* (A.36) *is asymptotically stable if there exist a scalar* $\hat{\alpha} > 0$ *and a real symmetric matrix* P *such that*

$$\begin{pmatrix} PA_0 + A_0^T P + \hat{\alpha} P & PA_1 \\ A_1^T P & -\hat{\alpha} P \end{pmatrix} < 0 \tag{A.39}$$

If $A_0 = 0$ then Proposition A.34 cannot be directly applied since this does not produces a feasible solution. Thus, more general time-delay-dependent conditions need to be applied. Time-delay-dependent stability conditions produce less conservative stability criteria. The following theorem presents a set of time-delay-dependent conditions to ensure asymptotic stability of (A.36) (Corollary 5.8 in [55]).

Theorem A.35 *The system* (A.36) *is asymptotically stable if there exist a real symmetric matrix* P *and real scalars* $\hat{\alpha}_0 > 0$, $\hat{\alpha}_1 > 0$ *such that*

$$\begin{pmatrix} M & -PA_1A_0 & -PA_1^2 \\ -A_0^T A_1^T P & -\hat{\alpha}_0 P & 0 \\ -(A_1^2)^T P & 0 & -\hat{\alpha}_1 P \end{pmatrix} < 0 \tag{A.40}$$

where

$$M = \frac{1}{\tau}\big[P(A_0 + A_1) + (A_0 + A_1)^T P\big] + (\hat{\alpha}_0 + \hat{\alpha}_1)P \tag{A.41}$$

An alternative approach based on Lyapunov–Krasovskii stability theory is less conservative than the Lyapunov–Razumikhin approach reviewed above. The trade-off in using Lyapunov–Krasovskii theory is in its analytical and computational complexity. This theory depends on Lyapunov functionals, which are structurally more complicated than traditional Lyapunov functions.

Appendix B
List of Notations

B.1 Notations and Symbols

Notations for signals and noise

Location	Notation
Transmitter	u_i—signal power of channel i
	$\mathbf{u} = [u_1 \cdots u_m]^T$—vector signal power
	n_i^0—noise power of channel i
Receiver	$\mathbf{n}^0 = [n_1^0 \cdots n_m^0]^T$—vector noise power
	p_i—signal power of channel i
	$\mathbf{p} = [p_1 \cdots p_m]^T$—vector signal power
	n_i—noise power of channel i
Link l	$\mathbf{n} = [n_1 \cdots n_m]^T$—vector noise power
	$\mathbf{y} = [y_1 \cdots y_m]^T$—vector OSNR
	$u_{i,l}$—signal power of channel i at the input
	$n_{i,l}^{\text{in}}$—noise power of channel i at the input
	$p_{i,l}$—signal power of channel i at the output
	$n_{i,l}^{\text{out}}$—noise power of channel i at the output
	$\mathbf{u}_l = [u_{1,l} \cdots u_{m,l}]^T$—vector signal power at input of link l
	$\mathbf{p}_l = [p_{1,l} \cdots p_{m,l}]^T$—vector signal power at output of link l
	$\bar{\mathbf{u}}$—vector signal power at input of all links
	$\bar{\mathbf{u}} = [\mathbf{u}_1^T \cdots \mathbf{u}_L^T]^T$
	$\mathbf{y}_l = [y_{1,l} \cdots y_{m,l}]^T$—vector OSNR
Span s on Link l	$p_{i,l,s}$—signal power of channel i at the output
	$n_{i,l,s}$—noise power of channel i at the output

L. Pavel, *Game Theory for Control of Optical Networks*,
Static & Dynamic Game Theory: Foundations & Applications,
DOI 10.1007/978-0-8176-8322-1, © Springer Science+Business Media New York 2012

Notations for parameters

$\mathcal{L}$	Set of links
$\mathcal{M}$	Set of channels
$\mathcal{R}_i$	Route of channel i
$\mathcal{M}_l$	Set of channels transmitted on link l
$\widehat{P}_l$	Total power target on link l
$L_{l,s}$	Fiber loss coefficient at span s on link l
$G_{i,l,s}$	Gain coefficient of channel i at span s on link l
$ASE_{i,l,s}$	ASE noise power on channel i at span s on link l
$h_{i,l,s}$	Span transmission for channel i at span s on link l
$T_{l,i}$	Link transmission for channel i on link l
$G_{i,l}$	Gain value for channel i on the spectral shape of the optical amplifier on link l
$G = [G_{i,s}]$ $\delta_{l,s}$	Filter loss of the optical amplifier at span s on link l
Γ	Network system matrix
Γ_l	Link system matrix

B.2 List of Symbols

H_∞ Hardy space of all complex-valued functions of a complex variable s which are analytic and bounded in the open RHP.

$\bar{C}_+$ closed RHP of the complex plane.

$\hat{t}$ slow time-scale continuous-time variable.

$\dot{v} = \frac{dv}{dt}$ dot notation represents the derivative with respect to the slow time-scale, $\hat{t}$.

$\lambda_{\min}(A)$, $\lambda_{\max}(A)$ maximum and minimum eigenvalues, respectively, of matrix A.

$\bar{\sigma}(A)$ largest singular value of a matrix A.

$\underline{\sigma}(A)$ smallest singular value of a matrix A.

$\mathcal{C} = \mathcal{C}([-\tau, 0], \Re^n)$ set of all continuous functions mapping $[-\tau, 0]$ to $\Re^n$.

$x_t \in \mathcal{C}$ set of values $x(q)$ for all $q \in [t-\tau, t]$.

$\mathcal{C}_{\hat{h}}$ space of continuous functions, ϕ, mapping $[-\tau, 0]$ to $\Re^n$ such that $\|\phi\|_c < \hat{h}$, where $\hat{h}$ is a positive real number.

$OSNR_i$ OSNR for the ith channel.

$G_{\ell,k,i}$ or $G_{\ell,i}$ channel dependent gain of the OA in the kth span on the ℓth link for the ith channel.

$L_{\ell,k}$ wavelength independent loss coefficient for the kth span on the ℓth link.

$ASE_{\ell,k,i}$ ASE noise power introduced by the OA in the kth span on the ℓth link for the ith channel.

$h_{\ell,k,i}$ optical span transmission, or the total power transmitted by the kth span on the ℓth link for the ith channel.

$T_{\ell,i}$ link transmission, or the total transmitted power across all spans for the ℓth link and the ith channel.

λ_i optical wavelength corresponding to channel i.

ν_i optical frequency corresponding to channel i.

u_{-i} vector of elements u_i with the ith entry deleted.
$u_{\max}$ largest possible input power, i.e., $u_i \le u_{\max}$ for all i.
J_i Nash game cost function for channel i.
$\mathcal{U}_i$ utility function for the Nash game or channel i.
α_i channel cost associated with the input power u_i.
β_i utility weight associated with U_i.
a_i a channel dependent design parameter for channel i.
$\tilde{\Gamma}$ Γ matrix such that the diagonal elements are replaced with $\mathrm{diag}(a_i)$.
μ scalar channel price computed by the link algorithm.
K number of iterations of the time step, n, before the link algorithm is updated.
$\bar{k}$ iterative time step on the slow time-scale of the link algorithm.
$\tau^f_{i,j}$ forward propagation delay experienced by the signal from source j to OSNR output i.
τ^b_i backward propagation delay experienced by the signal from OSNR output i to the channel source i.
$\tau_{i,j}$ time-delay from source j to OSNR output i to the source i.
τ^f_j forward propagation delay from source j to the single OSNR output of the single-sink network configuration.
τ^f forward propagation delay from source to OSNR output in the single-link configuration.
τ^b backward propagation delay from the OSNR output to the source in the single-link configuration.
τ round trip time-delay, from source to OSNR output and back, in the single-link configuration.
ρ_i tunable control parameter for the ith channel algorithm.
$\mathrm{eig}(A)$ set of eigenvalues of the matrix A.
$\mathrm{Re}[s]$ real component of the complex value s.
$\mathrm{Im}[s]$ imaginary component of the complex value s.
s complex domain variable.
1_{row} a row vector with all elements equal to 1.
t continuous-time variable on the "fast" time-scale of the channel algorithm.
ε relates $\hat{t}$ to t by $\hat{t} = \varepsilon t$, where $\varepsilon = \frac{1}{K}$.
$\hat{u}_i$ composition $u_i \circ t$.
β column matrix with elements β_i.
z_i shifted u_i variable, $z_i = \hat{u}_i - u_i^*$.
x shifted channel price parameter, μ, i.e. $x = \mu - \mu^*$.
$\tilde{\Gamma}_a$ a matrix with elements $\frac{\tilde{\Gamma}_{i,j}}{a_i}$.
ρ diagonal matrix with elements ρ_i, i.e. $\rho = \mathrm{diag}(\rho_i)$.
$\hat{z}(\hat{t})$ coordinate shift defined as $\hat{z}(\hat{t}) = z(\hat{t} - \hat{\tau}^f) - h(x(\hat{t}))$.
$\hat{x}(\hat{t})$ solution of the reduced system (11.46).
z composition defined as $z = (\hat{y} \circ \hat{t})$.
$\hat{\tau}$ time-delay in the slow time-scale which relates to τ via ε as $\hat{\tau} = \varepsilon\tau$.
$\check{z}$ solution of the boundary-layer system (11.47).
A_1 boundary-layer system matrix, $A_1 = -\rho\tilde{\Gamma}_a$.

ζ represents a characteristic polynomial.
Ω_{Γ} uncertainty set for the system matrix, Γ.
Γ_{Δ} perturbed system of the nominal system matrix Γ, i.e., $\Gamma_{\Delta} \in \Omega_{\Delta}$.
$\Delta\Gamma$ additive uncertainty matrix for the nominal system matrix Γ.
$\Delta\Gamma_{i,\text{row}}$ an $(n \times n)$ matrix with all elements equal to zero except the ith row of $\Delta\Gamma$.
$\hat{q}_i$ independent uncertainty bound on the 2-norm of $\Delta\Gamma_{i,\text{row}}$.
Ω_j uncertainty set for the input u_j.
$u_{\Delta j}$ uncertain input signal, i.e., $u_{\Delta j} \in \Omega_j$.
δ_j input multiplicative uncertainty on the input signal u_j.
$\bar{q}_j$ independent uncertainty bound for the input uncertainty, i.e., $\|\delta_j\|_2 \leq \bar{q}_j$.
Ω_{ni} uncertainty set for the transmitter noise $n_{0,i}$.
$n_{\Delta,0,i}$ uncertain transmitter noise, i.e., $n_{\Delta,0,i} \in \Omega_{ni}$.
δ_{ni} transmitter noise uncertainty.

References

1. Acemoglu, D., Ozaglar, A.: Costs of competition in general networks. In: Proceedings of the 44th IEEE Conference on Decision and Control, December, pp. 5324–5329 (2005)
2. Agrawal, G.P.: Fiber-Optic Communication Systems, 3rd edn. Wiley, New York (2002)
3. Alpcan, T.: Noncooperative games for control of networked systems. Ph.D. thesis, University of Illinois at Urbana–Champaign, Illinois, USA (2006)
4. Alpcan, T., Basar, T.: A game-theoretic framework for congestion control in general topology networks. In: Proceedings of the 41st IEEE Conference on Decision and Control, December, pp. 1218–1224 (2002)
5. Alpcan, T., Basar, T.: Global stability analysis of an end-to-end congestion control. In: Proceedings of the 42nd IEEE Conference on Decision and Control, December, pp. 1092–1097 (2003)
6. Alpcan, T., Basar, T.: A hybrid systems model for power control in multicell wireless data networks. Perform. Eval. **57**(4), 477–495 (2004)
7. Alpcan, T., Basar, T.: Distributed algorithms for Nash equilibria of flow control games. Adv. Dyn. Games: Appl. Econ. Finance Optim. Stoch. Control Ann. Dynam. Games **7**, 473–498 (2005)
8. Alpcan, T., Basar, T., Srikant, R., Altman, E.: CDMA uplink power control as a noncooperative game. In: Proceedings of the 40th IEEE Conference on Decision and Control, December, pp. 197–202 (2001)
9. Alpcan, T., Fan, X., Basar, T., Arcak, M., Wen, J.T.: Power control for multicell CDMA wireless networks: a team optimization approach. Wirel. Netw. **14**(5), 647–657 (2008)
10. Altman, E., Altman, Z.: S-modular games and power control in wireless networks. IEEE Trans. Autom. Control **48**(5), 839–842 (2003)
11. Altman, E., Basar, T.: Multiuser rate-based flow control. IEEE Trans. Commun. **46**(7), 940–949 (1998)
12. Altman, E., Basar, T., Jimenez, T., Shimkin, N.: Routing into two parallel links: game-theoretic distributed algorithms. J. Parallel Distrib. Comput. **61**(9), 1367–1381 (2001)
13. Altman, E., Basar, T., Srikant, R.: Nash equilibria for combined flow control and routing in networks: asymptotic behavior for a large number of users. IEEE Trans. Autom. Control **47**(6), 917–930 (2002)
14. Altman, E., Boulogne, T., El-Azouziand, R., Jimenez, T., Wynter, L.: A survey on networking games in telecommunications. Comput. Oper. Res. **33**(2), 286–311 (2006)
15. Altman, E., Solan, E.: Constrained games: the impact of the attitude to adversary's constraints. IEEE Trans. Autom. Control **54**(10), 2435–2440 (2009)
16. Arrow, K.J., Debreu, G.: Existence of an equilibrium for a competitive economy. Econometrica **22**(3), 265–290 (1954)

L. Pavel, *Game Theory for Control of Optical Networks*,
Static & Dynamic Game Theory: Foundations & Applications,
DOI 10.1007/978-0-8176-8322-1,

17. El Azouzi, R., Altman, E.: Constrained traffic equilibrium in routing. IEEE Transactions on Automatic Control **48**(9), 1656–1660 (2003)
18. Bala, K., Brackett, C.: Cycles in wavelength routed optical networks. IEEE J. Lightwave Technol. **14**(7), 1585–1594 (1996)
19. Banerjee, D., Mukherjee, B.: A practical approach for routing and wavelength assignment in large wavelength-routed optical networks. IEEE J. Sel. Areas Commun. **14**(5), 903–908 (1996)
20. Basar, T., Olsder, G.J.: In: Dynamic Noncooperative Game Theory. SIAM Series Classics in Applied Mathematics, 2nd edn. (1999)
21. Basar, T., Srikant, R.: Revenue-maximizing pricing and capacity expansion in a many-users regime. In: Proceedings of the 21st IEEE Conference on Computer Communications (INFOCOM), June 2002, pp. 294–301 (2002)
22. Bazarra, M.S., Sherali, H.D., Shetty, C.M.: In: Nonlinear programming: theory and algorithms, 3rd edn. Wiley-Interscience, New York (2006)
23. Bellare, M., Goldreich, O., Sudan, M.: Free bits, PCPs and non-approximability—towards tight results. SIAM J. Comput. **27**, 804–915 (1998)
24. Bertsekas, D.P.: Nonlinear Programming, 2nd edn. Athena Scientific, Nashua (1999)
25. Bertsekas, D.P., Tsitsiklis, J.N.: Parallel and Distributed Computation: Numerical Methods. Prentice Hall, New York (1989)
26. Bilo, V., Flammini, M., Moscardelli, L.: On Nash equilibria in non-cooperative all-optical networks. In: Diekert, V., Durand, B. (eds.) STACS 2005. LNCS, vol. 3404, pp. 448–459. Springer, Berlin (2005)
27. Bilo, V., Flammini, M., Moscardelli, L.: On Nash Equilibria in Non-cooperative All-Optical Networks. Optical Switching and Networking (2010)
28. Bilo, V., Moscardelli, L.: The price of anarchy in all-optical networks. In: S'ykora, R.O. (ed.) SIROCCO 2004. LNCS, vol. 3104, pp. 13–22. Springer, Heidelberg (2004)
29. Borella, M., Jue, J., Banerjee, D., Ramamurthy, B., Mukherjee, B.: Optical components for WDM lightwave networks. Proc. IEEE **85**(8), 1274–1307 (1997)
30. Boyd, S., Vandenberghe, L.: Convex Optimization. Cambridge University Press, Cambridge (2004)
31. Brown, G.W.: Activity Analysis of Production and Allocation. Wiley, New York (1951)
32. Charnes, A., Eisner, M., Kortanek, K.O.: On weakly balanced games and duality theory. Cah. Cent. Etud. Rech. Oper. **12**, 7–21 (1970)
33. Chraplyvy, A.: Limitations on lightwave communications imposed by optical-fiber nonlinearities. IEEE/OSA J. Lightwave Technol. **8**(10), 1548–1557 (1990)
34. Chraplyvy, A., Nagel, J., Tkach, R.: Equalization in amplified WDM lightwave transmission systems. IEEE Photonics Technol. Lett. **4**(8), 920–922 (1992)
35. de Meyer, B.: Repeated games, duality and the central limit theorem. Math. Oper. Res. **21**(1), 237–251 (1996)
36. de Meyer, B., Marino, A.: Duality and optimal strategies in the finitely repeated zero-sum games with incomplete information on both sides. In: Cahiers de la Maison des Sciences Economiques, pp. 1–7 (2005)
37. Debreu, G.: A social equilibrium existence theorem. Proc. Natl. Acad. Sci. USA **38**(10), 886–893 (1952)
38. Dubey, P.: Inefficiency of Nash equilibria. Math. Oper. Res. **11**(1), 1–8 (1986)
39. Erlebach, T., Jansen, K., Kaklamanis, C., Persiano, P.: Optimal wavelength routing on directed fiber trees. Theor. Comput. Sci. **221**(1–2), 119–137 (1999)
40. Facchinei, F., Fischer, A., Piccialli, V.: On generalized Nash games and variational inequalities. Oper. Res. Lett. **35**(2), 159–164 (2007)
41. Falomari, D., Mandayam, N., Goodman, D.: A new framework for power control in wireless data networks: games utility and pricing. In: Proceedings of the Allerton Conference on Communication, Control, and Computing, pp. 546–555 (1998)
42. Fanelli, A., et al.: Game theoretical issues in optical networks. In: Proceedings of ÊICTON 2006, Sep. (2006)

43. Fiedler, M.: Special Matrices and Their Applications in Numerical Mathematics. Martinus Nijhoff Publishers, Leiden (1986)
44. Fishman, G.S.: Monte Carlo: Concepts, Algorithms, and Applications. Springer, Berlin (1996)
45. Forghieri, F., Tkach, R.W., Favin, D.L.: Simple model of optical amplifier chains to evaluate penalties in WDM systems. IEEE/OSA J. Lightwave Technol. **16**(9), 1570–1576 (1998)
46. Foschini, G., Miljanic, Z.: A simple distributed autonomous power control algorithm and its convergence. IEEE Trans. Veh. Technol. **42**(4), 641–646 (1993)
47. Friedman, D.: Evolutionary games in economics. Econometrica **59**, 637–666 (1991)
48. Fudenberg, D., Kreps, D.M.: Learning mixed equilibria. Games Econ. Behav. **5**, 320–367 (1993)
49. Fudenberg, D., Levine, D.K.: Theory of Learning in Games. MIT Press, Cambridge (1998)
50. Fudenberg, D., Tirole, J.: Game Theory, 2nd edn. MIT Press, Cambridge (1991)
51. Garey, M.R., Johnson, D.S., Miller, G.L., Papadimitriou, C.H.: The complexity of coloring circular arcs and chords. SIAM J. Algebr. Discrete Methods **1**(2), 216–227 (1980)
52. Gargano, L., Vaccaro, U.: Routing in all-optical networks: algorithmic and graph theoretic problems. In: Numbers, Information and Complexity, pp. 555–578. Kluwer Academic, Dordrecht (2000)
53. Georgakopoulos, G.F., Kavvadias, D.J., Sioutis, L.G.: Nash equilibria in all-optical networks. Discrete Math. **309**(13), 4332–4342 (2009)
54. Greenwald, A., Friedman, E.J., Shenker, S.: Learning in network contexts: experimental results from simulations. Games Econ. Behav. **35**, 80–123 (2001)
55. Gu, K., Kharitonov, V.L., Chen, J.: Stability of Time-Delay Systems. Birkhäuser, Boston (2003)
56. Harker, P.T.: Generalized Nash games and quasi-variational inequalities. Eur. J. Oper. Res. **54**(1), 81–94 (1991)
57. Harsanyi, J.C.: Games with randomly disturbed payoffs: a new rationale for mixed-strategy equilibrium points. Int. J. Game Theory **2**, 1–23 (1973)
58. Hofbauer, J., Sigmund, K.: Evolutionary Games and Population Dynamics. Cambridge University Press, Cambridge (1998)
59. Horn, R.A., Johnson, C.R.: Matrix Analysis. Cambridge University Press, Cambridge (1999)
60. Huang, M., Malhame, R.P., Caines, P.E.: Nash certainty equivalence in large population stochastic dynamic games: connections with the physics of interacting particle systems. In: Proceedings of the 45th IEEE Conference on Decision and Control, December, pp. 4921–4926 (2006)
61. Ji, H., Huang, C.: Non-cooperative uplink power control in cellular radio systems. Wirel. Netw. **4**(3), 233–240 (1998)
62. Jia, X., Du, D., Hu, X., Lee, M., Gu, J.: Optimization of wavelength assignment for QoS multicast in WDM networks. IEEE Trans. Commun. **49**(2), 341–350 (2001)
63. Johari, R., Mannor, S., Tsitsiklis, J.N.: Efficiency loss in a resource allocation game: a single link in elastic supply. In: Proceedings of the 43rd IEEE Conference on Decision and Control, December, pp. 4679–4683 (2004)
64. Johari, R., Tsitsiklis, J.: Efficiency loss in a network resource allocation game. Math. Oper. Res. **29**(3), 407–435 (2004)
65. Kandukuri, S., Boyd, S.: Simultaneous rate and power control in multirate multimedia CDMA systems. In: Proceedings of the 6th IEEE International Symposium on Spread Spectrum Techniques and Applications, September, pp. 570–574 (2000)
66. Kandukuri, S., Boyd, S.: Optimal power control in interference-limited fading wireless channels with outage-probability specifications. IEEE Trans. Wirel. Commun. **1**(1), 46–55 (2002)
67. Kelly, F.: Charging and rate control for elastic traffic. Eur. Trans. Telecommun. **8**, 33–37 (1997)
68. Kelly, F.P., Maulloo, A.K., Tan, D.: Rate control for communication networks: shadow prices, proportional fairness and stability. J. Oper. Res. Soc. **49**(3), 237–252 (1998)
69. Khalil, H.K.: Nonlinear Systems, 3rd edn. Prentice Hall, New York (2002)

70. Kinderlehrer, D.: An Introduction to Variational Inequalities and Their Applications. Academic Press, San Diego (1980)
71. Korpelevich, G.M.: The extragradient method for finding saddle points and other problems. Èkon. Mat. Metody **12**, 747–756 (1976) [translated as Matecon.]
72. Koskie, S., Gajic, Z.: A Nash game algorithm for SIR-based power control in 3G wireless CDMA networks. IEEE/ACM Trans. Netw. **13**(5), 1017–1026 (2005)
73. Koutsoupias, E., Papadimitriou, C.H.: Worst-case equilibria. In: Proceedings of the 16th Annual Symposium STACS, pp. 404–413 (1999)
74. Krasovskii, N.N., Subbotin, A.I.: Game-Theoretical Control Problems. Springer, Berlin (1988)
75. Krishna, V., Sjostrom, T.: On the convergence of fictitious play. Math. Oper. Res. **23**, 479–511 (1998)
76. Kunniyur, S., Srikant, R.: End-to-end congestion control schemes: utility functions, random losses and ECN marks. IEEE/ACM Trans. Netw. **11**(5), 689–702 (2003)
77. Laplante, P.A.: Comprehensive Dictionary of Electrical Engineering. Springer, Berlin (1999)
78. Libman, L., Orda, A.: The designer's perspective to atomic noncooperative networks. IEEE/ACM Trans. Netw. **7**(6), 875–884 (1999)
79. Loja, A., et al.: Inter-domain routing in multiprovider optical networks: game theory and simulations. In: Proceedings of Next Gen. Internet Networks, May, pp. 157–164 (2005)
80. Low, S., Paganini, F., Doyle, J.: Internet congestion control. IEEE Control Syst. Mag. **22**(1), 28–43 (2002)
81. Low, S.H., Lapsley, D.E.: Optimization flow control-I: basic algorithm and convergence. IEEE/ACM Trans. Netw. **7**(6), 861–874 (1999)
82. Luo, Z., Pang, J.: Analysis of iterative waterfilling algorithm for multiuser power control in digital subscriber lines. EURASIP J. Appl. Signal Process. **2006**, 1–10 (2006)
83. Marbach, P., Berry, R.: Downlink resource allocation and pricing for wireless networks. In: Proceedings of the 21st IEEE Conference on Computer Communications (INFOCOM), June, pp. 1470–1479 (2002)
84. Marden, J.R., Arslan, G., Shamma, J.S.: Joint strategy fictitious play with inertia for potential games. In: Proceedings of the 44th IEEE Conference on Decision and Control, December, pp. 6692–6697 (2005)
85. Mecozzi, A.: On the optimization of the gain distribution of transmission lines with unequal amplifier spacing. IEEE Photonics Technol. Lett. **10**(7), 1033–1035 (1998)
86. Milis, I., Pagourtzis, A., Potika, K.: Selfish routing and path coloring in all-optical networks. In: Proceedings of CAAN2007, pp. 71–84 (2007)
87. Mitra, D.: An asynchronous distributed algorithm for power control in cellular radio systems. In: Proceedings of the 4th WINLAB Workshop on Third Generation Wireless Information Networks, November, pp. 249–257 (1993)
88. Mokhtar, A., Azizoglu, M.: Adaptive wavelength routing in all-optical networks. IEEE/ACM Trans. Netw. **6**(2), 197–206 (1998)
89. Monderer, D., Shapley, L.S.: Potential games. Games Econ. Behav. **14**(1), 124–143 (1996)
90. Morgan, J., Romaniello, M.: Generalized quasi-variational inequalities and duality. J. Inequal. Pure Appl. Math. **4**(2), Article 28, 1–7 (2003)
91. Morgan, J., Romaniello, M.: Generalized quasi-variational inequalities: duality under perturbations. J. Math. Anal. Appl. **324**(2), 773–784 (2006)
92. Mukherjee, B.: WDM optical communication networks: progress and challenges. IEEE J. Sel. Areas Commun. **18**(10), 1810–1824 (2000)
93. Myerson, R.: Game Theory: Analysis of Conflict. Harvard University Press, Cambridge (1991)
94. Mync, H.: Nonnegative Matrices. Wiley, New York (1988)
95. Nash, J.: Equilibrium points in n-person games. Proc. Natl. Acad. Sci. USA **36**(1), 48–49 (1950)
96. Nash, J.: Non-cooperative games. Ann. Math. **54**(2), 286–295 (1951)

97. Nisan, N., Roughgarden, T., Tardos, E., Vazirani, V. (eds.): Algorithmic Game Theory. Cambridge University Press, Cambridge (2007)
98. Owen, G.: Game Theory, 3rd edn. Academic Press, San Diego (1995)
99. Ozdaglar, A., Bertsekas, D.: Routing and wavelength assignment in optical networks. IEEE/ACM Trans. Netw. **11**(2), 259–272 (2003)
100. Paganini, F., Wang, Z., Doyle, J.C., Low, S.H.: Congestion control for high performance, stability, and fairness in general networks. IEEE/ACM Trans. Netw. **13**(1), 43–55 (2005)
101. Pan, Y.: A game theoretical approach to constrained OSNR optimization problems in optical networks. Ph.D. Thesis, University of Toronto, Toronto, Canada (2009)
102. Pan, Y., Alpcan, T., Pavel, L.: A distributed optimization approach to constrained OSNR problem. In: Proceedings of the 17th IFAC World Congress, July (2008)
103. Pan, Y., Alpcan, T., Pavel, L.: Effects of parameters on Nash games with OSNR target. In: Proceedings of the 2nd International Workshop on Game Theory in Communication Networks (GameComm), October (2008)
104. Pan, Y., Alpcan, T., Pavel, L.: A system performance approach to OSNR optimization in optical networks. IEEE Trans. Commun. **58**(4), 1193–1200 (2010)
105. Pan, Y., Pavel, L.: OSNR optimization in optical networks: extension for capacity constraints. In: Proceedings of the 2005 American Control Conference, June, pp. 2379–2385 (2005)
106. Pan, Y., Pavel, L.: Games with coupled propagated constraints in optical networks: the multi-link case. In: Proceedings of the 46th IEEE Conference on Decision and Control, December, pp. 3443–3449 (2007)
107. Pan, Y., Pavel, L.: Global convergence of an iterative gradient algorithm for the Nash equilibrium in an extended OSNR game. In: Proceedings of the 26th IEEE Conference on Computer Communications (INFOCOM), May, pp. 206–212 (2007)
108. Pan, Y., Pavel, L.: Iterative algorithms for Nash equilibrium of an extended OSNR optimization game. In: Proceedings of the 6th International Conference on Networking (ICN), April (2007)
109. Pan, Y., Pavel, L.: OSNR optimization with link capacity constraints in WDM networks: a cross layer game approach. In: Proceedings of the 4th IEEE Conference on Broadband Communications, Networks and Systems (BroadNets), September (2007)
110. Pan, Y., Pavel, L.: A Nash game approach for OSNR optimization with capacity constraint in optical links. IEEE Trans. Commun. **56**(11), 1919–1928 (2008)
111. Pan, Y., Pavel, L.: Games with coupled propagated constraints in general topology optical networks. In: Proceedings of the 1st International Conference on Game Theory in Networks (GameNets), May, pp. 549–558 (2009)
112. Pan, Y., Pavel, L.: Games with coupled propagated constraints in optical networks with multi-link topologies. Automatica **45**(4), 871–880 (2009)
113. Parrilo, P.A.: Polynomial games and sum of squares optimization. In: Proceedings of the 45th IEEE Conference on Decision and Control, December, pp. 2855–2860 (2006)
114. Pavel, L.: Dynamics and stability in optical communication networks: a system theory framework. Automatica **40**(8), 1361–1370 (2004)
115. Pavel, L.: Power control for OSNR optimization in optical networks: a noncooperative game approach. In: Proceedings of the 43rd IEEE Conference on Decision and Control, December, pp. 3033–3038 (2004)
116. Pavel, L.: An extension of duality and hierarchical decomposition to a game-theoretic framework. In: Proceedings of the 44th IEEE Conference on Decision and Control, December, pp. 5317–5323 (2005)
117. Pavel, L.: A nested noncooperative game formulation for OSNR optimization in distributed optical links. In: Proceedings of the 44th IEEE Conference on Decision and Control, December, pp. 6958–6965 (2005)
118. Pavel, L.: A noncooperative game approach to OSNR optimization in optical networks. IEEE Trans. Autom. Control **51**(5), 848–852 (2006)
119. Pavel, L.: OSNR optimization in optical networks: modeling and distributed algorithms via a central cost approach. IEEE J. Sel. Areas Commun. **24**(4), 54–65 (2006)

120. Pavel, L.: An extension of duality to a game-theoretic framework. Automatica **43**(2), 226–237 (2007)
121. Pavel, L.: A nested noncooperative OSNR game in distributed WDM optical links. IEEE Trans. Commun. **55**(6), 1220–1230 (2007)
122. Perakis, G.: The price of anarchy when costs are non-separable and asymmetric. In: Proceedings of the 10th International Integer Programming and Combinatorial Optimization (IPCO) Conference, June, pp. 46–58 (2004)
123. Perakis, G.: The "price of anarchy" under nonlinear and asymmetric costs. Math. Oper. Res. **32**(3), 614–628 (2007)
124. Ramamurthy, B., Datta, D., Feng, H., Heritage, J., Mukherjee, B.: Impact of transmission impairments on the teletraffic performance of wavelength-routed optical networks. IEEE J. Lightwave Technol. **17**(10), 1713–1723 (1999)
125. Ramaswami, R., Sivarajan, K.: Routing and wavelength assignment in all-optical networks. IEEE/ACM Trans. Netw. **3**(5), 489–500 (1995)
126. Ramaswami, R., Sivarajan, K.N.: Optical Networks: A Practical Perspective, 2nd edn. Academic Press, San Diego (2002)
127. Robinson, J.: An iterative method of solving a game. Ann. Math. **54**, 296–301 (1951)
128. Rosen, J.B.: Existence and uniqueness of equilibrium points for concave n-person games. Econometrica **33**(3), 520–534 (1965)
129. Roughgarden, T.: Selfish Routing and the Price of Anarchy. MIT Press, Cambridge (2005)
130. Roughgarden, T., Tardos, E.: How bad is selfish routing? J. ACM **49**(2), 236–259 (2002)
131. Rouskas, G., Perros, H.: A tutorial on optical networks. In: Advanced Lectures on Networking, vol. 2497, pp. 155–193. Springer, Berlin (2002)
132. Sandholm, W.H.: Population Games and Evolutionary Dynamics. Cambridge University Press, Cambridge (2009)
133. Saraydar, C., Mandayam, N.B., Goodman, D.J.: Pricing and power control in a multicell wireless data network. IEEE J. Sel. Areas Commun. **19**(10), 1883–1892 (2001)
134. Saraydar, C., Mandayam, N.B., Goodman, D.J.: Efficient power control via pricing in wireless data networks. IEEE Trans. Commun. **50**(2), 291–303 (2002)
135. Scutari, G., Barbarossa, S., Palomar, D.P.: Potential games: a framework for vector power control problems with coupled constraints. In: Proceedings of IEEE International Conference on Acoustics, Speech and Signal Processing (ICASSP) (2006)
136. Selten, R.: Re-examination of the perfectness concept for equilibrium points in extensive games. Int. J. Game Theory **4**, 25–55 (1975)
137. Shamma, J.S., Arslan, G.: Dynamic fictitious play, dynamic gradient play and distributed convergence to Nash equilibria. IEEE Trans. Autom. Control **50**(1), 312–326 (2005)
138. Shandbag, U., et al.: Nash equilibrium problems with congestion costs and shared constraints. In: Proceedings of the Conference on Decision and Control, December (2009)
139. Shen, H., Basar, T.: Differentiated Internet pricing using a hierarchical network game model. In: Proceedings of the 2004 American Control Conference, June, pp. 2322–2327 (2004)
140. Slusarek, M.: Optimal online colouring of circular arc graphs. Inform. Theor. Appl. **29**(5), 423–429 (1995)
141. Smith, J.M.: The theory of games and the evolution of animal conflicts. J. Theor. Biol. **7**, 209–221 (1974)
142. Smith, J.M., Price, G.R.: The logic of animal conflict. Nature **246**, 15–18 (1973)
143. Srikant, R.: The Mathematics of Internet Congestion Control. Systems & Control: Foundations & Applications. Birkhäuser, Boston (2004)
144. Stefanovic, N.: Robust power control of optical networks with time-delays. Ph.D. thesis, University of Toronto, Toronto, Canada (2010)
145. Stefanovic, N., Pavel, L.: Link level power control of optical networks with time-delay. In: Proceedings of the 46th IEEE Conference on Decision and Control, December, pp. 5198–5203 (2007)
146. Stefanovic, N., Pavel, L.: A Lyapunov–Krasovskii stability analysis for game-theoretic based power control in optical networks. In: Proceedings of GameComm 2008, October (2008)

147. Stefanovic, N., Pavel, L.: Robust power control of optical links with time-delay. In: Proceedings of European Control Conference, Budapest, Hungary, August (2009)
148. Stefanovic, N., Pavel, L.: Robust power control of single sink optical networks with time-delays. In: Proceedings of the 48th IEEE Conference on Decision and Control, Shanghai, China, December, pp. 2034–2039 (2009)
149. Stefanovic, N., Pavel, L.: A stability analysis with time-delay of primal-dual power control in optical links. Automatica **45**(5), 1319–1325 (2009)
150. Stefanovic, N., Pavel, L.: A Lyapunov–Krasovskii stability analysis for game-theoretic based power control in optical links. J. Telecommun. Syst. **47**(1), 19–33 (2011)
151. Stein, N.D., Ozdaglar, A., Parrilo, P.A.: Separable and low-rank continuous games. In: Proceedings of the 45th IEEE Conference on Decision and Control, December, pp. 2849–2854 (2006)
152. Strand, J., Chiu, A., Tkach, R.: Issues for routing in the optical layer. IEEE Commun. Mag. **39**(2), 81–87 (2001)
153. Tan, C., Palomar, D., Chiang, M.: Solving nonconvex power control problems in wireless networks: low SIR regime and distributed algorithms. In: Proceedings of the 48th IEEE Global Telecommunications Conference (GLOBECOM), November (2005)
154. Tonguz, O., Flood, F.: Gain equalization of EDFA cascades. IEEE J. Lightwave Technol. **15**(10), 1832–1841 (1997)
155. Tonguz, O., Flood, F.: EDFA-based DWDM lightwave transmission systems with end-to-end power and SNR equalization. IEEE Trans. Commun. **50**(8), 1282–1292 (2002)
156. Turkcu, O., Subramaniam, S.: Blocking in reconfigurable optical networks. In: Proceedings of the 26th IEEE Conference on Computer Communications (INFOCOM), May, pp. 188–196 (2007)
157. van Damme, E.: Stability and Perfection of Nash Equilibria, 2nd edn. Springer, Berlin (1987)
158. Vincent, T.L., Brown, J.S.: Evolutionary Game Theory, Natural Selection, and Darwinian Dynamics. Cambridge University Press, Cambridge (2005)
159. von Neumann, J., Morgenstern, O.: Theory of Games and Economic Behavior. Princeton University Press, Princeton (1944)
160. Yaiche, H., Mazumdar, R., Rosenberg, C.: A game theoretic framework for bandwidth allocation and pricing in broadband networks. IEEE/ACM Trans. Netw. **8**(5), 667–678 (2000)
161. Yariv, A.: Signal-to-noise considerations in fiber links with periodic or distributed optical amplification. Opt. Lett. **15**(19), 1064–1066 (1990)
162. Yassine, A., et al.: Competitive game theoretic optimal routing in optical networks. In: Proceedings SPIE, May, pp. 27–36 (2002)
163. Yates, R.: A framework for uplink power control in cellular radio systems. IEEE J. Sel. Areas Commun. **13**(7), 1341–1347 (1995)
164. Zander, J.: Performance of optimum transmitter power control in cellular radio systems. IEEE Trans. Veh. Technol. **41**(1), 57–62 (1992)
165. Zang, H., Jue, J., Mukherjee, B.: A review of routing and wavelength assignment approaches for wavelength-routed optical WDM networks. Opt. Netw. Mag. **1**(1), 47–60 (2000)
166. Zhu, H., Mukherjee, B.: Online connection provisioning in metro optical WDM networks using reconfigurable OADMs. IEEE/OSA J. Lightwave Technol. **23**(10), 2893–2901 (2005)
167. Zhu, K., Mukherjee, B.: Traffic grooming in an optical WDM mesh network. IEEE J. Sel. Areas Commun. **20**(1), 122–133 (2002)
168. Zhu, Q., Pavel, L.: Enabling differentiated services using generalized power control model in optical networks. IEEE Trans. Commun. **57**(9), 2570–2575 (2009)

Subject Index

L. Pavel, *Game Theory for Control of Optical Networks*,
Static & Dynamic Game Theory: Foundations & Applications,
DOI 10.1007/978-0-8176-8322-1, © Springer Science+Business Media New York 2012

MIX
Papier aus verantwortungsvollen Quellen
Paper from responsible sources
FSC® C105338

If you have any concerns about our products,
you can contact us on
ProductSafety@springernature.com

In case Publisher is established outside the EU,
the EU authorized representative is:
Springer Nature Customer Service Center GmbH
Europaplatz 3, 69115 Heidelberg, Germany

Printed by Libri Plureos GmbH
in Hamburg, Germany